SCIENCE AND ENGINEERING OF DROPLETS

SCIENCE AND ENGINEERING OF DROPLETS

Fundamentals and Applications

by

Huimin Liu

spraysoft.com
Detroit, Michigan

NOYES PUBLICATIONS
Park Ridge, New Jersey, U.S.A.

WILLIAM ANDREW PUBLISHING, LLC
Norwich, New York, U.S.A.

Library of Congress Catalog Card Number: 98-27324
ISBN: 0-8155-1436-0
Printed in the United States

Published in the United States of America by
Noyes Publications / William Andrew Publishing, LLC
Norwich, New York, U.S.A.

10 9 8 7 6 5 4 3 2 1

Library of Congress Cataloging-in-Publication Data

Liu, Huimin, 1981–
Science and engineering of droplets : fundamentals and applications / by Huimin Liu.
p. cm.
Includes bibliographical references and index.
ISBN 0-8155-1436-0
1. Spraying. 2. Atomization. I. Title.
TP156.S6L58 1999
620'.43--dc21 99-27324
CIP

Foreword

Dew droplets scatter the light of the early morning sun like thousands of diamonds. Billions of passing rain droplets, refracting and internally reflecting the sunlight, form the rainbow, stretching across the sky and competing in its magnificence with the red fire ball of the sunset. Droplets are encountered not only in nature but also in a wide range of engineering applications, such as spray combustion, spray drying, spray cooling, etc., etc. The related areas span automotive, aerospace, metallurgy, material, chemicals, pharmaceuticals, paper, food processing, agriculture, meteorology, and power generation. It was these applications that stimulated the fundamental research and development of the technologies associated with discrete droplets. In the past century, droplet-related phenomena have been a subject of considerable scientific interest and practical importance. In recent years, strong research efforts have led to significant strides in theoretical and experimental studies of droplet processes. The drastic developments in high speed, large memory, parallel processing computer systems and advanced computational and measurement techniques have enabled the direct numerical modeling and *on-line*, *in situ* measurements of certain droplet properties, and thus largely advanced the design and optimization of droplet processes. These advances have significantly improved our fundamental understanding of the phenomena and processes associated with discrete droplets in general. It is therefore desirable to review these results in a timely manner.

This book presents an updated, systematic review of the recent advances in fundamental science and engineering applications of discrete droplet processes. The objective of this book is to familiarize the reader with the scientific and engineering aspects of discrete droplet processes, and to provide experienced researchers, scientists and engineers in academic and industry communities with the latest developments in this field. The scope of this book encompasses the fundamental science and engineering applications of discrete droplet processes of *both* normal liquids (i.e., aqueous or oil-based liquids) and melts (i.e., molten metals, alloys, ceramics or polymers). It starts with a systematic survey of various processes and techniques for droplet generation, along with their applications and associated materials systems. It is followed by a through description of the fundamental phenomena and principles involved in droplet processes, with emphasis on the mechanisms of droplet formation and deformation in various processes. The empirical and analytical correlations, theoretical calculations and numerical modeling of droplet properties are outlined and some approaches to intelligent process control are presented along with discussions of recent developments.

There have been books on droplet-related processes. However, the present book is probably the first one that encompasses the fundamental phenomena, principles and processes of discrete droplets of *both* normal liquids and melts. The author has attempted to correlate many diverse mechanisms and effects in a single and common framework in an effort to provide the reader with a new perspective of the identical basic physics and the inherent relationship between normal liquid and melt droplet processes. Another distinct and unique feature of this book is the comprehensive review of the empirical correlations, analytical and numerical models and computer simulations of droplet processes. These not only provide practical and handy approaches for engineering calculations, analyses and designs, but also form a useful basis for future *in-depth* research. Therefore, the present book covers the fundamental aspects of engineering applications and scientific research in the area.

University of Bremen, Germany Prof. Dr.-Ing. Klaus Bauckhage
January, 1999

NOTICE

Contents

1

General Introduction

Droplets are encountered in nature and a wide range of science and engineering applications.[1]-[5] Natural droplets can be found in dew, fog,[6] rainbows,[7][8] clouds/cumuli,[9]-[11] rains, waterfall mists, and ocean sprays.[12] A dispersion of droplets in surrounding air can be produced by shower heads, garden hoses, hair sprays, paint sprays, and many other spray devices. A variety of important industrial processes involve discrete droplets, such as spray combustion,[1][13]-[15] spray drying,[2][16] spray cooling,[17] spray atomization,[4] spray deposition,[3] thermal spray,[18] spray cleaning/surface treatment, spray inhalation,[19] aerosol (mist) spray, crop spray, paint spray, etc. The related areas span automotive, aerospace, metallurgy, materials, chemicals, pharmaceuticals, paper, food processing, agriculture, meteorology, and power generation. Droplets are used to collect dust in charged droplet scrubbers[20] and to cool hot surfaces by droplet evaporation.[21] Liquid crystals with polymer-dispersed nematic droplets have found wide applications in optic display.[22]-[24] Droplet properties and deformation characteristics are of interest in liquid-liquid phase separation,[25][26] emulsion,[27] gas-liquid mass transfer, and ink-jet printing applications. In materials processing such as synthesis of some semiconductor thin films and composite materials, droplet formation may be a necessary step. For example,

GaAs microcrystals are grown by Ga droplet formation and successive As (arsenic) supply in low pressure metalorganic chemical vapor deposition.[28] Silicon carbide whiskers are synthesized by the formation of silicon-rich liquid droplets in the vapor-liquid-solid growth process.[29] Fullerene is formed by droplet nucleation from supersaturated carbon vapor.[30] In some other processes, however, droplet formation is undesired. For example, in pulsed laser ablation and deposition of thin metal films[31] and $YBa_2Cu_3O_7$ high temperature superconductor films,[32] droplet formation needs to be reduced by decreasing target etching rate. Further examples of science and engineering applications of discrete droplets of both normal liquids and melts are listed in Tables 1.1 and 1.2 for an overview.

Droplets may form by vapor condensation and deposition,[33]-[36] liquid breakup,[37]-[41] or solid particle/wire melting.[42][43] The generation of droplets in most engineering applications involves the breakup of a liquid, i.e., *atomization*. Atomization can be achieved by a variety of means, aerodynamically, mechanically, ultrasonically, or electrostatically, etc. For example, a liquid jet or sheet can be disintegrated by the kinetic energy of the liquid itself, or by exposure to a high-velocity gas, or by the mechanical energy applied externally through a rotating or vibrating device. The liquids of practical interest may be classified into two fundamental categories, i.e., *normal liquids* and *melts*. Normal liquids are typically aqueous or oil-based liquids at near room temperature. Melts are usually molten metals, alloys or ceramics at high temperatures.

Atomization of normal liquids has been used in many fields. The typical applications include spray combustion, spray drying, and aerosol spray, as illustrated in Figs. 1.1, 1.2, and 1.3, respectively. Spray combustion involves atomization and burning of liquid fuels.[1] Spray drying is a process in which a solution or suspension is atomized into a spray of droplets in hot gaseous surroundings to produce solid powder of the solute.[2] Aerosol spray is among the effective and reliable technologies for fire suppression and application of agricultural chemicals.

Table 1.1. Science and Engineering Applications of Droplets of Normal Liquids

Area	Application
Spray Drying	- Milk Powder Processing - Food Processing - Chemical Processing
Spray Combustion	- Diesel Engines - Spark Ignition Engines - Gas Turbines - Rocket Engines - Industrial Furnaces - Domestic Heating Boiler
Spray Cooling	- Cooling of Ingot in Continuous Casting - Cooling of Nuclear Cores - Cooling of Turbine Blades - Coke Quenching
Spray Inhalation	- Vaporization of Volatile Anaesthetic Agents and Medication
Aerosol (Mist) Spray	- Humidification/Air Conditioning - Fire Fighting - Drenching Operations - Lubricating - Pollution/Dust Control
Crop Spray	- Applying Agricultural Chemicals - Spray Irrigation
Paint Spray	- Surface Finishing - Surface Coating
Spray Cleaning	- Gas (Wet) Scrubbing - Gravel Washing - Vegetable Cleaning - Surface Treatment - Car Washing

Table 1.2. Science and Engineering Applications of Droplets of Melts

Area	Application
Spray Atomization	Powder Production: - Dental Amalgams (AgSnCu Powder) - Shot Blasting Grits (Steel or Iron Powder) - Solid Rocket Fuels (Al Powder) - Metallic Paints (Al, CuZn Powder) - Filters (CuSn Powder) - Solder Creams (PbSnAg, $Bi_{57}Sn_{43}$ Powder) - Strips for Diamond Synthesis (Co Powder) - Flares (Mg Powder) - Jewellery-Brazing Pastes (Pd, Au, Ag Powder) - Dense Media for Mineral and Scrap Separation (FeSi Powder) - Coatings for Welding Electrodes (Fe, FeMn, FeSi Powder) - Explosives (Al Powder) - Food Additives (Fe Powder) - Batteries (Zn Powder)
Spray Deposition	Net or Near-Net Shape Manufacturing: - Spray Forming - Spray Casting - Spray Rolling
Thermal Spray	Surface Coating and Free-Standing Components: - Plasma Spray - High Velocity Oxy-Fuel (HVOF)

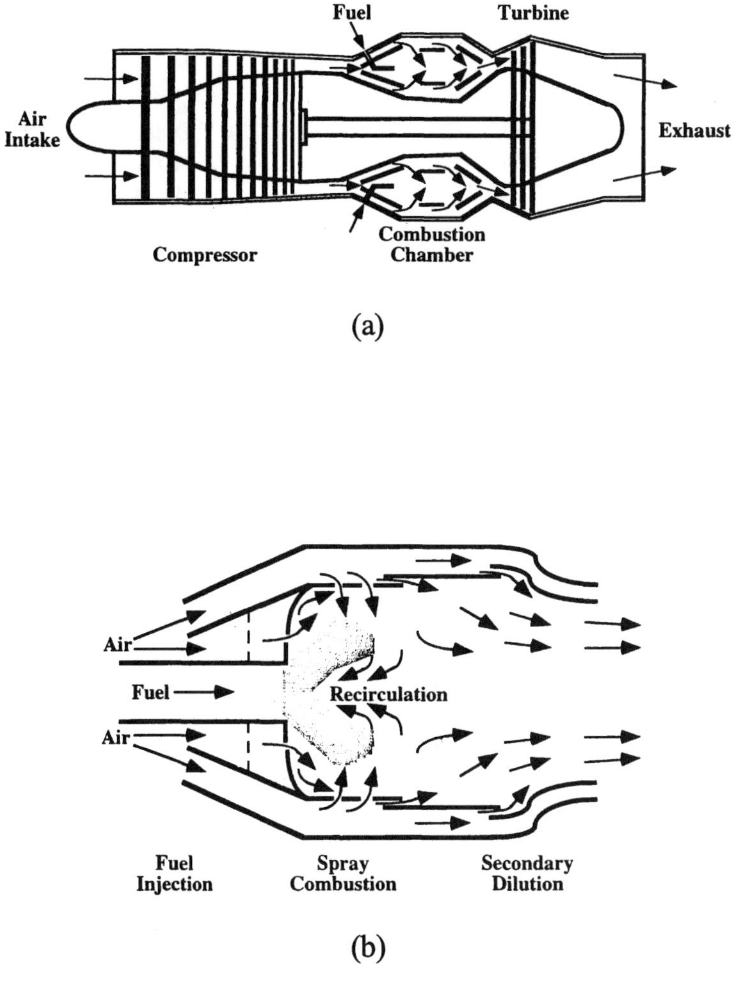

Figure 1.1. Schematic of spray combustion process: *(a)* annular combustion chamber in a single spool turbojet with an axial flow compressor; *(b)* fuel injection and droplet formation in combustion chamber.

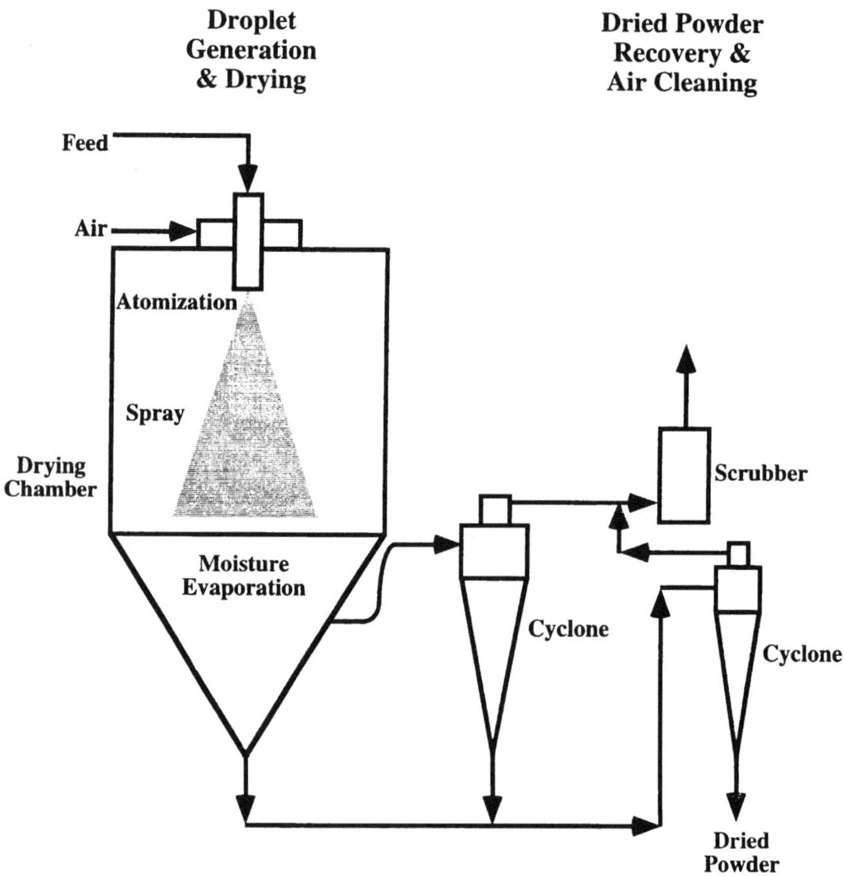

Figure 1.2. Schematic of spray drying process: open-cycle co-current layout.

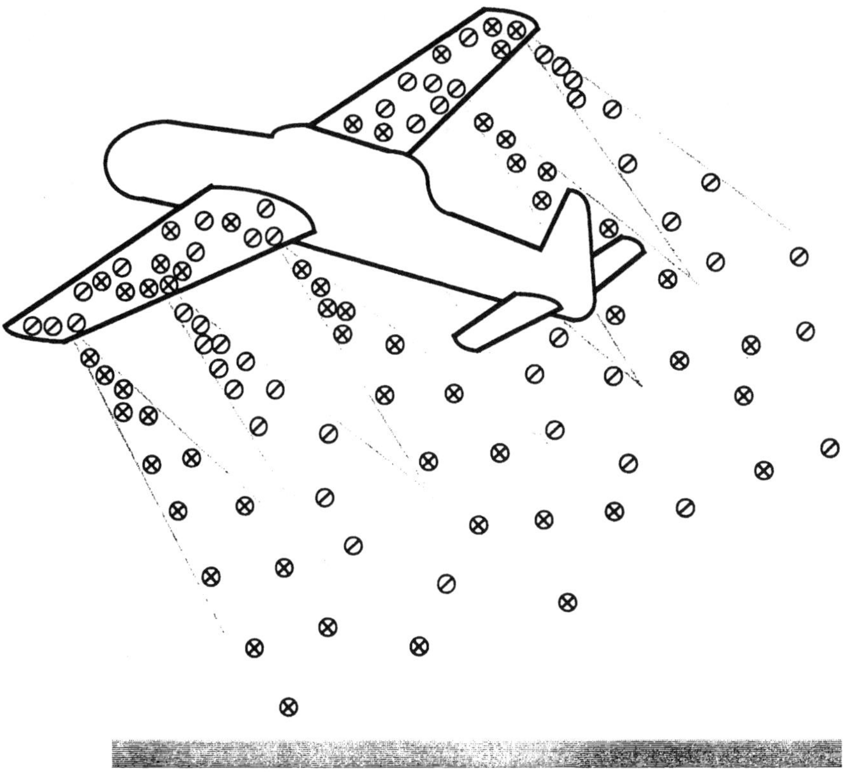

Figure 1.3. Schematic of aerosol spray process: electrostatic aircraft spray.

Atomization of melts has been used in two principal areas, i.e., powder production[4] and spray forming,[3] as shown in Figs. 1.4 and 1.5, respectively. It is not until recent years that the technologies for the atomization of melts have advanced sufficiently to ensure good yields of usable products over sustained periods of plant operation. In these two areas, some aspects such as melting of metals or alloys, basic atomizer designs, and atomization mechanisms are the same or similar. Metals or alloys melt at high temperatures to produce low viscosity but usually high surface tension liquids that are subsequently disintegrated into a dispersion of micro- to millimeter-sized droplets. Solidification of droplets always occurs simultaneously or subsequently, although a certain fraction of droplets may be in fully liquid state during impingement onto the deposition surface in a spray forming process. Rapid solidification that occurs in most atomization processes of melts is beneficial in promoting the metallurgical homogeneity and evenness of dendrite growth and distribution throughout the powders produced, leading to small grain size and elimination of macro-segregation in the end-products. Accordingly, the physical and mechanical properties of powder metallurgy (P/M) products or spray-deposited components are equivalent or usually superior to those of the parts manufactured *via* conventional routes. In addition, energy and cost savings are obvious advantages of P/M and spray atomization deposition processes. Hence, atomization of melts is a good choice for many manufacturing requirements.

In the applications associated with discrete droplets of both normal liquids and melts, large surface areas of droplets are utilized to achieve extremely high reaction rates, cooling and solidification rates, or evaporation rates. Therefore, droplet (particle) size and shape are the issues of most importance. As shown in Fig. 1.6, droplet (particle) size may vary from submicrometers, through tens of micrometers, to millimeters in different applications. Droplet (particle) shape may change from sphericity, through varying degrees of angularity, spongiform, to acicular, fibrous shapes, as illustrated in Fig. 1.7. By carefully selecting appropriate atomizers and atomization media, it is possible to obtain virtually any particular mean droplet (particle) size, size distribution, and any desired droplet (particle) shape. For example, high pressure gas atomization of melts with supersonic, close-coupled atomizers may generate fine metal droplets at high yields. Near-spheroidal droplets of metals can be obtained using gas atomization while water atomization of melts usually produces droplets of irregular shapes. In addition, efficient energy utilization, production of a narrow range of droplet (particle) size and shape, and droplet generation rate are some further issues of particular concern in the processes associated with discrete droplets.

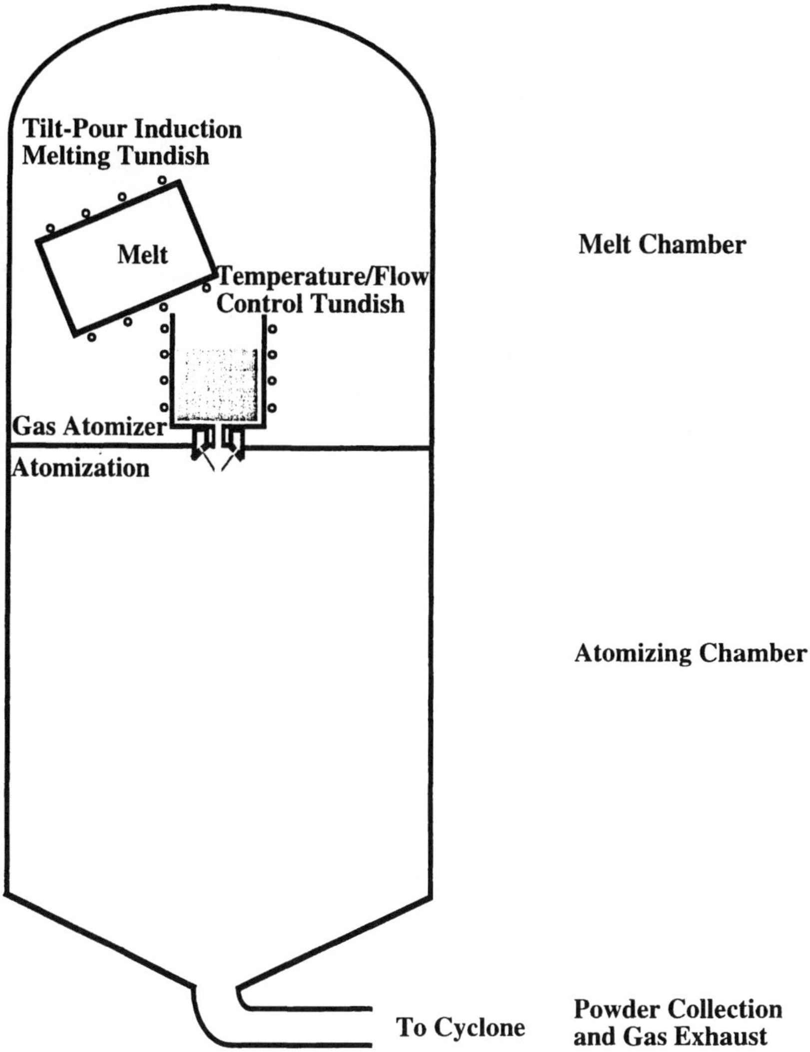

(a)

Figure 1.4. *(a)* Schematic of powder production process; *(b)* a facility for powder production *(Courtesy of Atomizing Systems Ltd., UK.)*

(b)

Figure 1.4. *(Cont'd.)*

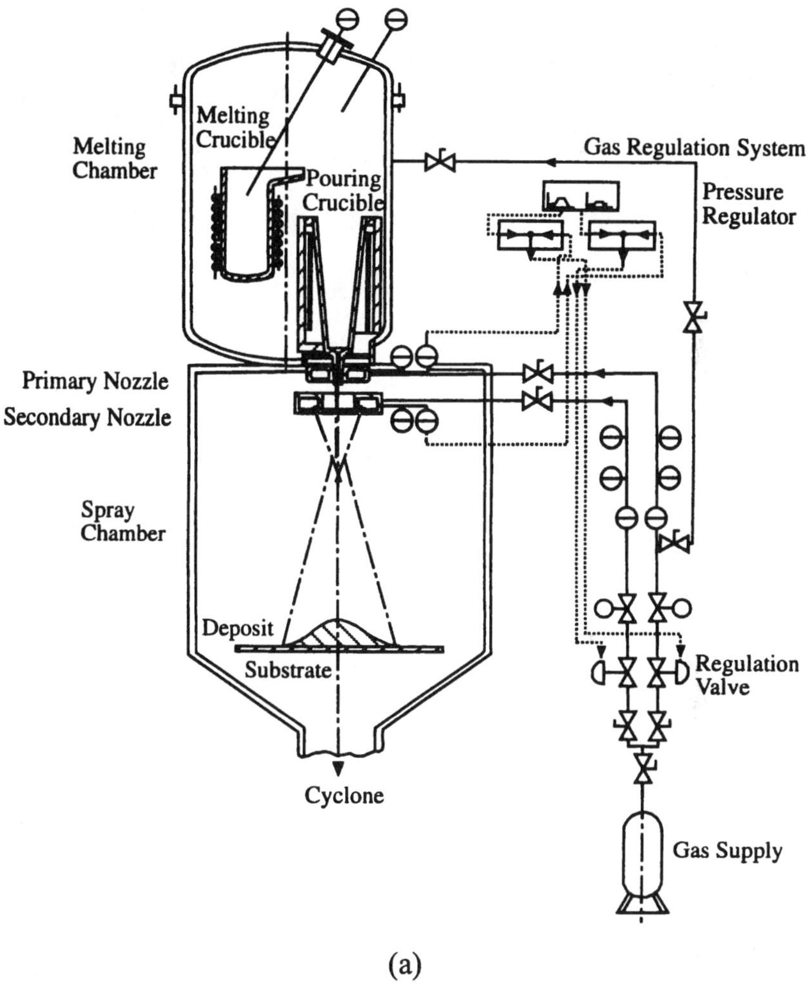

(a)

Figure 1.5. *(a)* Schematic of spray forming process; *(b)* a pilot-plant scale facility for spray forming at University of Bremen, Germany. *Top:* top view. *Bottom:* side view *(Courtesy of Prof. Dr.-Ing. Klaus Bauckhage at University of Bremen, Germany.)*

(b)

Figure 1.5. *(Cont'd.)*

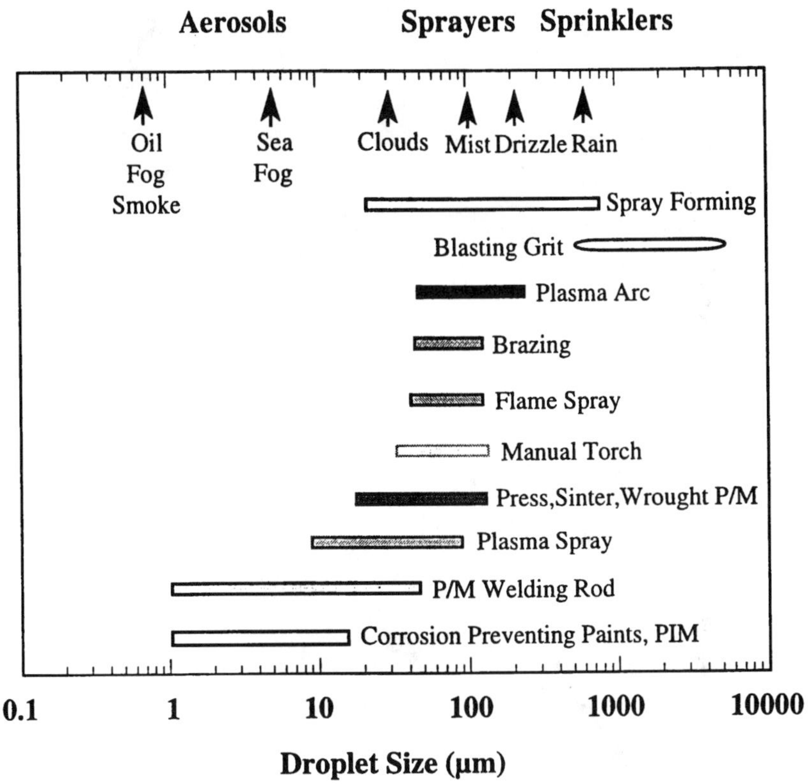

Figure 1.6. Size ranges of droplets/particles found in nature and generated by atomization of normal liquids and melts in aerosol spray, spray combustion, powder production, and spray forming processes.

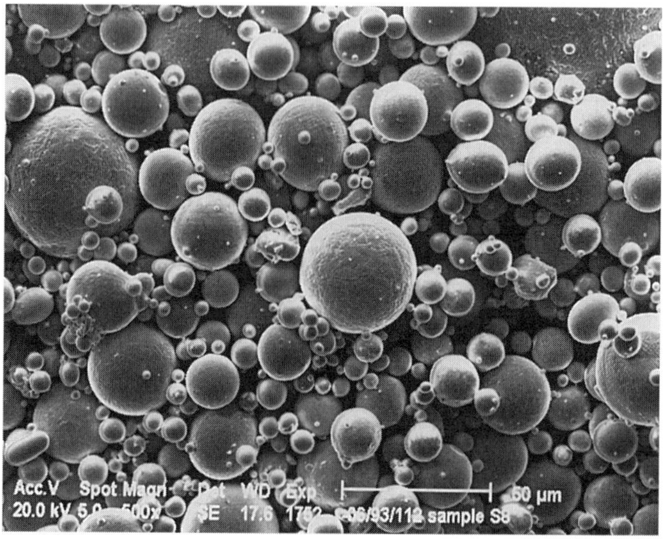

(a)

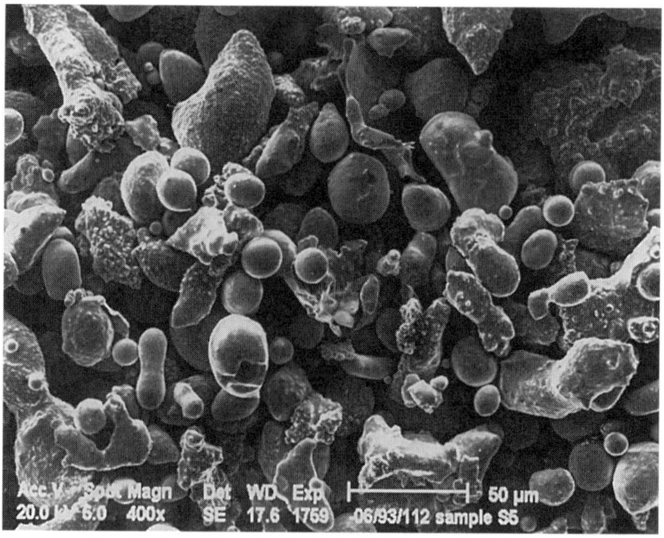

(b)

Figure 1.7. Shapes of solidified droplets (particles) generated in powder production and spray forming processes. *(a)* Spherical shape: gas-atomized gold alloy particles; *(b)* near-spherical and dendritic shapes: water-atomized bronze particles; *(c)* irregular and porous (spongiform) shapes: water-atomized zinc particles; *(d)* irregular aggregates: water-atomized copper particles *(Courtesy of Atomizing Systems Ltd., UK.)*

(c)

(d)

Figure 1.7. *(Cont'd.)*

Droplet deformation during impact onto a surface is an interesting subject in many scientific and engineering fields. The icing and erosion/ablation of aircraft surfaces may be caused by the impact of droplets in clouds and/or rains on the surfaces during flight.[44][45] The erosion of turbine blades[46] operating in wet steam and the erosion of terrestrial surfaces during rain are all related to droplet impact. The phenomena of droplet impinging and spreading on a surface are also encountered in a variety of engineering applications. These include, for example, spray combustion, spray cooling of surfaces, dispersed two-phase flow in once-through boilers, post-critical heat flux cooling, ink-jet printing,[47] metal welding/soldering, spray painting, lubrication, oil recovery from porous rocks,[48] and production of fine metal powders *via* impact atomization.[49] The spreading phenomenon is often accompanied by simultaneous heat transfer and solidification of droplets on deposition surfaces as observed in picoliter solder droplet dispensing for mounting of microelectronic components,[50] spray forming for near-net shape materials synthesis,[3] and thermal spray deposition for surface coating[18][51][52] (Fig. 1.8).

The extensive applications have stimulated the fundamental research and developments of the techniques associated with discrete droplets. In recent years, strong research efforts have led to significant strides in theoretical and experimental studies on droplet processes. The drastic developments in high speed, large memory, parallel processing computer systems and advanced computational and measurement techniques have enabled the direct numerical modeling and *on-line*, *in-situ* measurements of certain droplet properties, and largely advanced the atomizer design and the optimization of atomization and spray processes. These advances have significantly improved the fundamental understanding of the phenomena and processes associated with discrete droplets in general. Particularly, the design, testing and analysis of spray combustion in engine combustors have reached the highest levels in the overall field of spray science and technology.

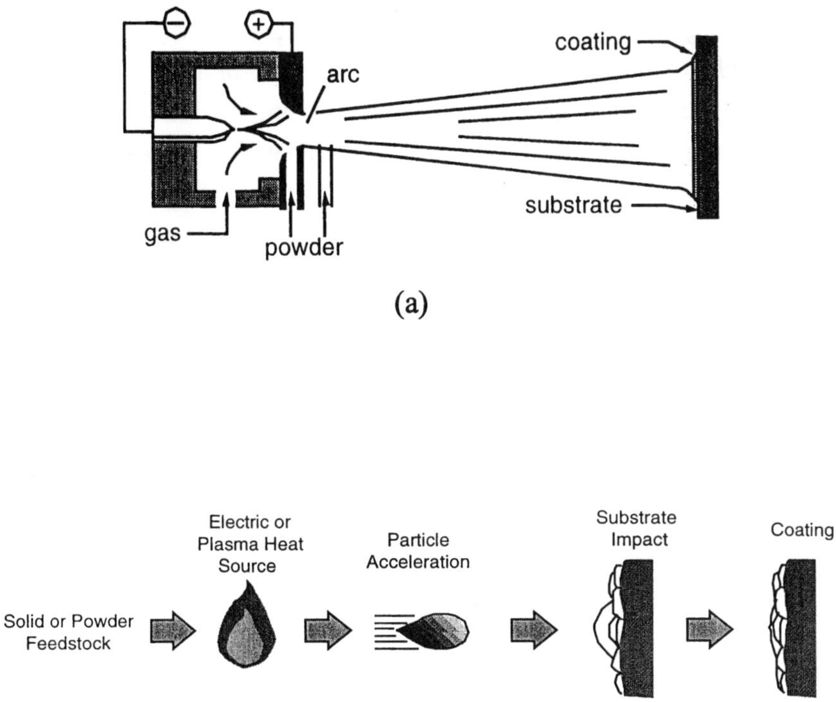

(a)

(b)

Figure 1.8. Schematic showing *(a)* plasma spray deposition process and *(b)* droplet formation, acceleration, surface impact, and coating formation.

Accordingly, this book is devoted to an updated, systematic review of recent advances in the area associated with discrete droplets. It is aimed at familiarizing the reader with the fundamental science and engineering applications of discrete droplets, and providing experienced researchers, scientists and engineers in academic and industrial communities with the latest developments in the field. Attempts are made to encompass the fundamental science and engineering applications of discrete droplets of both normal liquids and melts. It starts with a systematic survey of various processes and techniques for droplet generation, along with their applications and associated materials systems. It is followed by a thorough description of the fundamental phenomena and principles involved in droplet processes, with emphasis on the mechanisms of droplet formation and deformation in various processes. The empirical and analytical correlations, theoretical calculations and numerical modeling of droplet processes are discussed in detail to provide insight into the effects of process parameters on droplet properties and to summarize methodologies for analysis, design and optimization of droplet processes. Finally, the primary measurement techniques for droplet properties are outlined and the approaches to intelligent control over droplet processes are described along with discussions on recent developments. Detailed descriptions of the mechanisms involved in droplet condensation[33]-[35][53] and evaporation,[54][55] vaporization,[56]-[59] heat and mass transfer,[60]-[69] and chemical reactions in spray combustion, spray drying and other droplet processes are beyond the scope of the present book.

2

Processes and Techniques for Droplet Generation

Droplets can be generated by vapor condensation and deposition,[33][36][53][54][70]-[72] atomization and spray,[1]-[5] or plasma spray of solid particles/wires.[42][43] Atomization is perhaps the most widely used process for droplet generation. Atomization generally refers to the disintegration of a bulk liquid material *via* an atomizer into droplets in a surrounding gas or vacuum with or without a spray chamber. An atomizer can be a nozzle (or an array of nozzles) that ejects the liquid and atomizing medium (gas or liquid), a centrifugal device, or an ultrasonic vibrator. Nozzles are utilized in atomizing liquids in many industrial and domestic applications. In atomization of melts, bulk liquids are formed typically by melting solid materials. The atomized droplets usually experience rapid cooling before solidified into the end-products, i.e., powders, patterned matrix of dots, net or near-net shape preforms, or surface coatings. Droplets can also be generated by melting solid particles (powders), wires, fibers or other shapes of solid materials. A notable example is plasma spray atomization.[49] In this process, solid particles are melted in a plasma flame and projected against a target surface to break up into droplets of about one fifth of the original sizes.

In this chapter, various processes and techniques for droplet generation are described in detail, along with their applications and associated materials systems.

2.1.0 ATOMIZATION OF NORMAL LIQUIDS

Atomization is defined as the disintegration of a liquid into small drops or droplets. The resultant suspension of fine droplets or solid particles in a surrounding gas is termed *spray*, *mist*, or *aerosol*. Atomization of a liquid into discrete droplets can be brought about by a variety of means: aerodynamically, mechanically, ultrasonically, or electrostatically, etc. For example, the breakup of a liquid into droplets can be achieved by the impingement with a gas in two-fluid atomization, by centrifugal forces in rotary atomization, by ultrasonic vibrations utilizing a piezoelectric transducer in ultrasonic atomization, or by electrostatic/electromagnetic forces in an electric/magnetic field in electrostatic/electromagnetic atomization. Atomization processes may also be classified according to the energy used to produce the instability on a liquid. For example, pressure energy is used for pressure atomization, centrifugal energy for rotary atomization, gaseous/fluid energy for two-fluid atomization, and vibratory energy for ultrasonic or acoustic atomization.

Most practical atomization processes for normal liquids include: *(a)* pressure atomization, *(b)* two-fluid atomization, and *(c)* rotary atomization. Many other useful atomization processes for normal liquids have been developed in special applications, including: *(a)* effervescent atomization, *(b)* electrostatic atomization, *(c)* ultrasonic atomization, *(d)* whistle atomization, etc. These processes may be loosely classified into two major categories in terms of the relative velocity between the liquid being atomized and the surrounding ambience. In the first category, a liquid at high velocity is discharged into a still or relatively slow-moving gas (air or other gases). Notable processes in this category include, for example, *pressure atomization* and *rotary atomization*. In the second category, a relatively slow-moving liquid is exposed to a stream of gas at high velocity. This category includes, for example, *two-fluid atomization* and *whistle atomization*.

Atomization of normal liquids has been long studied in the fields of spray combustion and spray drying. The most widespread application of the atomization of normal liquids is in spray combustion processes. Generally, an injector or an atomizer is used to produce a spray of fuel inside a combustion chamber. With an *injector*, the fuel is introduced into the combustion chamber at high velocity, creating sufficient turbulence to generate a dispersion of fuel droplets. With an *atomizer*, the fuel is disintegrated into droplets either internally by spreading it on the surface of a swirl chamber or by an internal blast fluid, or externally by an external blast fluid, followed by the dispersion of the resultant droplets in the combustion chamber. Atomization of a fuel before injection into the combustion zone is critical to the combustion efficiency and performance of an engine or industrial furnace because atomization can promote the fuel evaporation and combustion by generating a very high surface-to-volume ratio in the liquid phase. A comprehensive theoretical and empirical background on the atomization of normal liquids has been established on the basis of the studies in these areas. A variety of atomizer designs, devices and techniques have been developed and tested for the atomization of normal liquids, the range of which is much wider than that for the atomization of melts.

The liquids involved in spray combustion, spray drying and other normal liquid processes are primarily hydrocarbon oils or aqueous liquids, although slurries, emulsions, and various non-Newtonian liquids have also been sprayed in some applications. The basic requirement in atomization applications is a narrow droplet size distribution, even a monosize distribution in some spray drying and aerosol spray applications. Small droplet size is desired in spray combustion for rapid heat transfer and vaporization. An ideal atomizer should possess the capability of providing energy-efficient and cost-effective atomization over a wide range of liquid flow rates and responding to changes in liquid flow rates. It should also be capable of scaling up for design flexibility, and not be susceptible to damages during manufacture, maintenance and installation. For spray combustion, an ideal atomizer should produce a uniform radial and circumferential fuel distribution, and be free of flow instabilities. It should not be susceptible to the gum formation by heat soakage and the blockage by contaminants or carbon buildup on the nozzle.

This section describes the atomization processes and techniques for droplet generation of normal liquids. A comparison of the features of various atomization techniques is summarized in Table 2.1. Emphasis is placed on the atomization processes used in spray combustion and spray drying from which many atomization processes have evolved. Advantages and limitations of the atomization systems are discussed along with typical ranges of operation conditions, design characteristics, and actual and potential applications. The physical properties of some normal liquids are listed in Table 2.2, where T is the temperature, ρ is the density, μ is the viscosity, σ is the surface tension, and P_v is the vapor pressure. Depending upon the priority levels of application requirements, many factors need to be considered for the selection of an appropriate technique. These include, for example, *(a)* the cost-effectiveness, *(b)* the ease of operation and maintenance, and *(c)* the capability of producing a monodisperse spray, covering the required range of droplet sizes, accommodating various liquid properties, and achieving sufficient production rates.

2.1.1 Pressure Jet Atomization

Pressure atomization is probably one of the most commonly used techniques in general application areas. In pressure atomization of a liquid, pressure is converted to kinetic energy to accelerate the liquid to a high velocity relative to surrounding ambience. Depending on the atomizer design and geometry, a liquid may be accelerated in different forms, such as a liquid jet in *pressure jet atomization*, a swirling liquid sheet in *pressure-swirl atomization*, or a fan liquid sheet in *fan spray atomization*. Pressure atomizers include: *(a)* plain-orifice atomizer used in pressure jet atomization, *(b)* simplex, duplex, dual-orifice and spill-return atomizers in pressure-swirl atomization, and *(c)* fan spray atomizer in fan spray atomization. Pressure atomization has been long studied by many investigators, for example, Hiroyasu and Kadota,[90] Mock and Ganger,[91] Carey,[92] Radcliffe,[93] and Dombrowski et al.,[94][95] among others.

Table 2.1. Comparison of Features of Various Atomization Techniques for Normal Liquids[1][5]

Method		Droplet Size (μm)	Application	Advantage	Limitation
Pressure Atomization	Plain-Orifice	25–250	Diesel engines, Jet engine afterburners, Ramjets	Simple, Rugged, Cheap	Narrow spray angle, Solid spray cone
	Simplex	20–200	Gas turbines, Industrial furnaces	Simple, Cheap, Wide spray angle	High supply pressure, Varying spray angle with pressure differential and ambient gas density
	Duplex	20–200	Gas turbine combustors	Simple, Cheap, Wide spray angle, Good atomization over a wide range of liquid flow rates	Narrowing spray angle with increasing liquid flow rate
	Dual-Orifice	20–200	A variety of aircraft and industrial gas turbines	Good atomization, Turndown ratio 50:1, Relatively constant spray angle	Poor atomization in transition range, Complexity in design, Susceptibility of small passages to blockage
	Spill Return	20–200	A variety of combustors, Good potential for slurries and fuels of low thermal stability	Simple, Good atomi-zation over entire range of liquid flow rates, Very large turndown ratio, Low risk of blockage due to large passages	Varying spray angle with liquid flow rate, Higher power require-ments except at maximum discharge
	Fan Spray	100–1000	High-pressure painting/coating, Annular combustors	Good atomization, Narrow elliptical spray pattern	High supply pressure

Table 2.1. *(Cont'd.)*

Method		Droplet Size (μm)	Application	Advantage	Limitation
Rotary Atomization	Spinning Disk	10–200	Spray drying. Aerial distribution of pesticides. Chemical processing	Good mono-dispersity of droplets. Independent control of atomization quality and liquid flow rate	Satellite droplets, 360° spray pattern
	Rotary Cup	10–320 [73][74]	Spray drying. Spray cooling	Capable of handling slurries	Possible requirement for air blast around periphery
Two-Fluid Atomization Air-Assist	Internal Mixing	50–500 [75][76]	Industrial furnaces. Industrial gas turbines	Good atomization. Low risk of clogging due to large passages. Capable of atomizing high-viscosity liquids	Possible liquid backup into air line. Requirements for external source of high pressure air & auxiliary metering device
	External Mixing	20–140 [77][78]	Industrial furnaces. Industrial gas turbines	Good atomization. Low risk of clogging due to large holes. Capable of atomizing high-viscosity liquids. No risk of liquid backup into air line	Requirements for external source of high pressure air. Limited liquid/air ratios
Two-Fluid Atomization Air-Blast	Plain-Jet	15–130 [79]-[82]	Industrial gas turbines	Simple, cheap, good atomization	Narrow spray angle. Atomizing performance inferior to prefilming air-blast type
	Pre-filming	25–140 [83]-[86]	Wide range of aircraft and industrial gas turbines	Good atomization especially at high ambient pressures. Wide spray angle	Poor atomization at low air velocities

Table 2.1. *(Cont'd.)*

Method	Droplet Size (μm)	Application	Advantage	Limitation
Effer-vescent Atomization	20–340 [87]	Combustion	Simple, reliable, very good atomization. Low risk of plugging due to large holes. Easy maintenance. Beneficial effects on reducing soot formation & exhaust smoke. Cheap.	Need for separate supply of atomizing air
Electrostatic Atomization	0.1–1000[88] 300–600[88] 100–250[126]	Paint spraying, Printing, Oil burner	Fine and uniform droplets	Very low flow rates, Strongly dependent on liquid electrical properties
Ultrasonic Atomization	Nebulizers 1–5 (55kHz, 0.12 l/min) 30–60 (50 kHz) 1–200[88]	Medical spray. Humidi-fication. Spray drying. Acid etching. Printing circuit. Combustion	Very fine and uniform droplets, Low spray rates	Incapable of handling high liquid flow rates
Whistle Atomization	~50 (10kHz, ≤75 l/min) 7 (>20 kHz, 0.125kg/min 0.33 MPa)	Atomization of liquid metals for powder production	Fine droplets, High gas efficiency	Broad droplet size distribution

Table 2.2 Physical Properties of Some Normal Liquids[1][88][89]

Liquid	T (K)	ρ (kg/m^3)	μ (Ns/m^2)	σ (J/m^2)	P_V (mm Hg)
Acetone	293	792	0.32×10^{-3}	0.0237	195
Aniline	288	1000	5.3×10^{-3}	0.04	243^a
Benzene	293	880	0.65×10^{-3}	0.029	80
Carbon tetrachloride	293	1588	0.97×10^{-3}	0.027	91
Castor oil	293	970	1	0.039	-
Chloroform	293	1489	0.58×10^{-3}	0.02714	-
Cottonseed oil	293	920	0.07	0.0354	-
Diesel Fuel	293	890	0.01	0.03	-
Ethanol	293	790	1.2×10^{-3}	0.022	47
Fuel oil (light)	313	916	0.047	0.023	-
Fuel oil (medium)	313	936	0.215	0.023	-
Fuel oil (heavy)	313	970	0.567	0.023	-
Gas oil	313	863	3.3×10^{-3}	0.023	-
Gasoline	293	680	0.35×10^{-3}	-	413
Glycerin	293	1258	1.49	0.063	10^{-4}
Heptane	300	684	0.38×10^{-3}	0.0194	1^b
Hexane	300	660	0.29×10^{-3}	0.0176	162
Kerosene	293	820	2.5×10^{-3}	0.025	7^c
Lube oil SAE10	293	900	0.1	0.036	-
Lube oil SAE30	293	900	0.3	0.036	1^c
Malathion (95%)	293	1230	0.045	0.032	4×10^{-5c}
Mercury	293	13550	1.53×10^{-3}	0.48	10^{-3}
Methanol	293	800	0.6×10^{-3}	0.022	100
n-Octane	300	703	0.5×10^{-3}	0.021	-
Turpentine	293	867	1.49×10^{-3}	0.027^d	3
Water	293	998	1×10^{-3}	0.0728	18

^{a} At 418k; ^{b} At233k; ^{c} At 303k; ^{d} At 283k

When passing a liquid of low viscosity through a small circular hole, i.e., a *plain-orifice*, the liquid can be readily atomized. At a low injection pressure, the liquid emerges at low velocity as a thin distorted-pencil-shaped stream. For injection pressures in excess of the ambient pressure by approximately 150 kPa, the liquid issuing from the orifice forms a high-velocity jet and disintegrates rapidly into droplets. The mean droplet size and size distribution depend on many variables, such as operation conditions, liquid properties, and nozzle geometry. Increasing the injection pressure can enhance the flow velocity of the liquid jet, leading to an increase in both the level of turbulence in the liquid jet and the aerodynamic drag forces exerted by the surrounding medium, and thereby promoting the disintegration of the liquid jet. A high injection pressure, low ambient pressure and/or small orifice diameter lead to small droplet sizes. An increase in liquid viscosity and/or surface tension usually hinders the liquid disintegration. High viscosity causes large droplet sizes, poor atomization, high spray jet penetration, low evaporation rates, heterogeneous mixing, and imperfect combustion. Surface tension opposes the distortion of the liquid surface. Liquid density affects the spray penetration as the kinetic energy of a liquid jet is one of the primary factors determining the spray behavior. The angle of the spray cone formed by a plain-orifice atomizer ranges typically from 5° to 15°. This angle is mainly affected by the turbulence and properties (viscosity and surface tension) of the liquid jet, and to a lesser extent by the diameter and diameter-to-length ratio of the orifice. An increase in the turbulence enlarges the cone angle by increasing the ratio of radial-to-axial velocity components of the liquid jet.

The flow of the surrounding gas or air has an important effect on pressure jet atomization with a plain-orifice. This is because the atomization process is not completed after a jet leaves the orifice. In fact, it continues in the surrounding medium until the droplet size falls below a critical value so that no further breakup may occur. For a given liquid, this critical size is dependent on the relative velocity between the liquid and the surrounding medium. If both the liquid and gas are flowing in the same direction, the penetration is augmented and atomization is retarded, leading to an increase in the mean droplet size. If the flows are in opposite directions, the penetration is decreased with a concomitant increase in the spray cone angle and the quality of atomization. The gas velocity affects the formation and development of a spray and the quality and degree of atomization. It should be noted that the effect of the gas flow on spray characteristics using a plain-orifice atomizer is relevant only to those operations in which the gas velocity is not high enough to change the nature of the atomization process. If the gas velocity is sufficiently high, the mechanism of liquid jet breakup may change to that of air-blast atomization.

Plain-orifice atomizers are widely used for injecting liquids into a flow stream of air or gas. The injection may occur in a co-flow, a contra-flow, or a cross-flow stream. The best known application of plain-orifice atomizers is perhaps *diesel injectors*. This type of injectors is designed to provide a pulsed or intermittent supply of fuel to the combustion zone for each power stroke of the piston. As the air in the combustion zone is compressed by the piston to a high pressure, a very high pressure (83–103 MPa) is required to allow the fuel to penetrate into the combustion zone and disintegrate into a well-atomized spray.

Another important application of plain-orifice atomizers is jet engine afterburner injectors. The fuel injection system typically consists of one or more circular manifolds supported by struts in a jet pipe. The fuel is supplied to the manifold by feed pipes in the support struts and sprayed into the combustion zone through the orifices in the manifold. Increasing the number of orifices and/or using a ring-like manifold may promote uniform distribution of liquid. To reduce the risk of blockage of orifices, a minimum orifice size of 0.5 mm is usually regarded as practical for kerosene-type fuels.

The third important application of plain-orifice atomizers is rocket injectors. The jets of liquid fuel and oxygen are designed to impinge each other out of the nozzles. At low injection velocities and large impingement angles, a well-defined sheet forms at right angles to the plane of the two jets and breaks up into droplets. The sheet becomes progressively less pronounced as the liquid velocities are increased and/or the impingement angle is reduced. Under typical operation conditions, atomization occurs *via* a process resembling the disintegration of a plain jet and by somewhat ill-defined sheet formation at the intersection of the two jets.

2.1.2 Pressure-Swirl Atomization

One of the limitations of plain-orifice atomizers is the narrow spray cone generated. For most practical applications, large spray cone angles are desired. To achieve a wide spray cone, a *simplex*, i.e., a *pressure-swirl atomizer*[96] can be used. A pressure-swirl atomizer consists of a conical swirl chamber with a small orifice at the vertex. During operation, a liquid is introduced into the swirl chamber through tangential ports and allowed to swirl. If the liquid pressure is sufficiently high, a high angular velocity is attained and an air-cored vortex is created. The swirling liquid then flows through the outlet of the swirl chamber and spreads out of the orifice under the action of both axial and radial forces, forming a tulip-shaped or conical sheet beneath the orifice. The sheet subsequently disintegrates into droplets. The liquid-air interaction, liquid surface tension and viscous forces are the primary factors governing the liquid breakup process. The spray cone angle may range from 30° to almost 180°, depending on the relative magnitude of the tangential and axial velocity components at the nozzle exit, and hence can be controlled by adjusting these variables. The droplet size is a function of liquid pressure and swirl chamber dimensions. The smaller the swirl chamber is, the finer the resultant droplets become. However, there exist quite restrictive flow limits for small droplets. Another undesirable feature is the instability to get sharp cutoff of spray due to dribbling.

Swirling flows are used extensively in atomizers and burners to generate recirculation zones with high shear rates. This type of atomizers makes use of centrifugal forces. The tangential and radial velocity components are of the same order or higher than the axial velocity component. The designs of pressure-swirl atomizers developed to date may be grouped into two basic types in terms of the spray pattern generated, namely, *solid-cone spray* and *hollow-cone spray*. In the first type of designs, droplets are distributed fairly uniformly throughout the volume of the spray cone, hence referred to as solid-cone spray. In the second type of designs, droplets are concentrated in the outer periphery of the spray cone, thereby termed hollow-cone spray.

A hollow-cone spray can be generated *via* a simplex atomizer. The spray pattern varies depending on the injection pressure. At very low pressures, liquid dribbles from the nozzle orifice. With increasing pressure, the liquid emerges from the orifice as a thin, distorted-pencil-shaped stream. A liquid cone may form at the orifice, but may contract into a closed bubble by surface tension forces. With further increasing pressure, the bubble opens into a hollow tulip-shape terminating in a ragged edge where the liquid disintegrates into large droplets. At high pressures, the curved surface of the tulip-shape straightens and a conical liquid sheet forms. As the sheet extends and its thickness diminishes, it becomes unstable and breaks up into ligaments that subsequently disintegrate into droplets in a well-defined hollow-cone spray. For low-viscosity liquids, the lowest injection pressure at which atomization can be achieved is about 100 kPa.

Solid-cone spray atomizers have been applied to many industrial processes such as gas scrubbing, gas washing, coke quenching, fire protection, chemical processing, foam breaking, dust suppression, gravel washing, vegetable cleaning, spray cooling, and drenching operations. The Delavan nozzle is a typical example of the solid-cone spray atomizers.

Various hollow-cone simplex atomizers (Fig. 2.1) have been developed for combustion applications, differing from each other mainly in the way that swirl is imparted to the issuing liquid jet. In these atomizers, swirl chambers may have conical slots, helical slots (or vanes), or tangential slots (or drilled holes). Using thin, removable swirl plates to cut or stamp the swirl chamber entry ports leads to economies of the atomization systems if spray uniformity is not a primary concern. Large simplex atomizers have found applications in utility boilers and industrial furnaces. Oil flow rates can be as high as 67 kg/min.

Solid-cone spray atomizers usually generate relatively coarse droplets. In addition, the droplets in the center of the spray cone are larger than those in the periphery. In contrast, hollow-cone spray atomizers produce finer droplets, and the radial liquid distribution is also preferred for many industrial applications, particularly for combustion applications. However, in a simplex atomizer, the liquid flow rate varies as the square root of the injection pressure. To double the flow rate, a fourfold increase in the injection pressure is required. This may lead to an unacceptably high value of the injection pressure. For very low flow rates, on the other hand, the injection pressure required may be very low so that the atomization quality may be extremely poor. This drawback of simplex atomizers has stimulated the development of various wide-range atomizers, such as duplex, dual-orifice, and spill-return atomizers.[91]-[94] These atomizers may readily accommodate the ratios of the maximum to the minimum flow rates in excess of 20 with the injection pressures up to 7 MPa. Thus, good atomization may be achieved over the entire operating range of liquid flow rates without resorting to impractical levels of pump pressure.

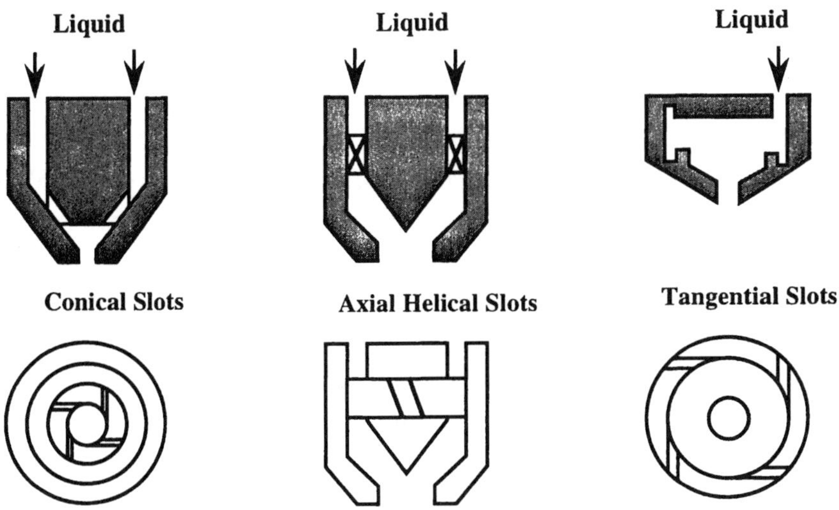

Figure 2.1. Schematic of various types of hollow-cone simplex atomizers.

In a *duplex atomizer* (Fig. 2.2), the swirl chamber consists of two sets of tangential swirl ports: primary and secondary ports. The primary ports are for low flow rates and the secondary ports are the main passage for high flow rates. During operation, the primary swirl ports are supplied first with a liquid from the primary manifold, while a spring-loaded pressurizing valve prevents the liquid from entering the secondary manifold. When a predetermined injection pressure is attained, the valve opens and the liquid flows through both the primary and secondary swirl ports simultaneously. Duplex atomizers exhibit superior performance to simplex atomizers, particularly at low flow rates. However, duplex atomizers tend to produce smaller spray cone angles in the combined flow range than in the primary flow range. In some designs, this problem is resolved by using a smaller tangent circle for the primary swirl ports than the secondary ports, giving rise to a constant spray cone angle. Duplex atomizers have been used in gas turbines for the atomization of fuels.

Dual-orifice atomizers have the same advantage as duplex atomizers, i.e., better atomization at low liquid flow rates. A dual-orifice atomizer (Fig. 2.3) comprises two simplex nozzles with the primary nozzle mounted inside the secondary nozzle concentrically. During operation, the primary spray does not interfere with either the secondary orifice or the secondary spray within the orifice. At low flow rates, a liquid flows entirely through the primary nozzle. Since the pressure required to force the liquid through the small ports in the primary swirl chamber is high, the atomization quality is good. With increasing flow rate, the injection pressure increases and eventually reaches a level at which the liquid is admitted to the secondary nozzle. Starting from this point and over a range of higher flow rates, the atomization quality is relatively poor due to the low secondary pressure. With further increase in the flow rate, the secondary pressure increases correspondingly, improving the atomization quality. In practice, one can employ a primary spray cone angle slightly wider than that of the secondary spray cone so that the two sprays coalesce and share energy within a short distance from the atomizer. However, this may alleviate the problem only to some extent. Therefore, the dual-orifice atomizer must be designed in such a way that the pressure for starting the secondary flow does not coincide with an engine operation point at which high combustion efficiency and low pollutant emission are primary requirements. Dual-orifice atomizers have been widely used in many types of aircraft and industrial engines.

A *spill-return atomizer* (Fig. 2.4) is essentially a simplex atomizer. The difference between the two types of atomizers is that the rear wall of the swirl chamber is solid in a simplex, while in a

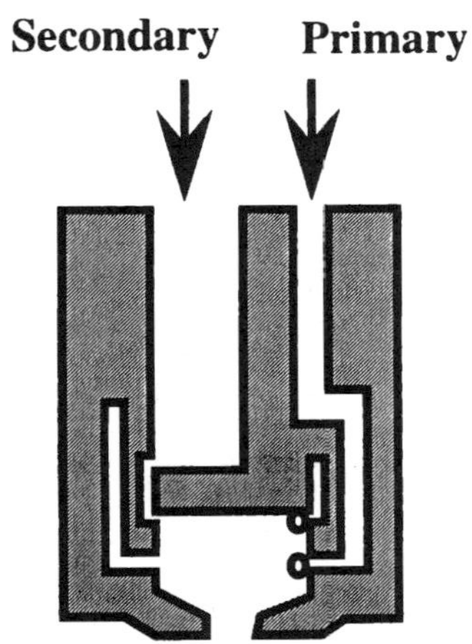

Figure 2.2. Schematic of a duplex atomizer.

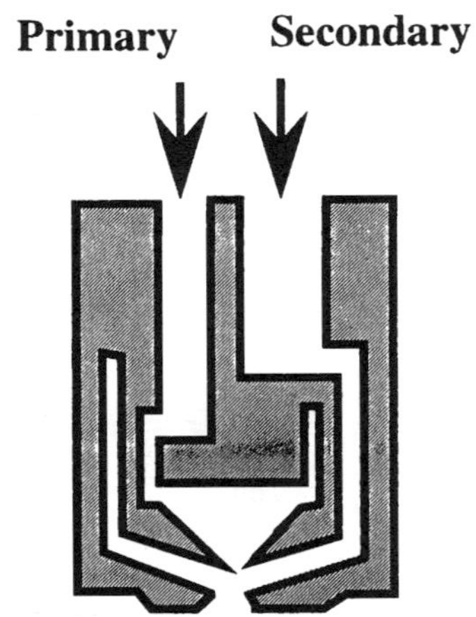

Figure 2.3. Schematic of a dual-orifice atomizer.

Figure 2.4. Schematic of a spill-return atomizer.

spill-return atomizer it contains a passage through which a liquid can be spilled away from the atomizer. The liquid is supplied to the swirl chamber always at the maximum pressure and flow rate. At the maximum operation capacity, a valve located in the spill line is closed, allowing the liquid to entirely emerge from the nozzle to form a well-atomized spray. Otherwise, the liquid will be diverted away from the swirl chamber by opening the valve, so that less liquid is left to pass through the nozzle. Satisfactory atomization may be achieved for flow rates down to 1% of the maximum value.[92] The atomization quality is high over a wide range of flow rates at a constant pressure and generally tends to improve as the flow rate is decreased. An even wider range of flow rates can be readily achieved if the pressure is also varied over a range. This is the most useful feature of spill-return atomizers. In addition, a spill-return atomizer is free from the blockage by contaminants in the liquid because the flow passages are designed to allow large flows during operation. A drawback associated with spill-return atomizers is the large variation in spray cone angle with change in flow rate. Lowering flow rate widens the spray cone angle due to the reduced axial velocity component and nearly unchanged tangential velocity component. The spray cone angle at a minimum flow rate may be 50° wider than at a maximum flow rate. Moreover, metering the flow rate is more complex than with other types of atomizers, and a large-capacity pump is required to handle the large recirculating flows. Therefore, spill-return atomizers are mainly used in large industrial furnaces, and the interest in the applications to gas turbines has declined. However, due to the excellent atomizing capability and freedom from blockage, spill-return atomizers are still an attractive alternative for spraying various fuels currently being considered for gas turbine applications, particularly those with high viscosity and aromaticity which may form gum deposition during operation.

The simplex atomizer has many other variants.[1] Designs are aimed at achieving good atomization over a wide range of flow rates by varying the effective flow area without the need for excessive hydraulic pressures. In practice, pressure-swirl atomizers used in gas turbine combustors are housed within an annular passage, through which a small amount of air flows. The airflow, usually less than 1% of the total combustor airflow, is discharged at the downstream end of the passage and flows radially inward across the nozzle face. This air, termed *shroud air* or *anticarbon air*, is used primarily to protect the nozzle from overheating by the flame and to prevent the deposition of coke that could interfere with the fuel spray and block the nozzle. In addition, the shroud air also exhibits some other beneficial effects. For example, swirled shroud air may improve atomization quality by reducing the mean droplet size and assisting in holding the spray tulip open at low fuel pressure while having no effect on the cone angle of the fully developed spray. This leads to an extension of weak-extinction limits and an improvement of light-up capability. Swirled shroud air tends to widen the spray cone angle by inducing radially outward flow, while unswirled shroud air tends to narrow the spray cone angle. Thus, the shroud air also provides a means for varying and controlling the spray cone angle. However, excessive amounts of unswirled shroud air may lead to a too-narrow spray cone angle and thus may result in combustion instability.

2.1.3 Fan Spray Atomization

It should be noted that the spray patterns generated with the plain-orifice and pressure-swirl atomizers are normally circular. To generate a narrow elliptical spray pattern, a *fan* or *flat spray atomizer* can be used. Flat spray atomizers may have different arrangements such as flood nozzle and flat nozzle. In a *flat spray flood atomizer*, a round liquid jet is arranged to impinge on a curved surface, producing a wide, flat spray pattern with a fairly uniform distribution of relatively coarse droplets. The nominal spray pattern is 120° or more. In a *flat spray atomizer*, the discharge orifice is the intersection of a V-shaped groove with a hemispheric cavity connecting with a cylindrical liquid inlet. The produced liquid sheet breaks up into a narrow elliptical spray pattern parallel to the major axis of the orifice. This is the most widely used type of fan spray atomizers. Using this type of atomizer, excellent atomization and spray patterns have been produced for viscous and non-Newtonian liquids. In addition, a fan spray can also be produced by cutting slots in plane or cylindrical surfaces and allowing a liquid to flow into the slots from two opposite directions. For a single fan spray injector, the liquid then issues from a shaped orifice and spreads out in the shape of a circle sector of about 75°.

The behavior of fan spray atomization has been studied by Dombrowski et al.,[95] Carr,[97] and others. The trajectory of the liquid sheet generated by a fan spray atomizer is found to be dependant on the injection pressure, sheet thickness and liquid surface tension, and less affected by liquid density.[95] The sheet thickness decreases with increasing distance from the orifice. Liquid surface tension tends to contract the edges of the sheet, leading to a curved boundary as the sheet expands. Liquid viscosity restrains the breakup of the edges of the sheet, resulting in relatively large droplet sizes. Increasing injection pressure and/or shroud air pressure can reduce mean droplet size. Ambient pressure has little effect on the spray angle produced by a fan spray atomizer. However, an increase in ambient pressure tends to increase the mean droplet size because, at a high ambient pressure, the liquid disintegration takes place closer to the nozzle exit where liquid sheet is thicker. For a given flow rate, fan spray atomizers normally generate coarser droplets than pressure-swirl atomizers. However, the Lucas fan spray fuel injection system[97] produces very fine droplets with a mean size less than 25 μm even for oils with viscosity as high as 8.5×10^{-6} m^2/s.

Fan spray atomizers have been widely used in the spray coating industry (Fig. 2.5), in some small annular gas turbine combustors, and in other special applications that require a narrow elliptical spray pattern rather than the normal circular pattern. In particular, fan spray atomizers are ideal for small annular combustors because they can produce a good lateral spread of fuel, allowing to minimize the number of injection ports.

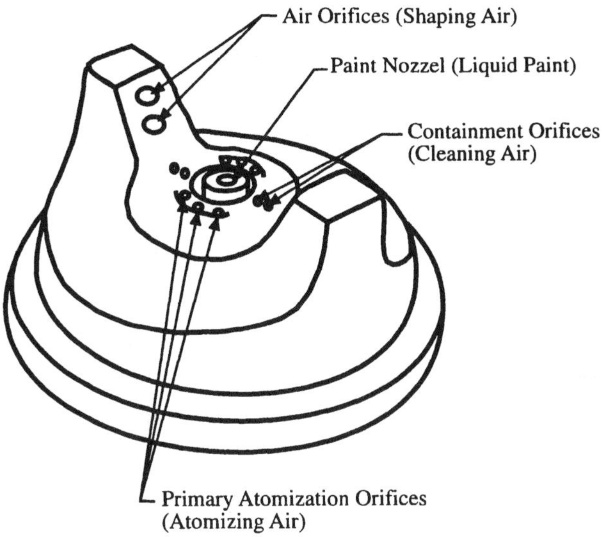

Figure 2.5. Schematic of a flat spray atomizer used for paint spray.

2.1.4 Two-Fluid Atomization

As mentioned in the previous section, a major drawback of the simplex atomizer is the poor atomization quality at the lowest flow rate due to too-low pressure differential if swirl ports are sized to allow the maximum flow rate at the maximum injection pressure. This problem may be resolved by using dual-orifice, duplex, or spill-return atomizers. Alternatively, the atomization processes at low injection pressures can be augmented *via* forced aerodynamic instabilities by using air or gas stream(s) or jet(s). This is based on the beneficial effect of flowing air in assisting the disintegration of a liquid jet or sheet, as recognized in the application of the shroud air in fan spray and pressure-swirl atomization.

Two-fluid atomization, sometimes also termed *twin-fluid atomization*, *two-phase atomization* or *aerodynamic atomization*, is one of the commonly used techniques in many areas. Two-fluid atomization may be divided into two categories: *air-assist* atomization and *air-blast* atomization.[98]-[102] In principle, the two atomization processes are exactly the same. In both processes, the bulk liquid to be atomized is first transformed into a stream/jet or sheet. The liquid stream/jet or sheet at a relatively low velocity is then exposed to high-velocity air (gas) stream(s). Air is used to improve and maintain the quality of atomization over a wide range of liquid flow rates. The kinetic energy of the high-velocity air stream(s) is used to shatter the liquid jet into ligaments that subsequently disintegrate into droplets. Large scale eddy structures in the air flow impact upon the liquid jet, causing destabilization, stretching, and flapping of the liquid jet. Due to the complex liquid-air interactions, the droplet sizes in aerodynamic atomization are very widely dispersed. The significant difference between the two processes is in the quantity and velocity of the air used. In air-assist atomization, the air flow rate is to be kept to a minimum, but there is no special restriction on air pressure. The air is supplied from a compressor or a high-pressure cylinder, passing through annular orifices surrounding the liquid jet, making a very high air velocity. In air-blast atomization, the air velocity corresponds to the pressure differential across the combustor liner and hence is limited to a certain value, usually up to 150 m/s. Thus, a large quantity of air (mass flow ratio of gas to liquid >1.0) is necessary to produce good atomization. In addition to atomization, the air also acts as a medium for the transportation and complete combustion of the droplets in the combustion zone. Therefore, air-assist atomization is characterized by a relatively small quantity of air with very high velocity, whereas air-blast atomization distinguishes itself by a large quantity of air with limited velocity.

In *air-assist atomization*, air is needed usually to augment the atomization process only at low liquid flow rates when the pressure differential is too low to produce satisfactory pressure atomization. In some designs, however, air assistance may be required over the entire range of operating conditions if the atomization quality achieved with a pressure atomizer alone is always poor. In an air-assist atomization process, the impingement of a low-velocity liquid stream by a high-velocity air stream may occur either within or outside the atomizer, referred to as *internal-mixing* (Fig. 2.6) and *external-mixing* (Fig. 2.7), respectively. In an internal-mixing atomizer, intense mixing of gas and liquid occurs within the mixing chamber, and the gas flow can be a single convergent jet or multiple Y-jets. An internal-mixing atomizer is well suitable for highly viscous liquids. Good atomization quality can be achieved for very low liquid flow rates. The spray cone angle can be controlled by air/gas flow. However, the aerodynamic and fluid dynamic flow patterns are highly complex. For example, in a Y-jet atomizer, a cylindrical liquid sheet may form within the atomizer due to shearing by high-velocity gas.[103] The sheet, filled with a cloud of stray droplets, is accelerated and thinned out by the gas as it travels at a velocity at least an order of magnitude lower than the gas, and disintegrates (outside the atomizer) into a spray more even than from a swirl atomizer. In an external-mixing atomizer, a high-velocity stream of gas is arranged to impinge at some angle onto a liquid jet or conical sheet at the center of the atomizer. An external-mixing atomizer can be designed to keep the spray cone angle constant for all liquid flow rates. The air utilization and energy efficiency of an external-mixing atomizer are less than an internal-mixing atomizer, and hence more energy input is required for atomization. However, an external-mixing atomization system is more robust in the sense that there is no danger of liquid flowing in the air line to interrupt the atomization operation.

A variety of designs of air-assist atomization systems have been developed and widely used in industrial gas turbines and oil-burning furnaces, as described by Gretzinger and Marshall,[102] Mullinger and Chigier,[103] Bryce et al.,[104] Sargeant,[105] and Hurley and Doyle.[106] Air-assist atomization systems are well suitable for large industrial engines particularly as the high-pressure air is needed only during engine light-up and acceleration. It is the need for an external supply of high-pressure air that prevents these systems from being applied to aircraft engines. For the atomization of coal-water slurries and other liquids that are difficult to atomize with a conventional pressure atomizer, the *Parker Hannifin atomizer* (Fig. 2.8) may be used in which air assistance is employed over the entire

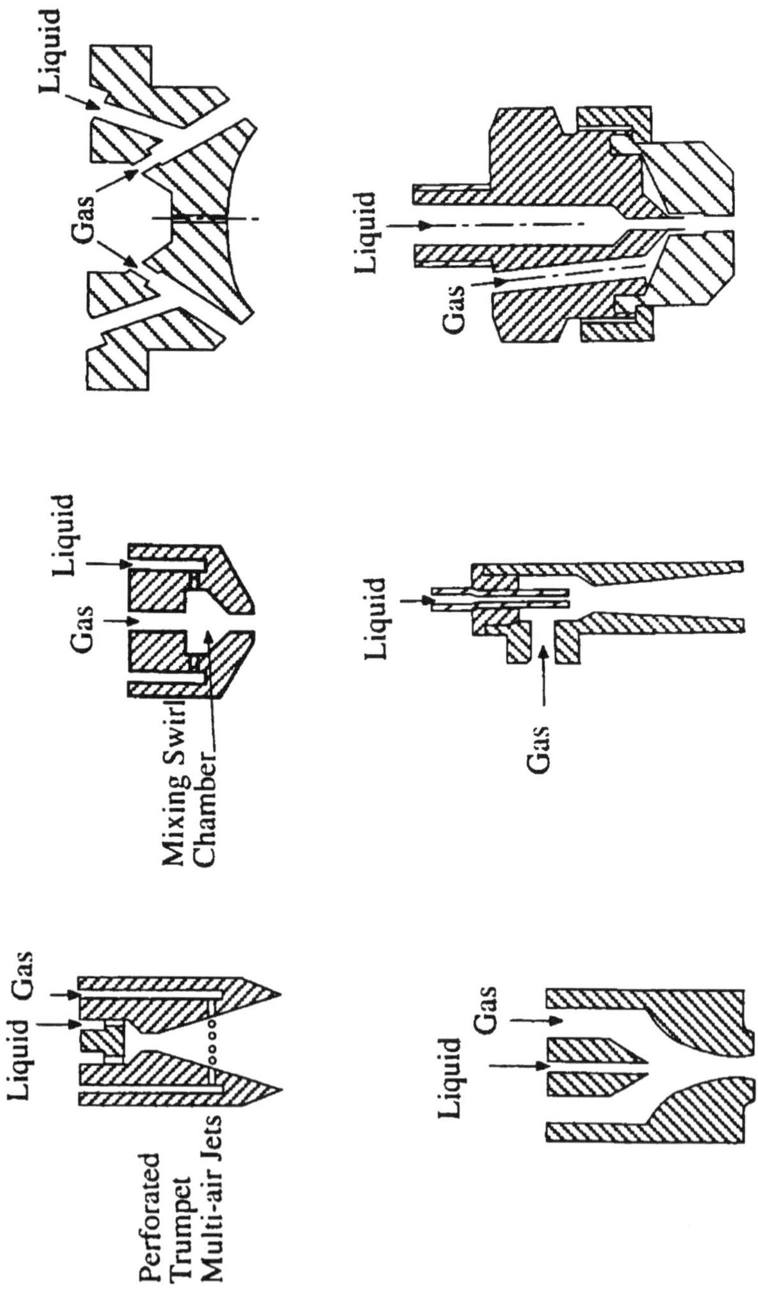

Figure 2.6. Schematic of internal-mixing two-fluid atomizers.

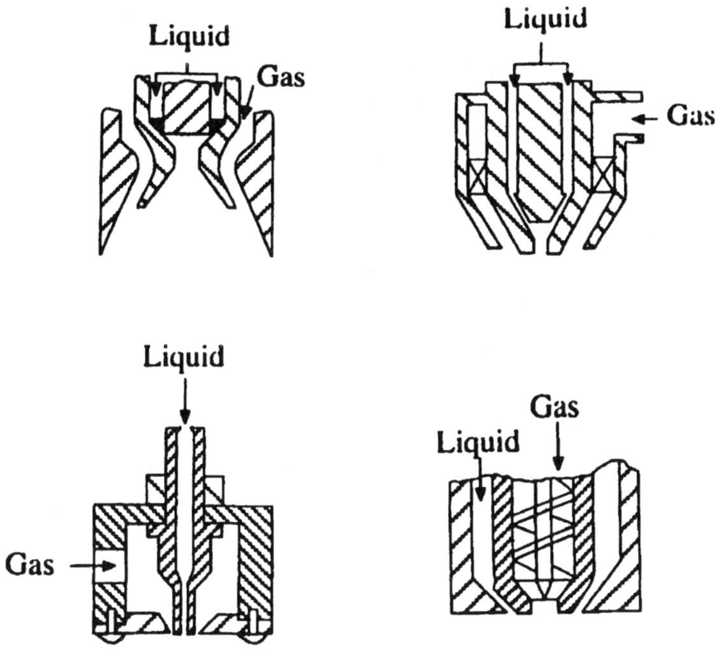

Figure 2.7. Schematic of external-mixing two-fluid atomizers.

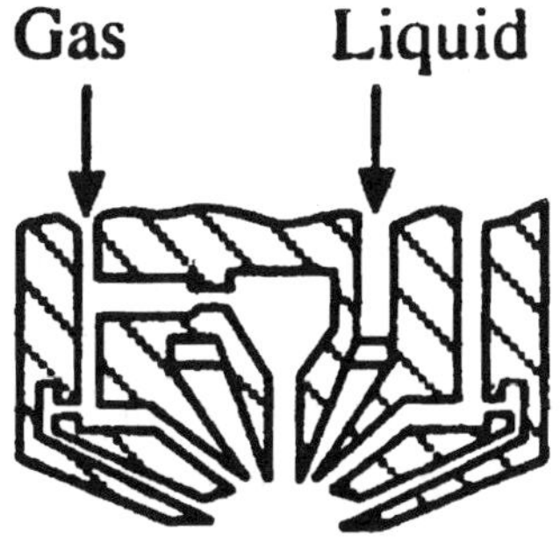

Figure 2.8. Schematic of Parker Hannifin atomizer designed for atomizing coal-water slurries and other difficult-to-atomize liquids. *(Courtesy of Parker Hannifin Corporation, USA.)*

range of operating conditions. In this atomizer, two clockwise swirled air streams, one inside and the other outside an annular, anticlockwise swirled liquid sheet, are used to shear the liquid sheet at the nozzle tip. The widely used types of air-assist atomizers include, for example, the National Gas Turbine Establishment atomizer,[101] the Y-jet atomizer[104][105] used in large oil-fired boiler plants, the combined pressure and twin-fluid atomizer,[107] and those designs of Delavan company used in industrial furnaces.

In *air-blast atomization*, a large amount of air is used as primary driving force of atomization. Fine droplets can be produced at low liquid pressures. Due to the thorough mixing of air and droplets, very low soot formation and a blue flame with low luminosity can be achieved in the ensuing combustion process, with concomitant decrease in flame radiation and exhaust smoke. In addition, air-blast atomizers are inherently simple in design. These advantages over pressure atomizers make air-blast atomizers particularly suitable for application to combustion systems operating at high pressures. Therefore, air-blast atomizers have been used in a broad range of aircraft, marine, and industrial gas turbines.

In air-blast atomization, a liquid can be injected into a high-velocity air stream in the form of one or more discrete jets. The jets disintegrate into droplets during flight in the air stream. An example of this type of design is the *plain-jet air-blast* atomizer used in gas turbines. Alternatively, a liquid is first arranged to form a continuous thin sheet before being exposed to the impact and/or shear by a high-velocity air. This is the commonly used type of air-blast atomizers, i.e., *prefilming* type. In this design, as illustrated in Fig. 2.9,[84] the liquid is introduced from equispaced tangential ports into a weir; through that, it flows over the prefilming surface as a thin film and then is discharged as a thin sheet at the atomizing lip with the help of two air streams. One air stream flows through a central circular passage, deflected radially outward by a pintle, and impacts the inner surface of the liquid sheet. The other air stream flows through an annular passage surrounding the main body of the atomizer and contacts the outer surface of the liquid sheet at a high velocity after leaving the exit of minimum flow area in the plane of the atomizing lip. After leaving the atomizing lip, the liquid sheet is stretched by the two surrounding air streams and breaks up into droplets. The two air streams then merge together and transport the atomized droplets into combustion zone. Swirling motion may be imparted to the air flows to optimize atomization. The atomization performance of prefilming atomizers is generally superior to that of plain-jet atomizers, especially under the adverse conditions of low mass flow rate ratios of air to liquid and/or low air velocities.[80][108] In addition, a prefilming air-blast atomizer can be designed to disperse fuel droplets throughout the entire combustion chamber, whereas a plain-jet air-blast atomizer tends to concentrate fuel droplets within a small region downstream of the nozzle exit. Therefore, most of the atomization systems now in service in spray combustion field are of the prefilming type. However, prefilming atomizers are fully effective only if both sides of the liquid sheet are exposed to atomization air. This introduces a complication in system design because it typically requires arrangement of two separate air flows through the atomizer. Thus, plain-jet atomizers are preferred in some applications. One of the major drawbacks of air-blast atomizers is the poor atomization at low air velocities during engine cranking. This problem may be overcome by using an additional simplex nozzle in a prefilming air-blast atomizer. The liquid is pressure-atomized through the simplex nozzle during engine cranking or flameout.

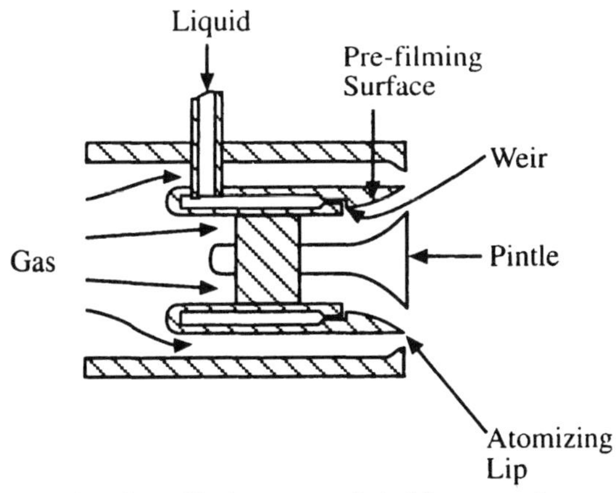

Figure 2.9. Schematic of prefilming type of air-blast atomizer.

2.1.5 Rotary Atomization

Rotary atomization is a centrifugal atomization method.[94][109] In a rotary atomization process (Fig. 2.10), a liquid is introduced onto a rotating surface at its center and then spreads out nearly uniformly under the action of centrifugal force. At low liquid flow rates, droplets may form directly near the edge of the surface, whereas at high liquid flow rates, ligaments or sheets may be generated at the edge and eventually disintegrate into droplets. The rotating surface may be a flat disk, vaned disk,[110][111] cup,[112] or slotted wheel, etc. Spinning disk and rotary cup atomizers are the widely used types. An air/gas flow around the periphery of rotating surface is sometimes used to shape the spray and to assist in transporting droplets away from the atomizer. Atomization is carried out usually in a cylindrical or conical chamber to accommodate the umbrella-like, 360° spray pattern created by rotating disk and downward air/gas streams. A symmetrical feed of liquid can be readily achieved using a cup. A conical-shaped cup is more usable for high flow rates than other shapes. The diameter of the rotating surface ranges from 25 mm to 450 mm. Small disks rotate at high speeds up to 1000 rps whereas large disks rotate at low speeds up to 200 rps. Higher speeds on the order of 2500 rps have been used when a pressurized liquid is fed onto a spinning disk.[88] Lower speeds on the order of 50 rps have been used when a coaxial air jet is used to assist liquid breakup. Atomization capabilities up to 84 kg/min can be achieved.[2] In the spray drying industry, advances have been made in handling high flow rates up to 2400 kg/min at high wheel peripheral speeds, yielding droplets finer than 20 μm.[2]

Rotary atomization processes in spray drying have been studied extensively by many researchers, for example, Kayano and Kamiya,[110] Tanasawa et al.,[111] Hinze and Milborn,[112] Christensen and Steely,[113] and Kitamura and Takahashi.[114] Details of the processes have been described and reviewed by Masters,[2] Dombrowski and Munday,[94] Matsumoto et al.,[109] Christensen and Steely,[113] Eisenklam,[115] and Fraser et al.,[116] among others.

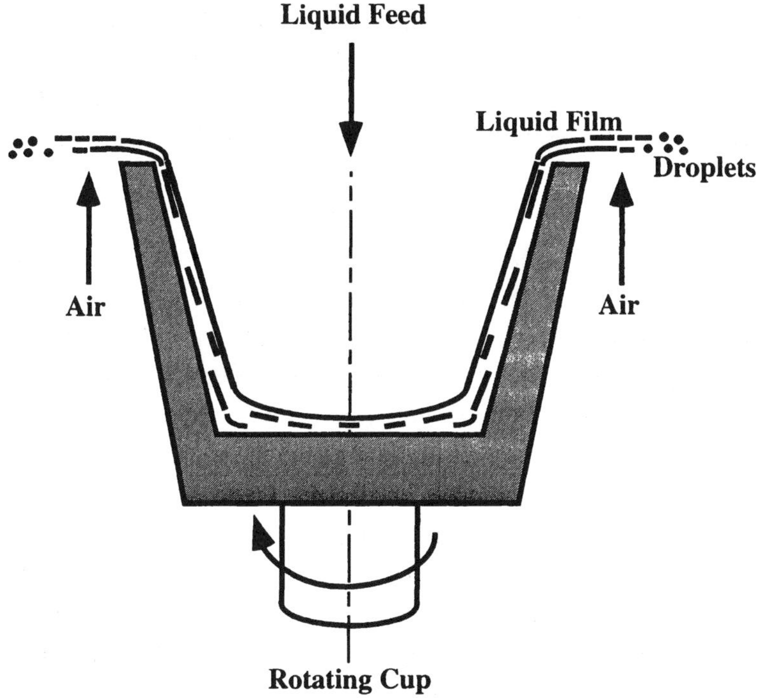

Figure 2.10. Schematic of rotary atomization process.

Generally, atomization quality may be improved by increasing the rotational speed of disk/cup, decreasing the liquid flow rate, and/or serrating the edge of the rotating element. A high liquid flow rate causes the liquid to leave the edge in the form of thin ligaments or sheets that eventually break up into droplets of various sizes, leading to wide size distribution. One way to increase the liquid flow rate is to increase the disk/cup diameter proportionally. This approach may require a large scale disk/cup that can cause difficulties in practical operation. Another way to increase the capacity is to design a system of stacked spinners with a large diameter.[88] Ligament formation can be made to produce much finer droplets by applying an electrostatic field between the rotating element and earth.[115] The thickness and uniformity of a liquid sheet can be controlled by varying the liquid flow rate and/or the rotational speed. In addition to these two process parameters, the generation of a uniform sheet thickness and thereby a uniform droplet size requires a large centrifugal force, a vibrationless rotation, a constant liquid flow rate, and a smooth disk/cup surface. Atomization quality for liquids of lower viscosity is usually higher than for those with higher viscosity.

The slippage between the liquid and the disk is a problem associated with flat disk atomizers. It may lead to an ejection velocity of the liquid from the disk edge much lower than the disk peripheral speed. This becomes a main drawback of flat disk atomizers, particularly at high rotational speeds. Therefore, conventional rotary disk atomization can generate a narrow spectrum of droplet sizes only when operating in the ligament mode at low liquid flow rates. In commercial atomizers, straight or curved radial vanes are used to prevent transverse flows of the liquid over the surface and to guide the liquid to the periphery. Thus, no slippage occurs after the liquid flows into the vanes, and the liquid ejection velocity may approach the disk peripheral speed. Another operation problem that seems to occur most often in spinning disk atomization is the production of satellite droplets. However, the satellite droplets can be separated dynamically from the larger primary droplets. This is usually done by using a separate air flow near the disk periphery. The velocity of the air flow is set to such a level that the satellite droplets can be dragged into the flow for collection and/or recirculation while the larger primary droplets can fly across it. Although this arrangement can eliminate the satellite droplets to a certain extent, it might be difficult to design such a secondary flow that would operate properly under variable practical conditions, particularly in agricultural aviation applications.[88]

An attractive feature of rotary atomization is the nearly uniform droplets produced with small disks at high rotational speeds and low liquid flow rates. Therefore, rotary atomization is probably the most generally successful method for producing moderately monodisperse sprays over a wide range of droplet sizes. The mean droplet size depends primarily on the disk rotational speed and diameter, the liquid density and surface tension, and to a lesser extent the liquid flow rate and viscosity, as well as the geometry parameters of vanes. For convenience, altering the rotational speed is a commonly used method of varying the droplet size. The average droplet size ranges approximately from 10 to over 200 μm. Droplet sizes from submicrometers up to 3000 μm have been produced for various liquids using spinning disks with different rotational speeds, disk diameters, and liquid flow rates. Geometric standard deviations of 1.05–1.72 have been reported.[88] Another useful feature is the flexibility in operation due to the fact that rotary atomizers allow independent variations of atomization quality, rotational speed, and liquid flow rate. Moreover, rotary atomization processes are extremely versatile and have been successfully applied to the atomization of liquids with a wide range of viscosities. The processes have a broad range of industrial applications such as spray drying and aerial distribution of pesticides, and have been widely used in the chemical processing industry for a century. As more reliable designs and operations are developed, rotary atomization will find increasing applications in spray drying industry.

One of the special rotary atomizers worth mentioning is the *windmill type* atomizer. In this atomizer, radial cuts are made at the periphery of a disk and the tips of segments are twisted, so that the disk is actually converted into a windmill that can rotate rapidly when exposed to an air flow at aircraft flight speed. The windmill type atomizer has been demonstrated[117] to be an ideal rotary atomizer for generating a narrow spectrum of droplet sizes in the range most suitable for aerial applications of pesticides at relatively high liquid flow rates.

2.1.6 Effervescent Atomization

In *flashing liquid jet* or *flashing injection atomization*,[118] a high-pressure liquid with dissolved gas flashes, shattering the liquid into small droplets in a fairly regular spray pattern. Flashing even small quantities of dissolved gas (<15% in mole fraction) can significantly improve atomization. To realize the beneficial effects, however, an expansion chamber is needed upstream of the discharge orifice. The bubble growth rate in dissolved gas systems is usually low, posing a limitation on the practical application of flashing injection atomization with dissolved gas systems.

To overcome the basic problems of flashing injection atomization with dissolved gas systems, an *effervescent atomization* method has been developed by Lefebvre et al.[1] In effervescent atomization, air or gas is injected through a plain-orifice nozzle into a bulk liquid within an atomizer at a location upstream of the discharge orifice. Unlike the above-described two-fluid atomization, the gas used in effervescent atomization is not intended to provide kinetic energy for breaking up the liquid stream, nor to be dissolved in the liquid as in the flashing injection atomization, but instead to penetrate into the liquid. The injection velocity is low and the pressure differential between the gas and the liquid is very small. The gas is supplied at nearly the same pressure as that of the liquid, just to allow the gas to penetrate the flowing liquid. The gas then forms bubbles in the liquid, and a two-phase flow condition is produced at the discharge orifice. During discharging, the liquid is squeezed by the gas bubbles into thin ligaments and/or shreds while the gas bubbles explode, an effect similar to that in flashing injection.[118] The ligaments and/or shreds are further shattered into droplets by the explosion of numerous small bubbles while leaving the discharge orifice.

Generally, an increase in the gas amount used may lead to an increase in the number of bubbles and a decrease in droplet size produced. However, atomization quality is very good even at very low injection pressures and low gas flow rates. Good atomization can be achieved at much lower liquid injection pressures than those typically used in pressure atomization. Mean droplet size is comparable to that produced in air-assist atomization for the same gas to liquid ratio. The large holes and passages in effervescent atomizers largely reduce the occurrence of nozzle plugging. This makes the effervescent atomizers ideal for combustion devices burning slurry and residual fuels. The presence of air bubbles is also beneficial to reducing soot formation and exhaust smoke in combustion applications. Moreover, this type of atomizer designs is simple, reliable, and of low cost.

2.1.7 Electrostatic Atomization

In *electrostatic atomization*, an electrical potential is applied between a liquid to be atomized and an electrode placed in the spray at a certain distance from liquid discharge nozzle. As a result of the mutual repulsion of like charges accumulated on the liquid surface, the surface becomes unstable and disrupts when the pressure due to the electrostatic forces exceeds the surface tension forces of the liquid. Droplets will be generated continuously if the electrical potential is maintained above a critical value consistent with liquid flow rate. Both DC and AC systems have been employed to provide high electrical potentials for generating fine droplets. Many configurations of electrode have been developed, such as hypodermic needles, sintered bronze filters, and cones.

A number of investigations have been made to explain the droplet formation mechanisms associated with electrostatic atomization.[119][120] It has been hypothesized that the dispersion of a liquid by electrostatic atomization occurs *via* the detachment of a single droplet from the capillary tip of the liquid. However, this mechanism has not been proven experimentally. Due to the complex physics involved, a generic theoretical model has not yet been established.

Generally, the droplet size generated in electrostatic atomization is a function of applied electrical potential, electrode size and configuration, liquid flow rate, liquid nozzle diameter, and liquid properties such as surface tension, dielectric constant and electrical conductivity.[121]-[124] When a low electrical potential is applied to a liquid, a stream of relatively uniform droplets will form below the liquid discharge nozzle. As the applied electrical potential is increased, the droplets produced become smaller, and the liquid velocity and droplet production rate both increase, with concomitant shortening of the distance between adjacent droplets.[121] With further increase in the electrical potential, the stream of droplets turns into a continuous liquid stream or jet issuing from the nozzle. When the electrical potential is increased beyond a certain value, the liquid no longer forms a jet, but instead disintegrates spontaneously into a mist of fine droplets at the nozzle exit.[121] For certain liquids, uniform droplets can be produced by electrostatic atomization. However, monodisperse sprays can not be obtained with a negative potential on the liquid. In some electrostatic atomization experiments of water, sugar solutions, lubricating oil and alcohol, it was observed that there is an upper limit of liquid electrical conductivity, beyond which atomization can not be achieved. Fine droplets can be generated for liquids of relatively low electrical conductivity, such as water, alcohol and dibutyl phthalate. However, some organic liquids of low dielectric constants such as benzene and carbon tetrachloride were found difficult to disperse using electrostatic atomization method.

Electrostatic atomization covers a variety of different processes with respect to electrical potential, liquid flow rate and droplet size. The electrical potential required to disintegrate a liquid depends on the flow rate and electrical properties of the liquid. Generally, however, the liquid flow rates in electrostatic atomization are very low. This drawback has limited its practical applications to electrostatic painting and non-impact printing. The process used in industrial electrostatic paintings involves a liquid flow rate of up to 3 l/min with a single nozzle and an electrical voltage on the order of 100 kV.[88] Painting surfaces are normally used as one electrode. Liquid paint is usually pressurized and fed into the atomizer. Droplet size distributions produced under these conditions are rather broadly dispersed. If a liquid is fed into the atomizer at the atmospheric pressure, the liquid flow rate is usually very low and thus the required electrical voltage is on the order of 10 kV. The average droplet sizes produced may range from 0.1 to 1000 μm with a relatively narrow size distribution.[88] In electrostatic atomization for manufacturing printed circuits in the electronic industry, liquid materials (usually molten metals) are atomized and deposited onto circuit boards normally under vacuum conditions. The liquid flow rate is extremely low (up to 6.7×10^{-5} l/min) and, correspondingly, the applied electrical voltage is typically on the order of 10 kV. The droplet sizes produced may range from 300 to 600 μm.[88] Significant work on electrostatic atomization has also been done for oil burner applications. In some designs, such as the spray triode invented by Kelly,[123] the electrostatic atomizer appears to have potential capability of handling the high flow rates used in common combustion devices. Detailed descriptions of experiment, theory and industrial applications of electrostatic atomization have been given by Okuda and Kelly.[125]

It should be noted that, to produce a reasonably monodisperse spray, the liquid flow rate should be maintained at an extremely low level, and thus the scaling up of such devices may pose some difficulties. It is also rather difficult to assess the liquid flow rate that can be achieved due to few quantitative studies and lack of comprehensive understanding of the underlying principles. Another drawback of the electrostatic atomization technique is that both the production and properties of droplets are significantly dependent on the electrical properties of the liquid, limiting the type of liquids that can be successfully atomized.

Recently, Sato et al.[126] proposed a new method for the production of uniformly sized insulating liquid droplets (such as kerosene or plastic monomer) in immiscible liquid media (distilled water) by means of an applied convergent electric field generated using AC or pulsed voltage. Video imaging and photographing of the disintegration of the liquid column showed that kerosene droplets with an essentially uniform diameter ranging from 100 to 250 μm were produced synchronously with the applied AC frequency using a nozzle of 100 μm in diameter. The droplet size can be controlled by varying the AC frequency, nozzle diameter, liquid flow rate, and velocity ratio between the oil-phase and co-flowing water. With increasing flow rate of the co-flowing liquid and increasing synchronous frequency, droplet size decreases due to the elongation of the liquid jet. The disintegration mechanism is most likely the forced oscillation of the liquid jet stimulated by each cycle change of the applied voltage.

2.1.8 Vibration Atomization

The breakup of a liquid into droplets can be brought about by periodical vibrations of the liquid. For example, droplets can be generated by periodically vibrating a liquid jet or liquid reservoir. The simplest form of *vibration atomization* is the vibrating capillary droplet generation that was originally developed to study the collision and coalescence of small water droplets. This simple droplet generator consists of a hypodermic needle vibrating at its resonant frequency and can generate a stream of uniform droplets. The size and frequency of droplet generation are dependent on the needle diameter, resonant frequency, liquid flow rate, and amplitude of oscillation of the needle tip.

The technique of periodical vibrations for droplet generation is based on the instability of a liquid jet emerging from a capillary tube or an orifice. If a liquid jet is emitted from an orifice under pressure, the jet is by nature unstable and will eventually disintegrate into droplets by the action of internal and/or external forces depending on the orifice diameter as well as flow conditions and properties of the liquid and surrounding medium. The disintegration of a liquid jet into fairly uniform droplets may be achieved by applying periodical vibrations of proper amplitude and frequency to the jet. Vibrations may be generated by using *(a)* a piezoelectric transducer, *(b)* an acoustic vibrator, or *(c)* a direct mechanical means. Accordingly, vibration atomization may be loosely divided into three categories: *(a)* ultrasonic atomization, *(b)* acoustic atomization, and *(c)* mechanical vibration atomization. Each of the three categories differs not only in design but also in droplet size range that can be generated.

In *ultrasonic atomization*, a liquid droplet is produced when powerful high frequency sound waves are focused onto the liquid. A liquid may be present in large volume and contained in a reservoir. A certain form of concave reflector such as a curved acoustic transducer made of barium titanate may be placed at the bottom of the bulk liquid to generate high frequency sound waves that propagate upwards into the bulk liquid. The intensive waves may cause the formation and subsequent collapse of cavities. The liquid over the acoustic transducer may also form a wavy layer. If the strength of the waves is sufficiently high to overcome the surface tension forces of the liquid, droplets will eject from the ripple crests on the top surface of the bulk liquid. Alternatively, a liquid may be introduced onto a rapidly vibrating solid surface over which it spreads into a thin film while capillary waves form in the film. When the amplitude of the vibrating surface increases to a level at which the wave crests in the film become unstable and collapse, the liquid film breaks up and ejects away from the surface into a mist of droplets.[127] The resultant droplet size is dependent on the ripple wavelength that in turn is determined by the vibration frequency.

The disintegration of a liquid jet/stream emerging from an orifice can be precisely controlled by periodically vibrating the orifice. The periodical vibrations can be achieved by implanting the orifice into a disk made of an electrostrictive material such as a piezoelectric crystal and applying electrical signals to the piezoelectric crystal. This is the design concept of ultrasonic atomization with an electrostrictive disk type generator.

An ultrasonic atomizer (Fig. 2.11) typically consists of a pair of piezoelectric transducer elements sandwiched between a pair of titanium horn sections. The horn, i.e., velocity transformer, is used to increase the amplitude of the vibrations at the atomizing surface. The common contact plane of the two piezoelectric transducer disks constitutes one of two electrical input terminals for high frequency electrical signals. The metal horn forms the other terminal. The disks will either expand or contract simultaneously, depending on the polarity of the input signals. The cyclic expansion and contraction generate a propagation of traveling pressure waves longitudinally outward in both directions along the nozzle axis, at the same frequency as the input signals. Normally, the length of the ultrasonic nozzle is designed to be equivalent to one wavelength of the pressure waves. Thus, a standing wave pattern is generated, a necessary condition for resonant operation. The interface between the two transducer disks is located at a nodal plane. Antinodes are located at both free ends, where the amplitude of the standing wave is a maximum. The amplitude at the atomizing end is much larger than at the other end due to the placement of a diameter step transition, a well-known method for increasing vibrational amplitude. A liquid is delivered through an axial tube inside the acoustically resonant device to the surface at the atomizing end, where the atomization of the liquid occurs at the maximum vibrational amplitude. Descriptions of a class of widely applicable ultrasonic nozzles have been given by Berger.[128]

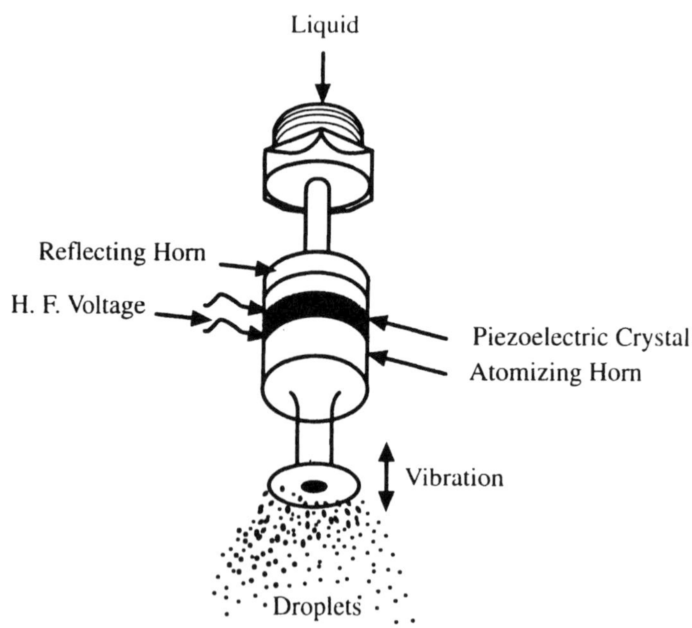

Figure 2.11. Schematic of an ultrasonic atomizer.

The droplet size generated by an ultrasonic atomizer depends primarily on the signal frequency and liquid flow rate. Ultrasonic atomizers may be of low-frequency (between 20 kHz and 100 kHz) or high-frequency (up to 3 MHz). More power can be supplied at the lower frequencies. The orifice diameter normally ranges from 3 to 20 μm.[88] The horn diameter is usually designed to be less than one quarter of the wavelength of vibration waves. This restricts the liquid flow rate. High-frequency ultrasonic atomizers are characterized by low throughput and fine droplet sizes, and are frequently used as nebulizers for inhalation in medical applications. The droplet sizes produced in the frequency range of 10 to 1000 kHz are approximately 3–50 μm.[88] Generally, the droplet sizes in ultrasonic atomization may range from submicrometers up to a few hundred micrometers.

Originally, ultrasonic atomizers were developed for use in small boilers for domestic heating and the emphasis of applications has been in the combustion area since then. Over the past decade, however, ultrasonic atomization has been extensively developed and applied to a variety of industrial areas such as humidification, medication, pharmaceutical coatings, semiconductor processing, spray drying, and vaporization of volatile anaesthetic agents.[19] One of the most useful features of ultrasonic atomization is its low spray rate due primarily to the low amplitudes of vibrations that ultrasonic transducers generate.[88] For example, a combination of frequency of 700 kHz and droplet size of 10 μm yields a flow rate of 2.2×10^{-5} l/min, while a combination of 10 kHz and 200 μm results in a flow rate of 2.5×10^{-3} l/min.[88] Thus, the droplets generated can be readily entrained and conveyed in a moving stream as a uniform mist. This feature makes ultrasonic atomization particularly suitable for applications such as coating, humidification, moisturizing, and spray drying. In addition, ultrasonic atomizers are capable of generating very fine droplets (1–5 μm)[129] at extremely low flow rates (< 0.1 l/min at a frequency of 55 kHz). This feature is highly desired in some pharmaceutical and lubrication processes. However, this also makes it difficult to successfully handle the high flow rates prevailing in most engine and furnace applications.

One of the approaches for increasing the liquid flow rate is to increase both the vibration frequency and liquid flow velocity, while keeping the wavelength according to Rayleigh's linear theory on the instability of a liquid jet,[37] as will be discussed in the following chapter. Another possible solution is to combine the effect of ultrasonic atomization with that of whistle atomization.[88] In this approach, the amplitude of signals produced initially by an ultrasonic transducer is further amplified by means of the resonant effects created by the hollow space of horn shape. While this design allows the capacity to be increased substantially, it will restrict the operating frequency of the transducer to one value. Since the droplet size generated by ultrasonic atomization is dependent on the transducer frequency, this arrangement is limited to producing one droplet size. Thus, a series of different sets are necessary to cover a required size range. Other designs may be used to increase the liquid flow rate. For example, many atomizers may be installed in parallel. This arrangement, however, may cause mutual interference of vibrations produced by individual transducers, leading to polydisperse droplets. One way to resolve the problem is to install each transducer in a separate chamber. Alternatively, multiple tubes/nozzles can be mounted in an atomization system and vibrated at a common frequency, rather than installing many individual units. It may also be feasible to design a large perforated plate mounted on a shallow liquid reservoir whose wall or bottom can be vibrated.

The ultrasonic atomizers made by the Delavan company and used in combustion, spray etching, and spray drying may operate at a frequency of 50 kHz, and generate droplets of 30 to 60 μm. Larger ultrasonic atomization devices made by Lechler GmBH, Germany, have a length of 14 cm and can generate water droplets of 30 μm at a flow rate of about 1 kg/min. The capability of various ultrasonic atomizers to atomize various liquids has been evaluated by Berger,[128] and the performance of a variety of ultrasonic atomizers has been examined by many investigators.[128]-[137] Designs of new piezoelectric droplet generators and devices[138]-[141] have been developed and evaluated, including those for liquid metal atomization.[142]-[144]

Instead of using electrical signals to an electrostrictive orifice in ultrasonic atomization, periodical vibrations can also be imparted onto a tube in the form of acoustic waves using audio signals. This is the design concept of acoustic atomization with vibrating tube using an audio speaker. In *acoustic atomization*, an oscillator generates signals that are amplified and transmitted to a speaker. A liquid jet is introduced through a tube and vibrations are transmitted to the tube *via* a metal rod that is connected to the speaker diaphragm. The liquid jet is then disintegrated into droplets of uniform sizes. The diameter of the capillary tube ranges typically from 50 to 1500 μm,[88] much larger than the orifice diameter used in ultrasonic atomization. The resultant droplet diameter lies between 100 and 3000 μm. The signal frequency is in the range of 0.3 to 30 kHz. Exceptionally good uniformity and predictability of droplet sizes have been reported by investigators.

In *mechanical vibration atomization*, periodical vibrations are applied to a liquid jet by using some mechanical means. For example, a fine whisker can be used to dip periodically into a liquid tube above a liquid reservoir.[88] The whisker is connected to a flat spring of silicon iron that is vibrated by an electromagnetic field. An AC current of 50 Hz is used to create the magnetic field. The whisker has a round shaped tip of 0.015 mm in diameter. When the whisker emerges above the liquid surface, a liquid thread is formed between the whisker tip and the liquid surface. As the whisker moves up further, this thread separates from both the whisker tip and the liquid surface under the action of gravitational and surface tension forces and/or by blowing with a traverse air flow, forming a droplet. The droplet is subsequently blown away by the air flow. The droplet size produced is dependent on the immersion depth of the whisker as well as the liquid surface tension and viscosity. In an alternative approach, a periodically rotating needle is employed that breaks up the liquid jet emerging from a capillary tube.

Disintegration of liquids can also be achieved by applying periodical virations to a liquid reservoir rather than to an orifice or tube. The reservoir wall or bottom may be made of piezoelectric crystals so that a pressurized liquid contained in the reservoir can be squirted out through an orifice or multiple orifices into droplets. This type of technique may be considered to be similar to those of vibrating tube/orifice in that individual droplets are periodically produced by externally controlled disturbances, although there are distinct differences between them in configuration and design.

There may be many other types of periodical vibrations for liquid disintegration or dispersion in various applications. The features, operation parameters, performance and applications of various experimental and commercially available ultrasonic and acoustic atomizers have been described and compared by Topp and Eisenklam.[129]

Vibration atomization techniques can generate droplets of a wide range of sizes (approximately 1–1000 μm). Droplets smaller than 50 μm can be best produced using an electrostrictive transducer such as piezoelectric crystals. Periodic acoustic signals generated by sound-speaker type vibrations are well suited for producing droplets larger than 50 μm. These two distinctively separate ranges are due primarily to the difference in the vibration frequencies used in the two techniques.

Droplets produced using periodic vibrations generally exhibit excellent monodispersity whether the vibrations are generated by a piezoelectric crystal, a sound speaker, or some sort of mechanical means. It has been reported that in atomization using an earphone-like vibrator, only 0.2–1.5% of the droplets produced have sizes different from the rest.[88] The converted standard deviation on a weight basis is 0.0008–0.0022; that is equivalent to a geometric standard deviation of 1.001–1.002. Geometric standard deviations of 1.01 and 1.005–1.08 have been achieved in atomization processes using a piezoelectric transducer and a whisker, respectively.[88] The droplets generated by the vibrating tube method at low flow rates appear to be more monodisperse than those produced by ultrasonic atomization techniques.

In vibration atomization, each individual droplet is produced one at a time by means of a periodic disturbance. Therefore, the resultant droplet size is not greatly dependent on the liquid properties. For a given liquid, the droplet size is practically determined only by liquid flow rate and vibration frequency.

Some experiments revealed that monodisperse sprays can be generated in the wavelength range of $3.5d_0$ to $7d_0$,[88] where d_0 is the liquid jet diameter. There always exists a certain minimum threshold frequency below which the droplets generated are not uniform. This threshold frequency was found to be 0.7 and 0.4 times the optimum frequency for high and low viscosity liquids, respectively. The amplitude of disturbances was found to have very little effect on the monodispersity of droplets. The optimum wavelength range is very narrow while uniform droplets can be generated over amplitudes of several orders of magnitude. However, high amplitudes may be necessary for highly viscous liquids. Since the amplitude of pulses can be readily varied for a standard pulse generator, it is an operating condition rather than design criterion.

2.1.9 Whistle Atomization

Whistle atomization is also known as acoustic or ultrasonic gas atomization. In a whistle atomizer,[88][131] a liquid jet is disintegrated into droplets by directing high-pressure gas on it, as schematically depicted in Fig. 2.12. The focusing gas flows create strong sound waves inside the nozzle, and hence it is called *whistle* or *stem-cavity* atomizer. This type of atomizer operates typically at a sound frequency of about 10 kHz and may generate droplets of about 50 μm at flow rates up to 75 l/min. In the Hartmann-whistle acoustic atomizer, a converging-diverging nozzle is used to generate a supersonic gas jet. Acoustic oscillations in shock wave pattern are produced when the jet impacts an open-ended chamber. This type of atomizer operates at an ultrasonic frequency (>20 kHz) and can generate droplets of about 10 μm. However, the droplet size distribution is quite broad with a standard deviation of about 2.5, and the droplet size cannot be readily controlled unless the nozzle dimension is changed. These are some of the drawbacks of whistle-type atomizers. The principle of Hartmann-whistle acoustic atomization can be applied to different design configurations in which a supersonic gas jet impacts a properly shaped solid surface and the liquid to be atomized is introduced into the shock wave zone. A design variation of the Hartmann atomizer has been developed for the atomization of liquid metal, as discussed in the subsequent section.

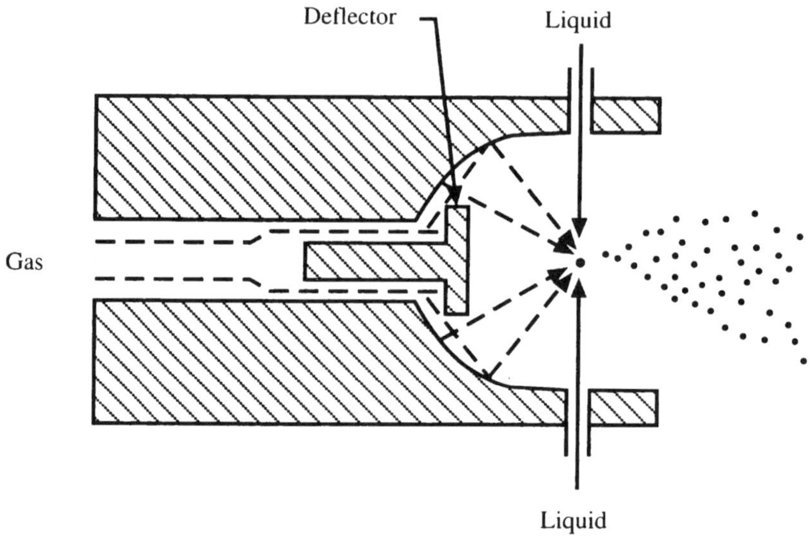

Figure 2.12. Schematic of a whistle atomizer. *(Reprinted from Ref. 88 with permission.)*

The Hartmann-whistle acoustic atomizer requires gas pressures in excess of 0.3 MPa and air to liquid mass ratios greater than 0.2. Flow rates as large as 1.7 kg/min (water or oil) can be handled with large atomizers. Water droplets as fine as 7 μm can be generated at a flow rate of 0.125 kg/min, a gas pressure of 0.33 MPa, and a spray angle of larger than 90° which is much larger than those in two-fluid atomization.[5] Liquid supply pressure is strongly coupled with the gas pressure. An increase in the gas pressure from 0.1 to 0.4 MPa, for example, requires an increase in the liquid supply pressure from 0.1 to 0.17 MPa for a constant water flow rate of 0.17 kg/min.[5]

Precise mechanisms of atomization in these whistle-type devices are still ambiguous, and the behavior and role of the shock waves are less clear. Plausibly, the shock waves may have certain beneficial effects on atomization. However, systematic studies[131] indicated that the sound field is not a critical factor in this type of atomization process. The degree of atomization is much more sensitive to gas pressure than resonant frequency.[115] These seem to imply that the liquid breakup is caused mainly by the aerodynamic forces of the gas, similarly to the interaction mechanisms prevailing in two-fluid atomization. It was also suspected that all the whistle atomizers operate simply on a principle similar to that in air-assist type of atomizers,[129] and the whistle atomizers can be considered to extend the range of normal internal-mixing twin-fluid atomizers without resorting to very small orifices.[5] In addition, it was found that the degree of atomization is almost independent of liquid viscosity.[115] This is a promising feature for the atomization of heavy fuel oils.

2.1.10 Vaporization-Condensation Technique

Vaporization-condensation technique involves atomization of a liquid, vaporization of the atomized droplets, and subsequent condensation of the vapor.[88] The original droplets are normally generated by pressure atomization. Therefore, the spray produced initially is typically polydisperse. The spray is then vaporized by heating it to above the boiling point of the liquid, usually using a combustor or an electrical heater. The vapor is subsequently mixed with a stream of hot air containing a regulated number of condensation nuclei. The mixture of the air and vapor passes through a section in which it is slowly cooled, becoming supersaturated and condensing uniformly upon the nuclei to form uniform droplets.

Vaporization-condensation technique has been applied to spraying certain insecticide chemicals. A pulse jet or a reciprocating engine has been used as a heating source. Liquid flow rates are up to 40 gallons per hour. The technique can also be used to produce a cloud of smoke for military applications. The desired droplet size is in the same range as the light wavelength for obscuration, i.e., 0.4 to 0.8 μm. Generally, the droplets produced by vaporization-condensation technique are in the size range of aerosols, i.e., smaller than 1 μm.[88]

When a vapor jet emerges into ambient air, a large amount of air is entrained into the jet due to drag and turbulent mixing. In the vapor-air mixing region, the vapor condenses on the nuclei that proceed to grow into droplets. As the distance from the vapor nozzle increases, the droplets encounter more air and the vapor pressure decreases until the growth rates of the droplets become negligible. The mean droplet size produced is primarily determined by the condensation rates that in turn are controlled by the heat content and velocity of the vapor, the liquid-air ratio, and the number of nuclei. The latter is generally known to be an important factor governing the droplet size in the condensation process. Since the droplet formation is based on the vapor condensation on nuclei, the droplet size is somewhat difficult to control. The process parameters that need to be carefully controlled include the number of nuclei, vapor temperature and pressure, and flow rates of vapor and air.

The droplet size distribution produced by vaporization-condensation technique is strongly dependent on the chemical composition and properties of the liquid. If well controlled on a small scale, vaporization-condensation technique can produce moderately monodisperse sprays with geometric standard deviations ranging from about 1.2 to 1.8.[88]

In an evaluation of various techniques for droplet generation,[88] periodic vibration of liquid jet, spinning disk and ultrasonic atomization techniques have been rated as the most appropriate methods for producing monodisperse sprays. These techniques were found to be very effective and appeared promising for refinement, modification and development into forms suitable for agricultural aviation applications in terms of the range of mean droplet size, monodispersity and capability of operating over a range of liquid properties. However, all of these techniques require some extension of their flow rate capabilities to match those required in agricultural aviation applications. Electrostatic atomization and vaporization-condensation techniques were found not to be adequate for agricultural aviation applications despite of their capabilities of generating nearly uniform droplets.

It should be noted that some problems may arise in the techniques or devices for producing monodisperse or near-monodisperse sprays. One of the problems is droplet coalescence. Initially uniform droplets may coalesce rapidly to create doublets or triplets, particularly in a dense and turbulent spray, deteriorating the monodispersity of the droplets. This problem may be lessened by using appropriate dispersion air around the spray.[88] Another problem is non-spherical droplet shapes that make estimations of monodispersity difficult.

2.1.11 Other Atomization Methods

In addition to the above-discussed techniques, some other atomization methods have also been developed for specific applications. These include, for example, solution atomization, liquefied gas atomization, and spark-discharge atomization,[88] etc.

Solution atomization involves dissolution of a relatively non-volatile liquid (solute) in a volatile solvent and atomization of the solution. During the atomization, the solvent material will evaporate in surrounding medium (air), leaving only nucleus droplets of the nonvolatile solute. The final droplet size is a function of the initial droplet size, the mass concentration of the solute, and the density ratio of the solution to the solute. The limitation of this technique lies in that it requires the dissolution of the liquid to be dispersed in a solvent.

A similar technique has been applied to the generation of monodisperse suspensions in water. This type of method was first used in medical field and then widely used to spray monodisperse solid particles such as polystyrene latex particles. Aerosols of solutes have also been produced by atomizing solutions of salt, sugar or methylene blue dye dissolved in water. In practical operations, a low concentration of solid particles in a solvent is recommended in order to avoid possible agglomeration of suspensions in the solvent.

Sprays of fine droplets can be generated by first mixing a liquid with liquefied gas under pressure and then expanding the mixture through a nozzle. This technique, referred to as *liquefied gas atomization*, has been used in many applications such as commercial aerosol cans. The mean droplet size generated with this technique is very small. In very few systematic studies, the measured droplet size distribution was found rather widely spread.[88] It is not clear, however, how the liquid amount, pressure, and nozzle design affect the mean droplet size and size distribution.

Fine droplets can also be generated by introducing electrodes into a liquid and applying a high potential across the electrodes. The disintegration of the liquid into droplets is due to spark discharge. The technique is therefore termed *spark-discharge atomization*. Small scale experiments have been performed to investigate the phenomena associated with the atomization process primarily for domestic burner application. The produced droplets exhibited a mass median diameter of about 100 μm, and the droplet size distribution was found quite widely dispersed.[88] No reliable or proven results seem to be available.

Some design concepts for generating uniform droplets have been proposed by Lee et al.[88] These include *(a)* centrifugal type chamber, *(b)* atomization by two opposing air-liquid jets, and *(c)* spinning disk coupled with an ultrasonic field. Some other conceptions include *(d)* rocket nozzle chamber, *(e)* frozen particles, *(f)* rotating brush, and *(g)* periodic vibrations using saw-tooth waves, etc.

2.2.0 ATOMIZATION OF MELTS

Atomization of melts has, in principle, some similarity to the atomization of normal liquids. The atomization processes originally developed for normal liquids, such as swirl jet atomization, two-fluid atomization, centrifugal atomization, effervescent atomization, ultrasonic piezoelectric vibratory atomization, and Hartmann-whistle acoustic atomization, have been deployed, modified, and/or further developed for the atomization of melts. However, water atomization used for melts is not a viable technique for normal liquids. Nevertheless, useful information and insights derived from the atomization of normal liquids, such as the fundamental knowledge of design and performance of atomizers, can be applied to the atomization of melts.

Atomization of melts has many applications and advantages for the production of powders and components compared to conventional techniques. Research and applications of atomization of melts have a fairly short history. Very little work had been done on this subject before the twentieth century. The main obstacle in the development of atomization of melts was a lack of suitable materials and techniques for handling molten metals. Over the past decades, atomization of melts has become the dominant mode for powder production due to its attractive features. First, high powder production rates make the technology viable and competent economically. Second, for certain atomization methods, it is possible to control powder size, size distribution, shape and surface morphology. Third, atomization offers a high degree of flexibility in alloying and control over impurity levels. Presently, it is the only viable fabrication technology for highly alloyed specialty alloy powders. The atomized powders are prealloyed and exhibit composition homogeneity. In addition, the production capacity may range from hundreds of grams to tonnage quantities. By the early 1980's, over 60% by weight of all the powders produced in North America were manufactured by atomization.[4] It was estimated that the worldwide atomization capacity is approaching 10^6 metric tons per year.

Atomization of melts has also been used in advanced materials processing and synthesis of net or near-net shape components without the formation of powders. Thermal spray, plasma spray deposition,[43] spray atomization deposition,[3] centrifugal spray deposition, and spray co-injection deposition are some examples of this category of applications. In thermal spray processes, a solid material in the form of powder, fiber or wire is directed into a plasma or combustion flame where it is rapidly molten into a spray of droplets and accelerated to a high velocity towards a target substrate to form a surface coating or a free-standing component. The powder feed is usually obtained from atomization of melts. Spray atomization deposition generally refers to a material synthesis approach that involves the energetic disintegration of a molten material by high-speed gas jets into micro-sized droplets, followed by the immediate deposition of solidified, semi-solid, and liquid droplets on a substrate to form a coherent, near fully dense material of desired shape. This process has also been termed *spray casting, spray rolling,* or *spray forming*, depending on whether or what post-processing steps follow the deposition. Various shapes, such as disk, tube, strip, and sheet, can be deposited by manipulating the substrate and/or changing the atomizer parameters. Refined microstructure and improved physical/mechanical properties can be achieved as a result of rapid solidification. Therefore, spray atomization deposition has attracted considerable attention as a viable processing alternative for structural materials over the past two decades.

Numerous atomization techniques have evolved for the production of metal/alloy powders or as a step in spray forming processes. Atomization of melts may be achieved by a variety of means such as aerodynamic, hydrodynamic, mechanical, ultrasonic, electrostatic, electromagnetic, or pressure effect, or a combination of some of these effects. Some of the atomization techniques have been extensively developed and applied to commercial productions, including *(a)* two-fluid atomization using gas, water, or oil (i.e., gas atomization, water atomization, oil atomization), *(b)* vacuum atomization, and *(c)* rotating electrode atomization. Two-fluid atomization has been the predominant technology and accounts for over 95% of the atomization capacity worldwide.[4] Some other atomization techniques have been evaluated in laboratory and pilot scale operations and may be considered as near-commercial methods, such as *(a)* ultrasonic gas atomization, *(b)* rotating disk atomization, *(c)* electron beam rotating disk atomization, and *(d)* roller atomization. A number of other atomization techniques do not appear to have established significant commercial applications but are of high scientific interest. These include, for example, *(a)* centrifugal shot casting atomization, *(b)* centrifugal impact atomization, *(c)* spinning cup atomization, *(d)* laser spin atomization, *(e)* Durarc® process, *(f)* vibrating electrode atomization, *(g)* ultrasonic atomization, *(h)* steam atomization, *(i)* electromagnetic and electrostatic atomization, and *(j)* impaction and pressure atomization. These methods demonstrate engineering innovation and may be used to produce high quality powders of a variety of advanced specialty materials. Some of them may be commercialized in the near future. The development of these techniques has not only advanced atomization technology in general but also improved our fundamental understanding of atomization mechanisms. As the need for advanced powder and particulate materials increases, more elaborate atomization techniques will be required.

Similar to the atomization of normal liquids, the basic requirements for atomization quality are a small droplet size and a narrow size distribution. In the atomization of melts, relatively fine droplets can be obtained using some techniques and, as a result, high cooling rates, ranging from 10^2 to 10^6 °C/s, can be achieved which allow the control of powder microstructure to a fine degree. Cooling rates of 10^5 to 10^6 °C/s are readily attainable in substrate quenching of fine liquid droplets using rotating discs, high speed rolls, or other metallic substrate systems. However, many atomization techniques are incapable of producing a narrow size distribution, i.e., the yields of useful powders are low, thus the production costs are high. Therefore, the requirement of a narrow size distribution is very important to reducing the costs, particularly of high alloy materials, and to reproducing microstructures of powders.

In addition to the difference in the size distributions of droplets/particles produced, the overall production costs are apparently the point of differentiation among all atomization techniques under consideration. As the criteria for the evaluation of a specific atomization technique/system, the following factors are of importance: high yields, minimum use of expensive gases, clean droplets/ particles, and high throughput.

In spite of the differences among the various atomization techniques for metallic melts, there are some features in common. First, the metal to be atomized is generally molten by high-frequency induction or resistance heating, fuel firing, or arc melting in a crucible (tundish). Melting equipment and procedures are similar. For reactive metals, induction melting is conducted in vacuum. With rare exceptions, the liquid metal is delivered through a refractory or ceramic nozzle. Refractory-free, clean melting systems used in casting processes (for example, electroslag refining system) may be employed for metal melting in atomization to reduce ceramic inclusions and defects. In the electroslag refining system (ERS), the liquid metal is transferred directly from the ERS pool to the atomization system *via* a ceramic-free copper funnel, i.e., a cold-walled induction guide. Second, in contrast to the atomization of normal liquids, the cooling of liquid metals occurs simultaneously during the atomization of melts, leading to an increase in liquid viscosity and surface tension and eventual solidification of atomized droplets. One of the issues to be considered in the atomizer design is therefore the possible blockage of the pouring system due to solidification and/or buildup of the liquid metal at the metal delivery nozzle tip caused by inappropriate flow pattern (typically recirculation flow) and rapid cooling. Cooling and solidification also affect critically the resultant droplet size distribution and determine, to a large extent, the droplet shape. Third, some post-processing and treatment systems for some atomization techniques may be similar. A comparison of the features of various atomization techniques is given in Tables 2.3, 2.4 and 2.5.

Table 2.3. Comparison of Features of Commercial Atomization Techniques for Melts[4][5]

Method	Droplet Size (μm)	Metal/Alloy	Cooling Rate (°C/s)	Throughput (kg/min)	Capacity (Metric Tons)	Advantage	Limitation
Gas Atomization	50–300 Standard deviation: 1.9–2.5 <10–50 at high gas pressures with close-coupled atomizer	Solder materials, Precious metals, Cu, Fe, Al, Mg, Co, Ti, Zn, Al-6Cr-3Fe-2Zr alloy, Low-alloy steels, High speed steels, Stainless steels, Specialty alloys, Ni-base superalloys, Alumina, Intermetallics	10^3–10^5	1–70	~6	Spherical & smooth particles, Cleanliness, Rapidly-solidified structures, Acceptable production rates	High cost, Low volume, Low energy efficiency (EE), Gas-filled porosity in particles
Ultrasonic Gas Atomization	<30	Low melting point alloys, Fe, stainless steels, Ni, Co alloys, Carbon steels	10^4–10^6	Annular: 2.4–15 Linear: much higher	Steel strip: ~2.4 Al strip: ~0.45	Fine smooth droplets, High gas efficiency	High cost
Water Atomization	~100 Standard deviation: 1.7–2.4	Fe, Cu, Cu alloys, Stainless steels, Tool steels, Ni alloys, Precious metals	10^4–10^6	≤400	15–60 Steel & iron 0.5–5 Stainless steel & Cu alloys	High volume, Low cost	Powder purity & shape,Low EE
Oil Atomization	50–70 Broad size distribution	High speed steels, Bearing steels	10^3–10^5	> 125	~3	Low oxidation	C pick-up, Medium EE, Shape
Vacuum Atomization	40–70 Broad size distribution	Ni, Co superalloys, Al, Cu, Fe alloys	~10^2	5–100	~0.9	Spherical, smooth,clean particles	Difficult to control, Medium EE
Rotating Electrode Atomization (REP, PREP)	~20 Standard deviation: 1.3–1.5	Armco Fe, Cu, Al, Zn, Co-Cr, Ti, Zr, Ni alloys, Low carbon steels	≤10^2	1–10	~0.04	Spherical, very smooth, ultraclean particles, Relatively high EE	High cost, Low capacity and volume, Relatively Coarse particles

Table 2.4. Comparison of Features of Near-Commercial Atomization Techniques for Melts[4][5]

Method	Droplet Size (μm)	Metal/Alloy	Cooling Rate (°C/s)	Throughput (kg/min)	Capacity (Metric Tons)	Advantage	Limitation
Rotating Disk Atomization (RSR)	~60 Standard deviation: 1.4–1.6	Ni-superalloys, Al, Be, Ti, Mo, Cu alloys, Steels, Silicide	10^5–10^7	3–18	≤0.9	Spherical, clean, smooth particles, Narrow size distribution, High EE	Relatively low capacity & throughput
Electron Beam Rotating Disk Atomization (EBRD)	<700 Narrow size distribution	Refractory, reactive metals and alloys such as Ti	$>10^3$	~5.5	~0.5	Relatively high EE	Coarse (flaky) particles, Relatively low capacity & throughput
Roller Atomization	220–680	Pb, Sn, Cu, Al, Steels	Relatively low	~4	—	Low energy requirements	Irregular, coarse particles

Table 2.5. Comparison of Features of Other Atomization Methods for Melts[4][5]

Method	Droplet Size (μm)	Metal/Alloy	Cooling Rate (°C/s)	Through-put (kg/min)	Capacity (Metric Tons)	Advantage	Limitation
Centrifugal Shot Casting Atomization (CSC)	150–1000 Standard deviation: 1.3–1.4	Fe, Co, Ni, Ti alloys, Alumina, Uranium monocarbide	10^2–10^5	≤1	—	Narrow size distribution	Coarse particles, Low EE
Centrifugal Impact Atomization	10–1000	Si,Stainless steels,Ni, Cu, $Fe_{81.5}B_{14.5}Si_3C_1$, Raney® type alloys		0.5–0.9	0.5–0.9 $\times 10^{-3}$	Fine particles, Rapid quenching	Low volume
Spinning Cup Atomization (RSC)	<10–300 Standard deviation: 1.5–1.7	Sn, Pb, Al, Cu, Zn alloys, Stainless and High speed tool steels, Superalloys	10^5–10^6	—	0.5–1.4$\times 10^{-3}$	Fine, clean, spherical particles, Narrow size distribution Small facility	Relatively low volume & productivity
Laser Spin Atomization	~100	Ni-Al-Mo, Ti-6Al-4V	~10^5	—	—	Spherical, homogeneous, clean particles	Low volume & productivity, Poor EE, Cost
Durarc® Process	≥250	Ti and Zr alloys	—	—	—	Spherical, clean particles, Less limitation on starting materials	Low volume
Vibrating Electrode Atomization (VEP)	300–500	Mild steel, Cr-Ni steel, Cu-Ni alloy, W	—	~0.2	—	Spherical, high-purity particles, Simple	Low volume & productivity

Table 2.5. *(Cont'd.)*

Method	Droplet Size (μm)	Metal/Alloy	Cooling Rate (°C/s)	Through-put (kg/min)	Capacity (Metric Tons)	Advantage	Limitation
Ultrasonic Atomization	30–50 USWA: >100 Medium size distribution	Solder materials, Welding electrodes, Ag-, Cu-base alloys, Dispersion-hardened alloys	USWA: ~10^7	0.02–0.8	Low	Spherical & high quality particles, Low cost, Small unit, High EE	Low volume & productivity
Steam Atomization	Coarse	Carbon-, low-alloy-, stainless steels, Co-, Ni-base superalloys	~10^3	—	—	Between gas and water atomization	Irregular particle shape
Pressure Atomization	50–500 Standard deviation: 1.4	Sn, Pb, Sb, Bi, In, Mg, Al, Zn	—	~17	—	Narrow size distribution High EE	Limited to low melting point metals
Vibrating-Orifice Atomization	10^3–10^4 Standard deviation: <1.05	Ca, Zn, Al	Low	Very low	Low	Very narrow size distribution, High EE	Coarse particles, Very low volume & productivity
Electro-Magnetic Atomization	600–850 Narrow size distribution	Sn, Pb	—	0.05–0.12	Low	Narrow size distribution	Low volume & productivity
Impaction Atomization	3×10^3–15×10^3	Fe, Steels, Cast iron, Ferroalloys, Mattes	Low	Large	Large	Simple, High EE, Large throughput	Very coarse particles
Electrostatic Atomization	—	—	—	Very low	Low	Potentially narrow size distribution	Difficulties in practical operation

This section describes the atomization processes and techniques for metal droplet generation. Advantages and drawbacks of the atomization systems are discussed along with typical ranges of operation conditions, design characteristics, and actual and potential applications. Commonly used atomization media and their thermophysical properties are also included.

2.2.1 Gas Atomization

Gas atomization of melt is a two-fluid atomization process involving the interaction of the melt and an atomizing gas. During gas atomization, a fine dispersion of droplets is generated by the impingement of the high energy atomizing gas on the melt stream as a result of the transfer of the impact kinetic energy from the atomizing gas to the melt. Gas atomization has been widely used in materials processing (such as spray forming)[3] and the powder metallurgy industry[4] to produce pre-alloyed droplets and powders of closely-controlled composition. Atomization is a key stage in these processes. It determines the size distribution and initial conditions of droplets, and influences droplet velocity, temperature, cooling rate, and eventually the microstructure and mechanical properties of the sprayed powders.

Early development of two-fluid gas atomization was stimulated by the applications in chemical processing industry for the atomization of various liquids such as oils, chemical solutions, emulsions, dispersions, slurries, and gels. Air atomization of non-ferrous metals such as lead and tin started in the early 1920's.[145][146] In this atomization process, a single spray nozzle was employed for the production of powders used in solder pastes. Using similar techniques, molten zinc and aluminum were atomized into powders for the production of explosives and the purification of zinc liquors, respectively. During World War II, a novel atomization process, i.e., the R.Z. Mannesmann gas atomization method, was invented by Mannesmann in Germany for iron powder production in which iron melt was atomized by air. In the 1960's, inert gas atomization was applied to a wide range of metals in aerospace applications. Over the past four decades, gas atomization techniques have been extensively developed and reached the mature stage. To date, gas atomization has been applied to a variety of materials, both metallic and non-metallic, such as solder materials, precious metals, copper, iron, aluminum, magnesium, cobalt, alloys of these metals, low-alloy steels, high speed steels, stainless steels, a wide range of ferrous and non-ferrous specialty alloys, nickel-base and other superalloys, intermetallic compounds, and ceramics (alumina). The viability of gas atomization for titanium-base alloys has also been demonstrated. General descriptions of gas atomization can be found in the reviews by Lawley,[4] Yule and Dunkley,[5] Gummeson,[145] Beddow,[146] Klar and Shafer,[147] Klar and Fesko,[148] Howells et al.,[149] and Reinshagen and Neupaver.[150]

Different gases, such as nitrogen, argon, helium, air, or a mixture of two gases, have been used for the atomization of melts. Nitrogen and air are commonly used in gas atomization, and nitrogen and argon are more often used in spray casting. Inert gases (such as argon and helium) are preferred in the atomization of reactive metals such as superalloys and titanium or in the applications that require low oxygen content. Reactive gases are used for in-situ reaction synthesis of oxides, nitrides, carbides, etc. Mixtures of gases provide a choice for producing desired powder characteristics at low cost. For example, gas additives or dopants can be added in atomization gas to passivate the surfaces of active droplets/powders. The gas used for atomization is either recycled or vented to the atmosphere. The gas used for purging the atomization chamber or tank is usually the same as the atomization gas. The chamber is normally made of stainless steel and its internal surfaces are polished to minimize powder contamination. The thermophysical properties of the commonly used atomization gases and metals/alloys are listed in Tables 2.6, 2.7, 2.8, 2.9, 2.10, and 2.11, respectively, for an overview.

A variety of atomizer designs have been developed in an effort to control the droplet size distribution and to increase the yield of fine powders. Gas atomizers used for the atomization of melts may be loosely classified into two primary categories in terms of the interaction mode between a liquid metal and an atomization gas during atomization, i.e., *(1)* internal mixing and *(2)* external mixing.

For low melting-point metals (for example, solder materials), the liquid metal and the atomization gas may be mixed internally inside the atomizer; for both low and high melting point materials, the two fluids can be mixed externally outside the atomizer in the near-nozzle region.

Table 2.6. Gas Properties as a Function of Temperature

Gas Property	Symbol	Equation
Thermal Capacity	c_{pG} (J/kg K)	$c_{pG} = a_0 + a_1 T_G - a_2 \times 10^6 T_G^{-2} + a_3 \times 10^{-6} T_G^2$
Thermal Conductivity	k_G (W/m K)	$k_G = k_{G0} \left(\frac{T_G}{273} \right)^n$
Density	ρ_G (kg/m^3)	$\rho_G = \frac{273 \rho_{G0}}{T_G}$
Viscosity	μ_G (kg/m s)	$\mu_G = \mu_{G0} \left(\frac{273 + c}{T_G + c} \right) \left(\frac{T_G}{273} \right)^{3/2}$

a_0, a_1, a_2, a_3 — Constants in equation for c_{pG}
c — Constant in relation for μ_G
k_{G0} — Thermal conductivity of gas at 273 K
n — Power index in relation for k_G
T_G — Temperature of gas
ρ_{G0} — Density of gas at 273 K
μ_{G0} — Viscosity of gas at 273 K

Table 2.7. Thermophysical Properties of Commonly Used Atomization Gases (N_2, Ar, He, Air), Gas Additive (O_2), and Steam[151]

Gas Property	N_2	Ar	He	Air	O_2	Steam
M_G (kg/kmol)	28	40	4	29	32	18
R (J/kg K)	297	208	2078	287	260	462
ρ_{G0} (kg/m^3)	1.250	1.784	0.188	1.293	1.429	0.586 [b]
μ_{G0} (10^{-5} kg/m s)	1.658	2.125	1.865	1.705	1.874	0.824
k_{G0} (10^{-2} W/m K)	2.401	1.634	14.2	2.44	2.451	2.452 [b]
c	118	140.43 [a]	60 [a]	114	138	673
n	0.8	0.8	0.7 [a]	0.82	0.87	1.48
γ	1.4	1.6	1.66	1.4	1.4	1.33
a_0 (J/kg K)	1021.29	521.	5193.	944.36	936.81	1697.98
a_1	0.135	0.	0.	0.186	0.131	0.572
a_2	1.794	0.	0.	0.	5.234	0.
a_3	0.	0.	0.	4.827	0.	0.

[a] Derived from regression analyses of the tabulated data in Ref. 151

[b] Values at 380 K; $k_G = k_{G0}\left(\dfrac{T_G}{380}\right)^n$, $\rho_G = \dfrac{380\rho_{G0}}{T_G}$

M_G — Molecular weight of gas

R — Gas constant, R=8313/M_G for compressible fluid flow calculation

γ — Isotropic factor of gas

Table 2.8. Metal Properties as a Function of Temperature

Metal Property	Symbol	Equation	Refs.
Wiedemann-Franz-Lorenz law for thermal conductivity	k_L (W/m K)	$k_L = 2.45\times10^{-8} T/\rho_e$	[152]
Surface tension	σ (N/m)	$\sigma = \sigma_m - d\sigma/dT(T - T_m)$	[153]
Density	ρ (kg/m^3)	$\rho = \rho_m - d\rho/dT(T - T_m)$	[153]
Arrhenius equation for viscosity	μ_L (mN s/m^2)	$\mu_L = \mu_0 \exp(\frac{E}{RT})$	[153]
Grosse's approach for viscosity constant in Arrhenius equation	μ_0 (mN s/m^2)	$\mu_0 = \frac{5.7\times10^{-2}(MT_m)^{1/2}}{V_m^{2/3}\exp(E/RT_m)}$	[152]
Grosse's approach for activation energy in Arrhenius equation	E (cal/mol)	$E = 1.21 T_m^{1.2}$	[152]
Effective molecular diameter of melt	d_m (m)	$d_m = (V_m/N)^{1/3}$	[152]
Iida-Guthrie's empirical expression for self-diffusivity in melt at T_m	D_{lm} (cm^2/s)	$D_{lm} = 3.5\times10^{-6}(\frac{T_m}{M})^{1/2} V_m^{1/3}$	[152]

M — Molecular weight of metal

N — Avogadro's constant, N=6.02×10^{23} (atoms/mol)

R — Gas constant in Arrhenius equation for viscosity, R=8.3144 (J/K mol)

T — Temperature

T_m — Equilibrium melting temperature of metal

V_m — Molar volume of melt at T_m

σ_m — Surface tension of liquid metal at T_m

ρ_e — Electrical resistivity of metal

ρ_m — Density of liquid metal at T_m

Table 2.9. Thermophysical Properties of Metals and Alloys in Atomization Processes

Property	Mg	Al	Cu	Ni_3Al[154]	Ni	Co	Fe	Refs.
T_m or T_l/T_s (K)	923	933	1356	1663/ 1658	1728	1768	1809	[152]
k_L (W/m K)	90	100	165	40	30	43 a	32 a	[155]
k_S (W/m K)	155.5	238	397	80.87	88.5	96	78.2	[153]
c_{pL} (J/kg K)	1360	1178	490	690.52	652	684	795[153]	[152]
c_{pS} (J/kg K)	1038	917	386	677.69	452	427	456	[153]
ρ_m (kg/m^3)	1590	2385	8000	6810 (l) 7178 (s)	7905	7760	7015	[153]
$d\rho / dT$ (kg/m^{3}K)	0.265	0.280	0.801	—	1.160	0.988	0.883	[153]
μ_o (mN s/m^2)	0.0245	0.149	0.301	0.162 ^{b}	0.166	0.255	0.3699	[153]
E (kJ/mol)	30.500	16.500	30.500	43.123 ^{b}	50.200	44.4	41.4	[153]
σ_m (N/m)	0.559	0.914	1.285	1.778 ^{c}	1.778	1.873	1.872	[153]
$d\sigma / dT$ (mN/m K)	0.35	0.35	0.13	0.38 ^{c}	0.38	0.49	0.49	[153]
D_{lm} (10^{-9} m^2/s)	5.63	4.87	3.97	3.90	3.90	3.77 ^{d}	4.16	[152]
d_m (10^{-10} m)	2.95	2.66	2.37	3.567	2.32	2.33	2.37	[152]
V_m (10^{-6} m^3/mol)	15.3	11.31	7.94	29.96	7.43	7.6	7.94	[152]
ΔH_m (kJ/kg)	361.96	387.67	204.60	751	292.16	262.37	246.60	[152]

^{a} Estimated from electrical resistivity according to Wiedemann-Franz-Lorenz law
^{b} Arithmetic average of properties of alloy elements based on mole fractions
^{c} Properties of Ni
^{d} Estimated with Iida-Guthrie's empirical expression

c_{pL}, c_{pS} — Thermal capacities of liquid and solid metal, respectively
ΔH_m — Latent heat of fusion,
k_L, k_S — Thermal conductivities of liquid and solid metal, respectively
T_l, T_s — Liquidus and solidus temperatures of alloy, respectively

Table 2.10 Thermophysical Properties of Refractory Metals and Alloys in Atomization Processes

Property	Ti	Pt	Cr	Mo	Ta-2.5W [156]	W	Refs.
T_m or T_l/T_s (K)	1998	2041	2133	2880	3306 /3299	3655	[152]
k_L (W/m K)	20	69 [a]	165 [a]	117 [a]	67 [a,b]	71	[157]
k_S (W/m K)	21.6	73.4	91.3	137	58 [b]	174	[153]
C_{pL} (J/kg K)	700 [157]	178	756	570	244	230 [157]	[153]
C_{pS} (J/kg K)	528	134.4	461	251	234	138	[153]
ρ_m (kg/m^3)	4110	19000	6280	9340	14862	17600	[153]
$d\rho/dt$ (kg/m^3 K)	0.702	2.9	0.3	0.50 [152]	0.710 [c]	1.500 [157]	[153]
μ_0 (mN s/m^2)	0.32 [157]	0.462 d	0.276 [d]	0.316 [d]	0.386 c,d	0.420	[153]
E (kJ/mol)	46.28 [152]	47.48 d	50.06 [d]	71.769 d	83.285 c,d	95.53 [152]	[153]
ρ_m (N/m)	1.650	1.800	1.700	2.250	2.15 [b]	2.500	[153]
$d\rho/dt$ (mN/m K)	0.26	0.17	0.32	0.30	0.25 [b]	0.29	[153]
D_{lm} (10^{-9} m^2/s)	5.11 [e]	2.46 [e]	4.53 [e]	4.17 [e]	3.335	3.42 [e]	[152]
d_m (10^{-10} m)	2.68 [f]	2.58 [f]	2.39 [f]	2.58 [f]	2.94	2.59 [f]	[152]
V_m (10^{-6} m^3/mol)	11.6	10.31	8.27	10.3	12.1	10.5	[152]
ΔH_m (kJ/kg)	304.80	113.85	401.92	371.07	171.4	190.97	[152]

[a] Estimated from electrical resistivity according to Wiedemann-Franz-Lorenz law
[b] Properties of Ta
[c] Arithmetic average of properties of alloy elements based on mole fractions
[d] Estimated using Grosse's approach
[e] Estimated with Iida-Guthrie's empirical expression
[f] Estimated with the equation for d_m

Table 2.11. Thermophysical Properties of Precious Metals and Solder Materials in Atomization Processes

Property	Ag	Au	Pd	Bi57-Sn43 [152]	Sn	Bi	Pb	Zn	Refs.
T_m (K)	1235	1337	1827	411.3	505	544	600	693	[152]
k_L (W/m K)	174.8	104.4	—	87.8	31.4	15.5	16.6	49.5	[153]
k_S (W/m K)	425	315.5	75.2	80	62.2	7.04	34	119.5	[153]
c_{pL} (J/kg K)	283	149	—	200.97	242	143	144	481	[153]
c_{pS} (J/kg K)	234	130	247	167.47	226	125	130	394	[153]
ρ_m (kg/m^3)	9346	17360	10490	8613	7000	10068	10678	6575	[153]
$d\rho/dT$ (kg/m^3 K)	0.907	1.5	1.266	-0.33 [a]	0.61	1.33	1.32	1.10	[153]
μ_0 (mN s/m^2)	0.453	1.132	0.347 [b]	1.49 [c]	0.538	0.446	0.464	0.413	[153]
E (kJ/mol)	22.2	15.9	41.57 [b]	—	8.88 [152]	6.45	8.61	12.7	[153]
σ_m (N/m)	0.903	1.140	1.500	0.452	0.544	0.378	0.468	0.782	[153]
$d\sigma/dT$ (mN/m K)	0.16	0.52	0.22	—	0.07	0.07	0.13	0.17	[153]
D_{lm} (10^{-9} m^2/s)	2.56	2.05 [d]	3.15 [d]	1.47 [d]	2.31 [158]	0.815 [153]	2.20 [158]	2.06	[152]
d_m (10^{-10} m)	2.68	2.66	2.56 [e]	3.20 [e]	3.05	3.26 [e]	3.18	2.52	[152]
V_m (10^{-6} m^3/mol)	11.6	11.3	10.14	19.76	17.0	20.80	19.42	9.94	[152]
ΔH_m (kJ/kg)	102.69	64.78	157.55	44.799	59.61	52.06	23.22	111.35	[152]

[a] Estimated from discrete data
[b] Estimated using Grosse's approach
[c] Value for μ at 673 K
[d] Estimated with Iida-Guthrie's empirical expression
[e] Estimated with the equation for d_m

In *internal mixing* atomization (for example centrifugal-pneumatic atomization),[159] the liquid metal and gas enter the swirl jet atomizer tangentially under pressure (Fig. 2.13).[159] The two fluids rotate, form a mixture, and accelerate in the confuser. Due to the strong centrifugal force, the liquid metal forms a film at the nozzle exit even without the presence of the gas. With the applied gas, the liquid film is atomized into a fine dispersion of droplets outside the nozzle.

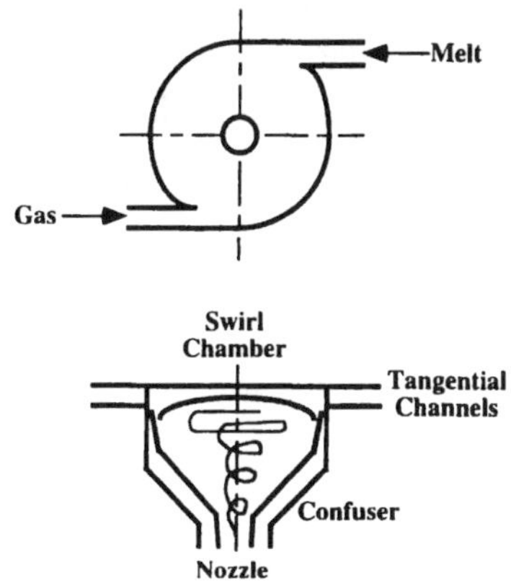

Figure 2.13. Schematic of an internal-mixing atomizer for atomization of melts.

In *external-mixing* atomization, the liquid metal stream issued from the metal delivery nozzle is impacted by an atomization gas outside the atomizer. In various atomizer designs, a multiplicity of gas jets are issued from various arrays or from an annular nozzle that may have a slit or discrete orifices distributed evenly in the circumference. An annular atomizer/nozzle with a slit is hereafter referred to as *annular-jet* atomizer/nozzle, and an annular atomizer/nozzle with discrete orifices is referred to as *discrete-jet* atomizer/nozzle. In terms of the geometry and configuration of atomizers, namely the geometrical location where the liquid metal and the gas interact, external mixing atomizers may be further divided into two major types, i.e., free-fall and close-coupled. In the *free-fall atomizer* (Fig. 2.14), the atomization gas and liquid metal stream meet away from the metal delivery nozzle. In the *close-coupled atomizer* (Fig. 2.15), the atomization gas interacts with the liquid metal in the immediate vicinity of the metal delivery nozzle at the base of tundish. The free-fall atomizer is most widely used in conventional gas atomization due to its stability, while the close-coupled atomizer is increasingly employed because of its relatively high atomization efficiency and powder yield. For each type of the atomizers, atomization can be subsonic or supersonic, depending on the gas pressure and the gas nozzle geometry and configuration applied. High-pressure gas atomization with supersonic close-coupled atomizers is emerging as an efficient production method for high yields of fine powders.

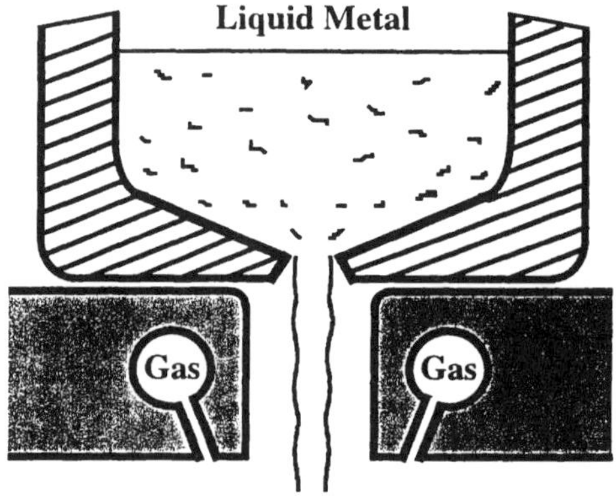

Figure 2.14. Schematic of a free-fall atomizer for atomization of melts.

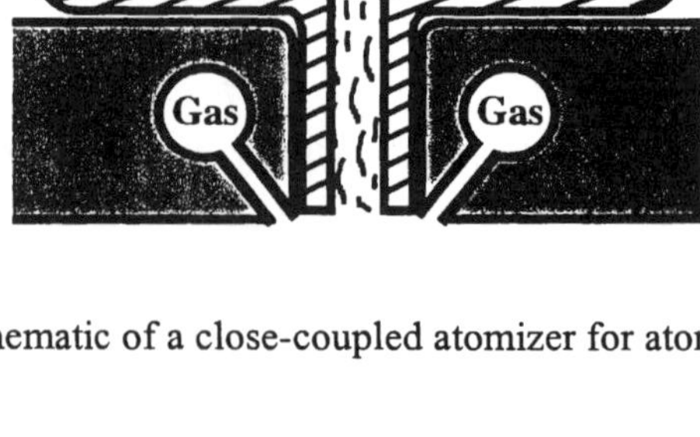

Figure 2.15. Schematic of a close-coupled atomizer for atomization of melts.

The close-coupled atomizer configuration has many variations (Fig. 2.16), such as the atomizers for high pressure gas atomization (HPGA),[160]-[162] SIGMA,[163][164] Ünal atomizer,[165]-[169] truncated plug atomizer,[170] NANOVAL atomizer,[171] annularized sheet atomizer, and ultrasonic gas atomizers.[154][172]-[174] The free-fall atomizer design may also have different geometry arrangements (Fig. 2.17). In some earlier designs, a single gas jet or multiple gas jets are arranged to horizontally impact a vertical, downward flowing metal stream. In most recent designs, multiple gas jets from discrete openings or a gas sheet jet from an annular nozzle form a converging cone concentric with the vertical metal stream. The atomized droplets solidify during flight downward in the chamber. To insure that all the droplet are completely solidified and cooled below certain temperature, the chamber must be sufficiently high (under certain conditions higher than 10 m) to prevent agglomeration by sintering at the base of the chamber. Thus, a vertical gas atomization facility may require a high building to house it. In this case, the horizontal gas atomization technique may be employed as an alternative. Due to the smaller dimension of a horizontal atomization facility with a typical chamber height of about 3 m, capital costs may be reduced. An additional solution may be filling the base of the chamber with water or liquid nitrogen to quench gas-atomized droplets for precious metals and tool steels. When water is used for quenching droplets, a subsequent powder drying process is then required similar to that in water atomization.

On average, current gas atomization processes have fairly low efficiency, perhaps 2–4%.[172] With expensive inert gases (for example, argon), production costs are high. The higher the gas velocity, the finer the powder size and the higher the yield of fine powders. To achieve the maximum efficiency in gas atomization and the maximum energy utilization of atomization gas, a laminar flow within the high velocity core of a gas jet is thus desired.[172] In other words, for the maximum efficiency, a liquid stream should be impacted by the gas at a maximum velocity, i.e., within a minimum distance for the gas flow from the gas nozzle to reach the metal stream, because the gas velocity and kinetic energy decay rapidly with distance downstream of the jet core.[175] Therefore, close-coupled atomizers are more efficient than free-fall atomizers, and internal mixing atomizers are even more efficient than close-coupled atomizers. Generally, the internal mixing atomization is more efficient relative to the external mixing atomization. An example worth mentioning is the internal mixing gas atomizer named *centrifugal-pneumatic atomizer*, proposed by Sheikhaliev and Dunkley.[159] This novel atomizer, although well known in the atomization of cold normal liquids, can produce metal droplets in the size range similar to those produced with the most efficient close-coupled atomizers but with much lower gas flow rates and lower gas pressures. This system has been used for the atomization of Sn, Zn, Cu-Zn, Al and Mg.

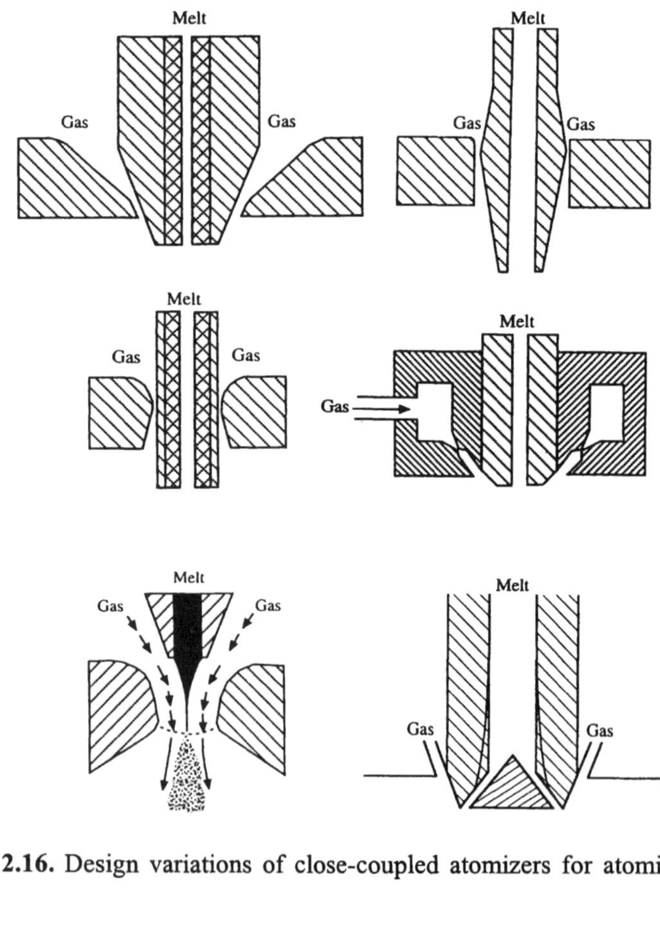

Figure 2.16. Design variations of close-coupled atomizers for atomization of melts.

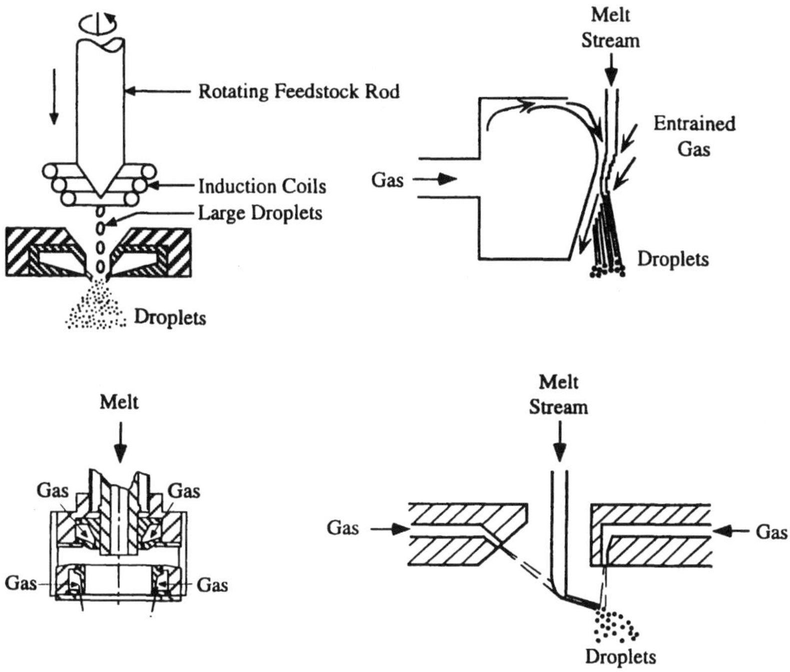

Figure 2.17. Different geometry arrangements of free-fall atomizers for atomization of melts.

In addition to the designs mentioned above, other atomizer configurations such as those used in spray combustion and spray drying processes may also be considered as an alternative in gas atomization of melts for special purposes after appropriate modifications.

Controlling variables in gas atomization of melts include nozzle geometry parameters and many process parameters. An exhaustive list of these parameters is given in Tables 2.12 and 2.13, respectively, along with typical values and/or ranges of the parameters.

The metal flow rates and capacity in gas atomization are lower than in water atomization. For example, in batch operations, metal flow rates are normally in the range of 1 to 70 kg/min, corresponding to a melt stream diameter of 2 to 15 mm. The flow rates for low melting-temperature metals range from 8 to 33 kg/min during continuous atomization operation. In commercial gas atomization processes, gas pressure is usually in the range of 0.5 to 9 MPa, corresponding to a gas flow rate between 0.02 and 0.24 m^3/s. Higher gas pressures, up to 18 MPa, in conjunction with close-coupled atomizer designs have been examined. The high-pressure gas atomization has been demonstrated to be able to generate fine droplets of mass median size below 10 μm. Gas velocity distribution depends on nozzle geometry and configuration, and may vary from several tens to hundreds meter per second. Melt superheat is normally between 75 and 150°C. The representative mass median particle size is within the range of 50 to 300 μm with a standard deviation of about 2. The cooling rates of gas-atomized droplets are lower than those of water-atomized droplets for a given droplet size.

Table 2.12. Geometry Parameters in Gas Atomization of Melts

	Parameter	Value/Range
Gas Atomizer	Nozzle Number	1, 2, 4, 12, 16, 18, 20, 24, or Annulus
	Nozzle Diameter (mm)	0.25, 0.5, 0.7, 0.8, 1, 3
	Relative Angle between Gas Nozzle and Metal Delivery Nozzle (°)	7–45 (Vertical) 90 (Horizontal)
	Configuration	*External Mixing*: Close-Coupled: SIGMA HPGA USGA Ünal Truncated Plug Annularized Sheet NANOVAL Free-Fall: Horizontal Osprey Ultrasonic *Internal Mixing*: Centrifugal-pneumatic
	Type of Nozzle	Straight Converging Converging-Diverging (Laval) Shock Wave (Ultrasonic)
Metal Delivery Nozzle	Diameter (mm)	1–15
	Configuration	Column Annularized Sheet

Table 2.13 Process Parameters in Gas Atomization of Melts

	Parameter	Value/Range
Atomization Gas	Pressure (MPa)	0.5–9 (1.4–8.2 in USGA)
	Flow Rate (m^3/s)	0.02–0.24
Metal/ Alloy	Superheat (°C)	75–150
	Flow Rate (kg/min)	1–70 (2.4–4.8 in USGA)
Melting Crucible (Tundish)	Material	Ceramics/Refractory
	Heating/Melting Method	Induction Resistance Fuel Firing Electroslag Remelting
Delivery Nozzle	Material	Ceramics/Refractory Steels
Spray Chamber	Pressure	Atmosphere or Subatmosphere

The relative angle between gas nozzle and metal delivery nozzle is an important factor influencing gas atomization. For an annular gas nozzle concentric with metal stream, too flat gas jets (large impingement angle) may cause a wide spray cone and a high back pressure against the metal stream. On the contrary, too steep gas jets (small impingement angle) may require a long distance for the gas flow from the gas nozzle to reach the metal stream, leading to the development of turbulence and poor gas efficiency. The impingement angle may vary from 15° to 80°, depending on the metal and atomization gas used.

Scale-up of gas atomizers is difficult and it requires the use of higher gas-to-melt mass flow rate ratio to maintain the same droplet size. The scale-up may also cause some complex phenomena to occur, such as the disappearance of the prefilming effect in close-coupled atomizers, the generation of turbulence in melt flow within delivery nozzle, and change in atomization mechanisms.

For the delivery of the atomization gas, different types of nozzles have been employed, such as straight, converging, converging-diverging, and ultrasonic nozzles. A straight tube nozzle is used in subsonic gas atomization, while converging and converging-diverging tube nozzles are employed in supersonic gas atomization at high gas pressure. In ultrasonic gas atomization (Fig. 2.18), an internal Hartman shock wave tube is utilized to modify the normal manifold gas pressure (1.4–4.1 MPa or 200–600 psi) to yield a gas velocity as high as Mach 2 with a frequency of 60 to 120 kHz at the nozzle exit. At such high, pulsed gas velocity, liquid metal is atomized into fine droplets, usually at a yield of 80 to 90% in the size range less than 30 μm. The ultrasonic waves are stimulated when the high-pressure gas is accelerated through resonance cavities. Except for gas nozzle geometry, the atomizer geometries and configurations as well as atomization facilities in subsonic, supersonic, and ultrasonic gas atomization are quite alike.

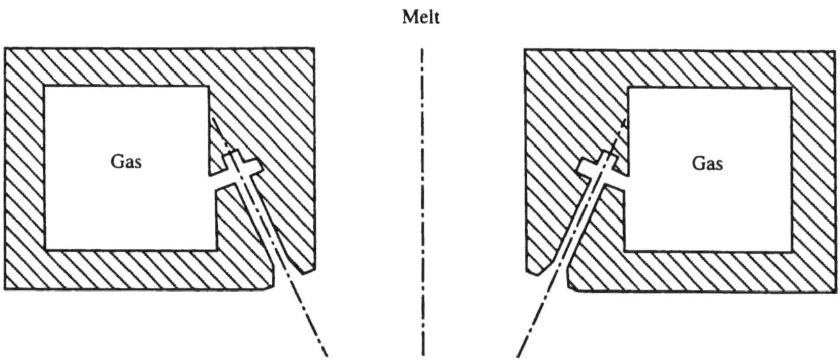

Figure 2.18. Schematic of a close-coupled, ultrasonic gas atomizer for atomization of melts.

Ultrasonic gas atomization, termed *USGA*, was developed at the Massachusetts Institute of Technology in the US by Grant[172] who radically modified the pulverization method invented in Sweden in the 1960's.[176] The method is conceptually akin to Hartmann-whistle acoustic atomization of normal liquids. Since the mid 1970's, USGA has been used in commercial scale for producing powders of some lower melting-temperature alloys (for example, aluminum). Both close-coupled and free-fall arrangements may be used in USGA. Only laboratory and pilot scale operations are reported for high melting-temperature alloys, such as stainless steels, Ni- and Co-base alloys, pure Co, Ni, and Fe. The important geometry parameters and process parameters in USGA are the same as those for conventional gas atomization with closed-coupled atomizer configuration.[177] These include metal delivery nozzle diameter, relative angle between gas nozzle and metal delivery nozzle, axial and radial distances between gas nozzle and metal delivery nozzle, melt superheat, and gas pressure. Argon and nitrogen have been used as atomization gas with a pressure up to 8.2 MPa. Liquid stream diameter is typically 1 to 5 mm, and mass flow rate is about 2.4–4.8 kg/min, depending on metal density, stream diameter, and overpressure. The throughput is limited by the metal stream diameter. Increasing the metal stream diameter can enhance the metal flow rate but the energy available for the atomization may be severely attenuated when the stream diameter is beyond about 5 mm. The upper limit of the flow rate is about 15 kg/min for an annular USGA atomizer. To overcome this limitation, a linear USGA atomizer has been developed.[178] In the linear USGA atomizer, the liquid metal stream has a rectangular cross-section with a high aspect ratio, and the atomization gas is issued from a rectangular slit nozzle surrounding the vertical, downward flowing metal stream. Increasing the longitudinal dimension of the atomizer can enhance the flow rate, while reducing the transverse dimension can minimize the attenuation of the pulsed gas energy with an attendant pre-filming effect.

Ultrasonic gas atomization has several inherent advantages, including high relative velocities between gas and liquid metal phases, very fine spherical droplets with smooth surfaces and few satellites, and enhanced gas cooling capability due to the expansion of the high pressure gas. As a result of these features, cooling rates of droplets are high and droplets solidify within a short time, so that the atomization chamber (or container) can be small. The height and diameter of the spray chamber in laboratory scale may be as small as 1.7 m and 1.3 m, respectively. The gas investment is low and the gas utilization efficiency is relatively high. The gas demand is roughly as low as 25% of that for subsonic gas atomization. In addition, due to the very fine droplet size, the cooling rates of droplets are approximately 10^5 °C/s, yielding highly refined microstructures in the solidified particles. An important application of USGA is the production of continuous strip by spray deposition using the linear USGA atomizer. It has been planned that a batch operation will have a capacity of about 2400 kg for carbon steel strip, and about 450 kg for aluminum strip.[4]

Although the primary goal in gas atomization of liquid metals is to produce smooth, spheroidal (spherical) particles, large variations from smoothness and roundness are usual in practice, particularly for aluminum-base alloys, zinc and copper-zinc alloys. This is due primarily to cooling and solidification of droplets. Fine satellites observed on coarser particles are formed due to collisions among droplets in the turbulent two-phase flow in the spray. This condition makes the separation of fine and coarse particles impossible, exerting negative effects on the microstructure and properties by incorporating powders with significantly different dendrite sizes and phase distributions into final products. The second major problem associated with gas atomization is the entrapment and dissolution of the atomization gas in droplets that eventually remains in solidified particles, i.e., powders.

2.2.2 Water Atomization

The earliest vestiges of *water atomization* technology can be recognized in the process invented by Marriott in England in 1872,[4] in which a steam injector was used to atomize metal melts. This process established the basis for the development of the technology and system for metal powder production from hundreds of grams to tonnage quantities in a wide range of chemical and metallurgical applications. With the development of the iron and steel industry and the growing need for shot to clean castings, water granulation with large capacity and pressures of 1–10 bar evolved for the production of shot in the size range between 0.5 and 5 mm. Following the idea of water granulation, high pressure water atomization was developed and applied first on copper and its alloys between the 1930's and 1940's, and later on steels in the 1950's. During this time period, Powder Metallurgy Ltd. in England developed a process for non-ferrous metals and alloys which established the basis for the modern commercial water atomization technology. This process involves an annular nozzle from which a high velocity conical jet of water is issued to impact the downward flowing metal stream. The process was further applied to high melting temperature alloys (iron, low alloy steels, stainless steels, high speed steels and superalloys) by B.S.A. Co. Ltd. in England and Höganäs in Sweden. To date, this technology has been used for making metal and alloy powders for a variety of metallurgical and chemical applications. It is the dominant method for the production of metal powders in terms of tonnage. The largest commercial application of water atomization is iron powder production for pressing and sintering. Water atomization is also used for the commercial production of copper, copper alloy, stainless steel, tool steel, and magnetic powders for press and sinter applications. In addition, water atomization is employed for the production of nickel alloy powders widely used in thermal spray coating and brazing, and precious metal powders used in brazing, sintering and dental amalgams.

Extensive reviews of water atomization processes have been made by Gummeson,[145] Klar and Fesko,[148] Lawley,[4] Beddow,[146] Yule and Dunkley,[5] and Reinshagen and Neupaver.[150] Similar to gas atomization, water atomization of melt is also a two-fluid atomization process involving the interaction of the melt and water. The significant difference is that water possesses a higher quench capability than gas due to its high thermal conductivity and capacity. A water atomization unit typically consists of a melting facility, a water pressurization system, an atomizer, an atomization chamber (tank) and a powder collection and drying system. In water atomization, metal or alloy charge is usually induction-melted in air, inert gas, or under vacuum, depending on the metal or alloy involved. Other melting techniques, such as arc melting and fuel heating, may also be used. In commercial-scale water atomization facilities, the molten metal is mostly poured into a tundish that provides a uniform and controlled temperature and flow of liquid metal through a circular delivery nozzle at the bottom of the tundish. The liquid metal exits from the delivery nozzle and falls a certain distance under gravity or pressure before impacted by water jets. The delivery nozzle serves as a controller of the diameter and shape, and to some extent, temperature and flow velocity of the melt stream. The melt stream is directed into the atomization zone where it is disintegrated into droplets by water jets. While the droplets fly downward in the atomization chamber they are rapidly cooled and solidified as a result of the interaction with water. The solidified powders are subsequently transported into the collection and drying system for post-processing. The overall process scheme of water atomization is similar to that of gas atomization with some minor differences. For example, the high-pressure gas supply in gas atomization is replaced by high-pressure water pump in water atomization, a cyclone particle separator and filter are replaced by a decant water line.

For an atomization unit, the atomizer is the most critical part. It controls the flow and interaction of water and liquid metal, and hence affects the atomization efficiency. The design of water atomizers is similar to that of gas atomizers with respect to nozzle geometry and configuration. A variety of atomizer designs have been developed and are currently in use, including both asymmetric or axisymmetric discrete multiple jets (such as opposed V-shaped jets) and annular slit nozzle concentric with the metal stream (cone jet). A fan spray pattern can be generated by two opposed, V-shaped jets. Annular nozzles are less flexible than V-shaped nozzles with two or more discrete jets distributed symmetrically about the metal stream. Therefore, annular nozzles are primarily used in large capacity plants for iron powder production. The melt is almost always in the free-fall configuration. The free-fall distance varies with designs but ranges typically from 10 to 30 cm.[4] A converging cone arrangement for water jets is the preferred configuration. Many atomizer designs are proprietary and some are described in patent literature.

In water atomization, a number of operation variables are to be considered in order to properly control the process. The variables include geometry parameters, process parameters, and thermophysical properties of metal/alloy and water. Each design and configuration of an atomization unit are unique and thus only some specific operation conditions may be employed. Many of the variables are interrelated. Therefore, there may exist more than one set of optimum variable combinations for a given atomization unit.

The geometry and process parameters of importance are not unlike those for gas atomization, as summarized in Tables 2.14 and 2.15, respectively. In conventional commercial water atomization, the water pressure is normally in the range of 5.5 to 21 MPa. The use of much higher water pressure from 50 to 150 MPa has been considered for the production of ferrous alloy powders in powder injection molding applications. Such high levels of water pressure can lead to very fine powders with a mass median size of 5 to 20 μm. In commercial water atomization, the powder size distributions are relatively broad (10–1000 μm) with a mass median size of about 100 μm and a standard deviation from 1.7 to 2.4. Based on the measurement of secondary dendrite arm spacing (SDAS) in particles, the cooling rates of droplets are estimated to be within the range of 10^4–10^6 °C/s, depending upon droplet size.

Table 2.14. Geometry Parameters in Water Atomization of Melts[145]

	Parameter	Value/Range
Water Atomizer	Nozzle Number	1, 2, 4, or Annulus
	Relative Angle between Water Nozzle and Metal Delivery Nozzle (°)	40–60 (Annular Jet) 15–35 (V Jets)
	Configuration	External Mixing & Free-Fall: Annular Cone Jet Symmetric Discrete Fan-Spray Jets (V Jets) Asymmetric Cross Jets (Horizontal)
Metal Delivery Nozzle	Diameter (mm)	~7
	Configuration	Column

Table 2.15. Process Parameters in Water Atomization of Melts[5][148]

	Parameter	Value/Range
Water	Pressure at nozzle exit (MPa)	5.5–21
	Velocity at nozzle exit (m/s)	70–230
	Mass Flow Rate (l/min)	110–380 (3000 for Annular Jet)
Metal/ Alloy	Superheat (°C)	75–150
	Mass Flow Rate (kg/min)	4.5–90 (400 for Annular Jet, 500 for Asymmetric Jet)
Heating/ Melting	Tundish Material	Ceramics/Refractory
	Heating/Melting Method	Induction Arc Melting Fuel Firing

Generally, the relative angle between water nozzle and metal delivery nozzle is much smaller in water atomization than in gas atomization. The efficiency of water atomization is between 4% and 5%. For achieving maximum efficiency, laminar flow of water jets is desired. The higher the water pressure, the finer the powder size and the higher the powder yield. For example, by increasing the water pressure about two times (from 60 to 125 atmospheres), the yield of mild steel powders of less than 500 μm can be enhanced from 60% to 80%. However, pressuring water to above 100 atmospheres, plus delivering it to atomization system and recycling it, is not an inexpensive operation. In addition, even intermediate levels of oxidizable elements in melt may form a refractory thin layer on liquid surface, preventing liquid from disintegrating and forming spheroids. Nevertheless, water atomization of liquid metals/alloys represents a far cheaper process than gas atomization with most gases except with air. Compared to the gas-atomized powders, the water-atomized powders are more irregular in shape with a rough dimpled surface texture (Fig. 1.7 b–d).

Water atomization is intrinsically a high volume, low cost process. Therefore, it is generally more cost-effective compared to other commercial atomization methods. However, powder purity, particularly with reactive metals and alloys, is a drawback associated with water atomization, although any metal or alloy melt that does not react violently with water may be processed with water atomization method.

2.2.3 Oil Atomization

In addition to water atomization, *oil atomization* using various hydrocarbons (oils) is the only other liquid atomization process used in the atomization of liquid metals. Due to the similarity of these two atomization processes, water atomizers can be readily adapted to oil atomization. Some problems inherent in water atomization, such as powder oxidation, can be avoided in oil atomization. The cost-effectiveness is superior to gas atomization, but not as good as water atomization.

Oil atomization process was developed by Sumitomo Metal Industries[179] for commercial production of powders of low alloy, high strength, high quality steels in the early 1980's. This is also the first time that Mn and Cr were successfully prealloyed in a P/M steel. In this process, an annular nozzle and a proprietary oil at a pressure of 14 MPa were used to produce injection-molding grade 4100 steel powder. Compared to water-atomized 4600 steel powders containing Ni and Mo, the oil-atomized, Mn and Cr prealloyed 4100 steel powders exhibited higher compressibility and hardenability, low level of carbon (0.002 w/o C) after annealing, and no significant oxidation (0.088 w/o O).

The drawback of oil atomization is that for high-melting-temperature alloys (for example, >800 °C), pyrolysis of the oil may occur leading to carbon pick-up so that powder must be decarburized by a wet hydrogen anneal followed by a dry hydrogen reduction to control the oxide level. Thus, the commercial application of oil atomization is limited to the production of steel powders with relatively high carbon content (~0.4 w/o).[4] Oil atomization process has the potential to be applied to high speed steels and bearing steels. This is because decarburization is not necessary for steels of >0.6

w/o C and a low temperature heat treatment is sufficient for removing the oil.

IPS Steel Powder AB in Sweden, probably the only operating plant using oil atomization, routinely produces steel powders using oil atomization. The oil pressure used in atomization is low (0.5 MPa) and thus the resultant powder size is coarse (~500 μm). The oxygen content is, however, very low (<0.01 w/o O). High-pressure oil atomization (up to 10 MPa) is currently under evaluation by this company in an attempt to produce finer powders.

The oil-atomized powders have typically a mass median size of 50–70 μm, with comparable size distribution and similar cooling rates to the water-atomized powders. However, oil atomization of some low-melting-temperature metals such as lead and tin generates finer powders than using water atomization, due to reduced cooling rate by nucleate boiling.[5] Finer powders are also possible by using higher oil pressures. In addition, the shape of oil-atomized particles is generally more spherical than water-atomized ones, particularly for the alloys which form tenacious oxide films such as NiCr, Al, Zn, etc.

2.2.4 Vacuum Atomization

Vacuum atomization is a commercial batch process.[180] The development of vacuum atomization started in the mid 1960's, concurrent with the development of inert gas atomization. In 1970, a patent for the vacuum atomization method was issued to Homogeneous Metals, Inc. Using vacuum atomization, this company routinely produces superalloy powders of fine size without great consumption of argon, giving powders free of inert gas filled porosity. Wentzell[180] has made detailed description of this proprietary process.

Vacuum atomization is a method conceptually similar to effervescent atomization. As schematically depicted in Fig. 2.19, a vacuum atomization facility consists of two chambers, one above the other. The overall dimension is about 18 × 4 m. In the lower chamber, metal is first induction-melted under vacuum and subsequently pressurized to 1–3 MPa with a mixture of inert gas (normally argon) and soluble gas (typically H_2). Thus, the volume of the lower chamber can be small but it must withstand vacuum conditions during melting and pressure conditions during pressurization. The upper chamber is under vacuum and serves as a spray and collection tank. During pressurization, the energy for the atomization is stored in the liquid metal. When the liquid metal supersaturated with the gas under pressure is suddenly exposed to vacuum by exhausting through a guide tube into the upper vacuum chamber, the gas expands and comes out of solution. The high velocity of the exploding melt and the gas desaturation lead to the atomization of the melt. Therefore, this process is sometimes also termed *soluble gas atomization* or *melt explosion atomization*. The considerable energy input in introducing gas into the flowing melt stream generates an internal-mixing nozzle effect.

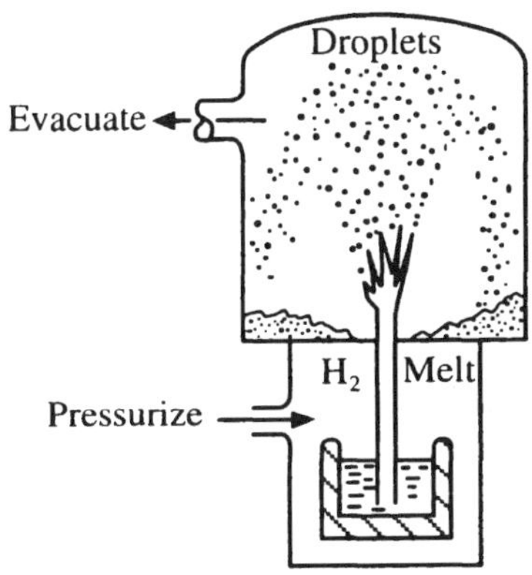

Figure 2.19. Schematic of vacuum atomization of melt.

Process parameters of importance include gas pressure, H_2 concentration in solution in liquid metal, diameter of the guide tube, and melt superheat. The concentration of H_2 gas in solution in the liquid metal ranges from 0.0001 to 0.001 w/o, proportional to the square root of H_2 partial pressure that is in turn controlled by the composition of the gas mixture. For a given H_2 partial pressure, the concentration of H_2 gas in the liquid metal is dependent on the metal type and superheat.

Normally, the vacuum-atomized droplets are clean. However, hydrogen may be entrapped or dissolved in the droplets. For powder metallurgy applications, powder treatment may be required to avoid porosity in final products. The vacuum-atomized droplets are normally spherical and smooth with few satellites. The cooling rates of droplets are lower than in gas atomization, typically in the order of magnitude of 10^2 °C/s. This is because the cooling of droplets is primarily by radiation heat transfer in vacuum atomization while it is by forced convection heat exchange with gas during gas atomization. Droplet size may range from 1 to 500 μm with a typical mass median diameter of 40–70 μm.

The arrangement of the melting and vacuum spray chambers is critical for guiding the liquid metal to eject into the vacuum chamber. Difficulties exist in precisely controlling the expulsion of the liquid metal into the vacuum chamber. Therefore, flaky droplets may be formed in vacuum atomization. Although vacuum atomization was developed mainly for the production of high-purity nickel and cobalt based superalloy powders, it is also applied to atomize the alloys of aluminum, copper and iron.

2.2.5 Rotating Electrode Atomization

Rotating electrode atomization is a commercial centrifugal atomization process. This process, designated REP (*Rotating Electrode Process*) or PREP (*Plasma Rotating Electrode Process*), was invented by Nuclear Metals, Inc. in 1963.[181] The original short rod design was modified to long rod design in 1974.[182] Since its invention, REP has been widely used for producing clean spherical powders of low carbon steels as carrier for carbon black in office copier applications. Over the past ten years, this atomization technique has been applied to the production of cobalt-chromium and titanium alloy powders for prosthetic devices. It has also been adapted to the production of ultra clean titanium alloy powders for potential aerospace applications.

Rotating electrode atomization process has been described in detail by many researchers.[183][184] As illustrated in Fig. 2.20, in the

REP, a rod of metal or alloy, referred to as a consumable electrode, is rotated at high speed about its longitudinal axis. Simultaneously, it is melted gradually at one of its ends by a heat source, such as an arc, a plasma, or an electron beam, etc. A thin film of the molten metal is detached from the rod end and ejected from the periphery of the rod by centrifugal force, forming spherical droplets. The atomization is conducted in an inert atmosphere, usually argon. Helium may be used to increase arc stability and convective cooling efficiency of droplets.

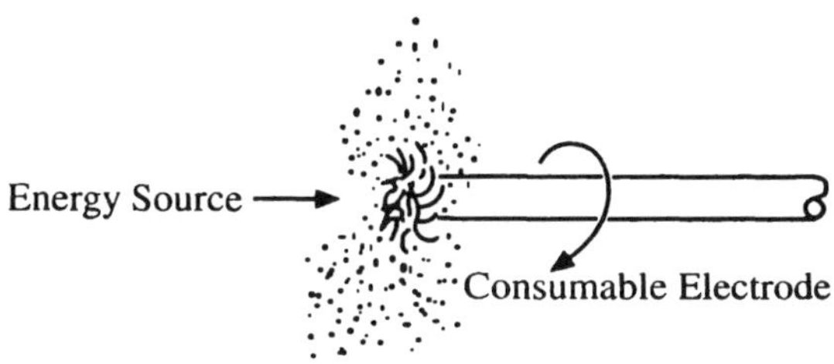

Figure 2.20. Schematic of rotating electrode atomization.

In the DC circuit, the rotating consumable rod serves as an anode and the transfer arc plasma torch or a cooled tungsten tipped device as the permanent stationary cathode. As the anode is consumed, it is moved into the spray chamber. The chamber diameter may be as large as 2440 mm.

The process parameters of practical importance are the diameter and rotational speed of the anode. In the short rod design, a consumable anode is typically 64 mm in diameter and 254 mm in length. Normally, more than 80% of the rod is converted into droplets. Due to the limited quantities of droplets generated, this design is suitable for small scale applications and for brittle alloys where an anode of high aspect ratio is not practical. In the long rod design, the diameter and length of an anode are 65 mm and 1524 mm, respectively, giving rise to an increased conversion efficiency. Normally, the anode rotational speed is about 1570 radians/s in both short and long rod designs, although it may be increased up to 2100 radians/s in experiments. Melt superheat is limited in the process since the liquid metal is removed by the centrifugal force from the rod as soon as it is melted. This makes it difficult to atomize the alloys with a wide melting range.

Compared to those intrinsically energy-intensive processes, such as gas atomization and water atomization, the energy used in REP (more generally, centrifugal atomization) is low and atomization occurs by means of centrifugal forces. In addition, REP requires no containment of liquid metals. This eliminates the contamination from refractory oxides. Therefore, the rotating-electrode-atomized particles are usually very smooth, spherical, almost satellite-free, and of high purity. This is the remarkable feature of REP and other centrifugal atomization methods operating in inert gas or vacuum at low rates. This is also an indication that the atomization occurs in the *Direct Droplet* regime, as detailed in the next chapter. The droplet size is relatively coarse, typically ranging from 100 to 400 μm with a mass median diameter of about 200 μm. The powder yield is high, i.e., the droplet size distribution is narrow. The cooling rates of droplets are up to 10^2 °C/s, depending on the gas used in the chamber and the droplet size.

Rotating electrode atomization may be applied to almost all metals and alloys since it does not require a crucible for melting and/or pouring. In particular, high melting-temperature metals and alloys, such as Ti and Zr, are well suited for the process. However, the production cost is still a drawback associated with the process, since electrode production is generally more expensive than a metal melt. In addition, production rates are relatively low compared to other atomization processes such as gas atomization and water atomization.

In a recently developed process, the consumable rods used in REP are replaced by disk-shaped electrodes, since such electrodes are easier to fabricate than long round rods. During atomization, a rotating consumable disk is melted at its periphery.

2.2.6 Rotating Disk Atomization

Rotating disk atomization is in principle a centrifugal atomization process. In this process (Fig. 2.21), a metal or alloy is usually induction-melted in a vacuum chamber. When the desired melt temperature (superheat) is attained, the chamber is back-filled with an inert gas (typically, helium), and a water-cooled disk is brought to speed up. Helium jets are turned on, while the molten metal is poured on the center of the rotating disk. When the stream of the molten metal impinges onto the surface of the rapidly spinning disk, the liquid metal ejects from the periphery of the disk under the action of centrifugal forces, forming droplets. Droplets rapidly cool and solidify during flight traversing the cross flow of the helium jets.

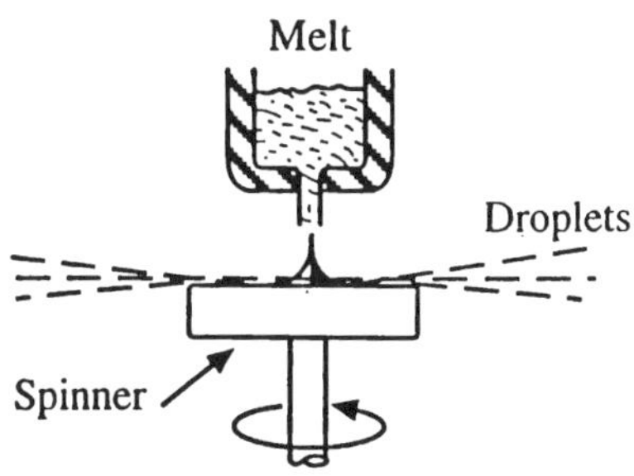

Figure 2.21. Schematic of rotating disk atomization of melt.

This process, originally designated as RSR (*rapid solidification rate*), was developed by Pratt and Whitney Aircraft Group and first operated in the late 1975 for the production of rapidly solidified nickel-base superalloy powders.[185][186] The major objective of the process is to achieve extremely high cooling rates in the atomized droplets *via* convective cooling in helium gas jets (dynamic helium quenching effects). Over the past decade, this technique has also been applied to the production of specialty aluminum alloy, steel, copper alloy, beryllium alloy, molybdenum, titanium alloy and silicide powders. The reactive metals (molybdenum and titanium) and silicide have been arc melted prior to atomization. The original RSR system had a melt capacity of 45 kg. It is increased to an intermediate size with 135 kg melt capacity and a spray chamber of about 2.3 m in diameter. The largest RSR facilities can handle batches of up to 900 kg with a spray chamber of about 5 m in diameter and a closed-loop helium recirculation system. The production rate varies from 3 kg/min to 18 kg/min for Ni-base superalloys.

In addition to mass flow rate and melt superheat, disk rotation speed is also one of the important process parameters in RSR. For the atomization of Ni-base superalloys, the process parameters typically take the following values: melt superheat 100 °C, melt flow rate 12 kg/min, helium gas velocity 170 m/s, helium gas flow rate 60 kg/min, helium back fill pressure 33 kPa, disk rotating speed 2500 radians/s.[4] Under these operation conditions, the droplet diameter is typically smaller than 200 μm and the mass median diameter can be as small as 60 μm, corresponding to a cooling rate of about 10^5–10^7 °C/s. The RSR-atomized droplets are normally spherical and have smooth surfaces without satellites. The ejection velocities of the droplets may range from 40 to 110 m/s, depending on the diameter of the rotating disk. As the centrifugal forces at the periphery of the rotating disk increase, the size of the ejected droplets decreases and correspondingly the cooling rates of the droplets are enhanced. The cooling rates are also significantly influenced by the melt superheat since it largely determines the droplet size. The melt superheat, in turn, may be changed either in the melting stage or by varying the heat flow to the rotating disk.

2.2.7 Electron Beam Rotating Disk Atomization

Electron beam rotating disk (EBRD) process[187] is a two-step centrifugal atomization process. A consumable metal ingot is melted at its bottom tip by an electron beam in high vacuum environment with some attendant refining effects. The molten metal drips onto a rapidly rotating container, either a disk or a cup, below the ingot. The liquid is ejected from the rim of the container under the action of centrifugal force and breaks up in flight into droplets. The ingot is either stationary or rotating at a low speed to develop conical tip geometry. Hence, the dimensional precision of consumables is not as critical as in REP, and rough and inhomogeneous ingots may be used as consumables. The droplets ejected off the rim are deflected downwards by a water-cooled copper shield. Thus, atomization chamber can be small, and cost and space requirements can be relatively low.

The details of the development of the EBRD process have been described by Pietsch et al.[187] There are two alternative operation modes in addition to the above continuous non-contact mode. The first one may be referred to as *continuous contact atomization*. In this mode, liquid metal contacts the bottom surface of the container instead of melt dripping, and then flows continuously from the center to the rim of the container. The second one may be termed *discontinuous non-contact atomization*. In this mode, the container is first filled by dripping melt while it is rotating at a very low speed of about 3×10^{-3} radians/s. The rotating speed of the container is then enhanced to about 14 radians/s while the metal or alloy is remelted and atomized. More than one focused electron beam may be used to provide energy for melting metal.

The EBRD process has been demonstrated applicable to the atomization of a variety of reactive and refractory metals and alloys in addition to Ti alloys. The process is being further developed to utilize highly contaminated raw materials and to produce high quality powders for hot isostatic pressing application without additional treatment.

As described in detail by Pietsch et al.,[187] in the largest production plant planned, the atomization chamber will have a diameter of 16 m which can accommodate consumables of 6 m in length and 0.165 m in diameter, corresponding to ~0.5 ton for Ti. The production rate may reach 5.5 kg/min at a rotating speed of 300 radians/s and <5 radians/s for the cup and ingot, respectively. Energy efficiency is relatively high, roughly 90%, and the energy requirement is about 2.5 MJ/kg of atomized particles.[4]

The EBRD-atomized particles may be of spherical or flaky shape. Droplet size is less than 700 μm, but much smaller droplets, 30 to 50 μm in diameter, have also been demonstrated. Cooling rates higher than 10^3 °C/s are achieved in fine droplets.

2.2.8 Roller Atomization

Roller atomization is a mechanical atomization process. It was invented in the mid 1970's.[188] In this process, as schematically depicted in Fig. 2.22, a stream of molten metal is fed into the gap between two counter-rotating rolls and forms a thin liquid sheet that subsequently disintegrates into droplets by the mechanical forces. In the original design, a pair of rollers of 100 mm in diameter are mounted in the same horizontal plane and rotate at speeds up to 1250 radians/s. The roll gap is about 50–100 μm, and the metal flow rate is up to 6 kg/min.

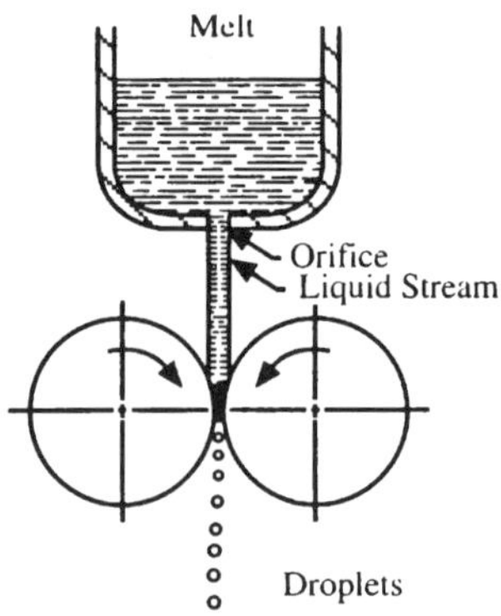

Figure 2.22. Schematic of roller atomization of melt.

To ensure that a liquid metal is completely disintegrated into droplets before the initiation of any solidification in the gap, the heat transfer between the liquid metal and the rolls must be minimized. Materials with low thermal conductivity and poor wetting properties have been found to have the most significant effect on reducing the heat transfer. Epoxy and phenolic resin coatings have been applied to the rolls to meet the heat transfer requirements for temperatures below about 300 °C, and epoxy-alumina powder mixture bonded to the surface of the rolls for higher temperatures. As an alternative, fluid barrier films such as light oils and releasing agents have been applied continuously to the roll surface to provide an effective thermal insulating layer between the liquid metal and rolls. Thus, the roller quenching effects required in producing rapidly solidified ribbons and splats can not be achieved in the roller atomization process.

Depending on operation conditions and metal properties, the shapes of the atomized particles may be spheroidal, flaky, acicular, or irregular, but spherical shape is predominant. The spheroidal particles are coarse. For example, roller-atomized Sn particles exhibited a mass median diameter of 220 to 680 μm. The large particle sizes and highly irregular particle shapes suggested that the disintegration process may be arrested either by the premature solidification or by the formation of a thick, viscous oxide layer on the liquid surface. The particle size distributions were found to closely follow a log-normal pattern even for non-uniform particle shapes.

Process parameters of primary importance include roll speed, differential roll speed, roll gap, metal flow rate, metal stream velocity, and melt superheat. The mass median diameter of particles diminishes exponentially as the roll speed increases. It is possible to obtain a smaller mass median diameter when one of the rolls is kept stationary rather than rotating the two rolls at the same speed. Metal flow rate seems to have a negligible effect on the mass median diameter. However, the mass median diameter increases with increasing metal stream velocity, suggesting that the relative velocity of the metal stream to the periphery of the rolls may be a fundamental variable controlling the mass median diameter. The size distribution is approximately constant for the conditions studied.

The roller atomization process has been applied to the atomization of many metals and alloys, such as lead, tin, aluminum, copper and steel. The production rate is potentially high, and the energy requirement is much lower than in commercial gas and water atomization processes. The energy efficiency is typically ~30%, as compared to ~3% and ~4% in gas and water atomization, respectively. The process also shows considerable potential for scale-up and development into a continuous process.

2.2.9 Centrifugal Shot Casting Atomization

Centrifugal shot casting (CSC) atomization[189] is conceptually similar to REP in the melting stage and to EBRD in the atomization stage. In this process, a stationary or slowly revolving consumable ingot (an as-cast, rolled or extruded electrode, normally as cathode) is positioned above a water-cooled, rapidly rotating container (normally a copper crucible, as anode) which usually has a skull of the same material as that to be atomized. The electrode may be made from powder compacted materials or from scrap. In an atmosphere of argon or helium, the electrode is melted at its bottom tip by an electronic arc struck between it and the crucible, and dripped into the crucible. By the action of centrifugal forces, the molten metal flows up to the rim of the container and breaks up into droplets while ejecting off the rim. Droplets freeze in flight into powder.

CSC atomization was developed by AEA Harwell Laboratories in the UK in the early 1970's. Initially, the CSC process was used for the atomization of refractory and oxide materials such as alumina, plutonium oxides, and uranium monocarbide in nuclear fuel applications. Since it is well-suited to the atomization of reactive metals/alloys or those subject to segregation, the CSC process has been applied to a variety of materials such as iron, cobalt, nickel, and titanium alloys and many refractory metals. The process also has potential to scale up to a continuous process.

The pressure of argon gas during atomization ranges from 0.03 to 0.1 MPa. The crucible diameter is 75 mm and its rotating speed is up to 400 radians/s. This combination of conditions gives a production rate up to 1 kg/min.[189] The CSC-atomized particles are either spherical or flaky. Spherical particles usually have smooth surfaces with a wide size range (100–1500 μm). The mass median diameter of droplets may be somewhere between 150 and 1000 μm with a standard deviation of about 1.3 to 1.4. From the microstructure of solidified particles, the cooling rates of droplets are estimated to be 2×10^2 to 7×10^4 °C/s, varying with droplet size and operation conditions.

2.2.10 Centrifugal Impact Atomization

In *centrifugal impact atomization*, a molten metal stream (2–3 mm in diameter) and a volatile liquid coolant are introduced simultaneously to a rapidly rotating impeller. The cone of the impeller disintegrates a small portion of the liquid metal into millimeter-sized drops and then the vanes strike these drops into small droplets. Most of the liquid metal forms a liquid film on the leading face of the vanes, accelerates to the same speed of the vanes, and then leaves the vanes, disintegrating into small spherical droplets. While the droplets are generated, the coolant vaporizes, taking heat from the droplets and leading to rapid cooling and solidification of the droplets. The majority of the droplets are solidified before contacting the shroud. Some liquid/droplets adhere to the vanes and continue to build up mass until centrifugal forces cause their removal. Detailed mechanisms for the droplet generation and rapid solidification in this process have been discussed in Ref. 190.

The centrifugal impact atomization process was developed in the early 1980's with patents granted to several researchers in Dow Corning Corporation.[190] In the original design, the atomization facility was capable of processing up to 0.9 kg batches and the propeller rotating speed was up to about 1250 radians/s. The process has been applied to the atomization of silicon, stainless steels (303, 304 and 316), Raney® type alloys (Ni/Al, Cu/Al and Mn/Al), nickel, copper and a glass forming alloy ($Fe_{81.5}B_{14.5}Si_3C_1$). Coolants include water, liquid ammonia, methanol and hexane. The flow rate of coolant is typically 6×10^{-4} to 2×10^{-3} m³/min. The droplet size ranges from <1 μm to >1 cm. Both fine spherical and coarse irregular-shaped droplets/particles have been generated with a size range of about <1–150 μm and <1–1000 μm, respectively. The droplet size distribution depends on many variables, but the propeller rotating speed and alloy composition are the most significant. Increasing the propeller rotating speed can reduce the droplet size and the agglomeration of particles. Although this process has not yet been scaled up for important industrial applications, it offers some flexibility in operating with different coolants, and high energy efficiencies and cooling rates.

2.2.11 Spinning Cup Atomization

In single-stage *spinning cup atomization* (Fig. 2.23), a small stream of molten metal is introduced into a layer of liquid quenchant that is held against the inner wall of a spinning cup. The liquid metal stream is sheared and broken up by the quenchant and the atomized droplets rapidly solidify within the quenchant. The solidified particles are collected on the inner wall of the cup and then separated from the liquid quenchant during post-processing. In two-stage spinning cup atomization, a liquid metal stream is first broken up into droplets (250–1000 μm) or fine ligaments (0.1–0.2 mm in diameter) *via* gas atomization or conventional centrifugal atomization. It is followed by the impact of the atomized droplets/ligaments onto the spinning cup containing liquid quenchant, leading to further breakup of the droplets/ligaments.

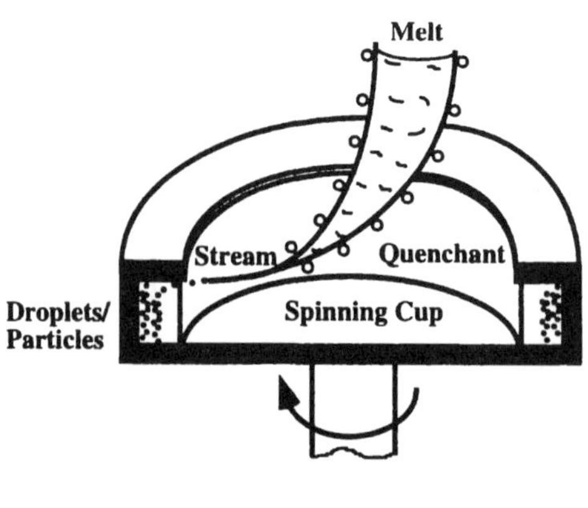

Figure 2.23. Schematic of single-stage spinning cup atomization of melt.

This process, also termed *rapid spinning cup* (RSC) process, was invented in the early 1980's contemporarily by Osaka University in Japan[191] and Battelle's Columbus Division in the US.[192] Unlike water atomization where water streams or droplets are used to disintegrate a molten metal, a coherent fast-moving liquid layer is used in the RSC process. Liquid quenchants include water, oil, glycerine, and other commercial quenching liquids. The materials atomized with the spinning cup method include a wide variety of metals and alloys such as tin, lead, aluminum alloys, copper alloys, iron alloys (stainless steels and high speed tool steels), zinc alloys and superalloys.[192]

The two-stage spinning cup method exhibits some advantages over gas and water atomization.[192] First, the design of the atomization facility is relatively simple and the dimension of the facility can be small because the atomization, cooling and solidification occur within the spinning cup. Second, the RSC process can generate very fine and clean spherical particles with narrow size range. The mass median size may range from <10 μm to ~300 μm with a standard deviation of 1.5 to 1.7. Third, cooling and solidification rates are high due to the intimate contact of droplets and the liquid quenchant. Fourth, some problems associated with nozzle system and post-processing inherent in gas or water atomization can be avoided in the RSC process. In addition, the process offers a high degree of flexibility in controlling particle characteristics such as shape, size and size distribution by changing process parameters. For example, the mean particle size can be reduced by increasing quenchant velocity, and the particle morphology can be changed in many instances from spherical to irregular to platelet shapes through proper selection of quenchant composition. Detailed descriptions of the RSC process, operation parameters, powder characteristics and applications have been given in Ref. 192. Although the RSC process has not yet established significant industrial applications, it is a method of interest.

2.2.12 Laser Spin Atomization

Laser spin atomization is conceptually a modification of REP. It was developed by Konitzer et al.[193] in the mid 1980's. In this process, a high power CO_2 laser beam is used to melt the top of a consumable metal/alloy ingot of ~10 mm in diameter while the consumable is spun at speeds up to 3650 radians/s by a high speed motor. Droplets are generated by spinning off the molten metal from the top of the consumable under the action of centrifugal force. Helium gas jets are introduced vertically to the streams of the droplets, giving rise to rapid quenching and solidification of the droplets. The evacuation of the unit prior to atomization minimizes the contamination of droplets.

The Laser-spin-atomized droplets are usually spherical, clean, and homogeneous in composition. A mass median diameter of ~100 μm has been obtained for a Ni-Al-Mo alloy. Cooling rates are estimated to be in the order of magnitude of 10^5 °C/s. Similarly to other centrifugal atomization techniques, droplet properties (shape, size, cooling rate, etc.) are dependent on the rotation speed, ingot diameter, superheat, and material properties.

2.2.13 Durarc® Process

Durarc® process is a centrifugal atomization process. In this process, the Durarc® electrode (a cylindrical metal/alloy of ~78 mm in diameter) is connected to a cylindrical field coil at its tip, and surrounded by several concentric cylinders which provide incoming and outgoing cooling water passages. The field coil is encapsulated within a removable, water-cooled copper tip. A magnetic field is generated by the coil with a large flux component parallel to the surface. When an arc is struck to the metal charge, held as a skull melt in the water-cooled copper crucible, the arc direction is to the melt. With the flux field parallel to the arc tip, it is normal to the arc current. Magnetic field interaction with the current causes the latter to rotate over the arc tip. With the liquid metal pool present, the arc is rotated rapidly. Electromagnetic forces also exist in the melt at the point of arc attachment and lead to the expulsion of the liquid metal from the melt pool and eventually disintegration of the liquid metal into droplets. The droplets solidify and are collected subsequently.

Durarc® process was developed for atomizing titanium alloys at the Research and Development Center of the Westinghouse Electric Corporation in the early 1970's.[194] The process utilizes the non-consumable arc melting technology for titanium alloys and involves melting of a material in a water-cooled copper skull crucible. The operation is carried out in a large chamber (2.6 m in height and 1.85 m in diameter in a pilot-scale facility), either under vacuum or in inert gas atmosphere (up to 0.92 MPa). This feature, along with the skull melting step, makes the Durarc® process attractive for reactive or high melting-temperature metals such as Ti, Zr, and others. In addition, the starting material can be scrap or contaminated feed stock and there is no limitation on its geometry. For Ti-6Al-2V-2Sn alloy, the atomized particles are spherical and relatively coarse (≥ 250 μm).

In summary, a variety of configurations of centrifugal atomization techniques have been examined for the atomization of metals/alloys and are undergoing further development. In centrifugal atomization techniques, the spinner can be a dish or a cup, a crucible or a flat disk, or some variations of these. Centrifugal forces can also be generated by rotating a consumable metal ingot. Generally, energy requirements of centrifugal processes are relatively low. Droplet shape can vary from spherical, flattened to elongated, and droplet size generally exhibits a narrow distribution with a standard deviation of 1.3–1.5. Some problems associated with centrifugal atomization include the erosion and/or dissolution of the spinner-atomizer device, and the creep or deformation of the spinner during the atomization of high melting-temperature metals/alloys. In an overflow spinning arrangement, control of the lip is critical for consistent product definition. The rim speed and diameter substantially determine the droplet size distribution. In addition, control over the atmosphere is also of importance. Since centrifugal atomization does not require an atomizing fluid, spray chamber could be in vacuum. However, the cooling rate in vacuum is low. Therefore, a fluid or metallic impact substrate is provided for higher cooling rates at increased cost to the atomization system.

2.2.14 Vibrating Electrode Atomization

Vibrating electrode atomization is a process similar to the rotating electrode atomization. The difference between them is that in the former an electrode vibrates while in the latter an electrode rotates during atomization. In the vibrating electrode process (VEP),[195] a wire (consumable electrode) is continuously fed by rollers into an atomization chamber. The rollers are attached to an electrodynamic oscillator so that vibrations are transferred to the wire electrode. In the chamber, a water-cooled rotating electrode made of copper or graphite is placed opposite of the end of the wire electrode. When an arc is struck between the rotating electrode and the end of the vibrating wire electrode, the end is molten and the liquid metal is ejected away, leading to atomization.

In the VEP, currents used are between 600 and 1200 A at potentials between 30 and 60 V. The vibration frequency of the wire electrode is up to 500 Hz. The materials atomized *via* VEP include mild steel, Cr-Ni steel, Cu-Ni alloy and tungsten. The VEP is carried out in an inert atmosphere (typically argon) for most alloys, but the arc is struck under water for tungsten wire. Wire diameter is 1–4 mm, and its feed rate is 1.7–4.3 m/min. The feed rate and current density must be determined properly according to the relationship between these two variables. At lower current densities, the wire electrode tends to stick to the rotating electrode. At higher current densities, the wire electrode becomes overheated, causing it to bend or even rupture.

The VEP-atomized particles are spherical. The mass median diameter of the particles ranges typically from 300 to 500 μm. Both the mass median diameter and size range of the particles reduce with decreasing wire diameter for a given vibration frequency. The narrowest particle size distribution is produced at the resonant frequency. The mass median diameter decreases slightly as the current density decreases, and the influence of the current density and vibration mode on the particle size distribution is little. Due to the simplicity and ease of use, the VEP is an attractive candidate for atomizing small quantities of high-purity, refractory metal powders of spherical shape.

2.2.15 Ultrasonic Atomization

Ultrasonic atomization is sometimes also termed *capillary-wave atomization.* In its most common form,[142] a thin film of a molten metal is atomized by the vibrations of the surface on which it flows. Standing waves are induced in the thin film by an oscillator that vibrates vertically to the film surface at ultrasonic frequencies. The liquid metal film is broken up at the antinodes along the surface into fine droplets once the amplitude of the capillary wave exceeds a certain value. The most-frequent diameter of the droplets generated is approximately one fourth of the wavelength of the capillary wave,[142] and thus decreases with increasing frequency.

The Ultrasonic-atomized droplets are spherical and of high quality if the atomization is conducted in an inert atmosphere. Droplet size is typically 30–50 μm, but finer droplets (10–20 μm) have also been produced with this method. Frequencies of up to 150 kHz have been used. The dimension of the chamber for ultrasonic atomization can be small (1–2 m in height and 1.5 m in diameter) due to the low droplet ejection velocities. During atomization, the vibrating surface needs to be heated to the melting temperature of the metal being atomized in order to avoid solidification of the melt on the surface. Thermal damage to the ultrasonic transducer and materials selection for the vibrating surface become severe problems as temperature rises to above a few hundred degrees. Thus, the process is only applicable to metals and alloys of low melting temperatures (up to about 1000 °C), such as solder materials,[144] welding electrodes, silver and copper-base alloys for electrical contacts, dispersion hardened alloys, and powder materials needed in the chemical and pharmaceutical industries. As a result of the high energy efficiency, the energy requirement is less than 100 J/kg of atomized particles. However, because of the interrelationship among scale, frequency and droplet size, the production rate is very low, typically of the order of 0.02–0.8 kg/min for droplet sizes less than 50 μm. This method is therefore not yet established as a near-commercial atomization process for powder production. Nevertheless, scale-up using arrays of vibrators or vibrating ribbons or other shapes is technically feasible.

Recently, a variant of the ultrasonic atomization process has been developed and demonstrated by Bauckhage et al.[143] at University of Bremen, Germany. This process, termed *ultrasonic standing wave atomization* (USWA), is in concept different from the earlier one. As schematically illustrated in Fig. 2.24, a molten metal stream (~2 mm in diameter) falls between two ultrasonic transmitters (sonotrodes) or one transmitter and one reflector which are positioned and tuned to set up an intense standing wave field. When the transmitters are positioned 1.5 sound-wavelengths apart, the liquid metal stream falling through a central, low sound intensity zone is peeled off at its edges, forming a flat spray. Atomization is conducted in a chamber filled with inert gas at up to ~1.6 MPa pressure. The mass median diameter of droplets may range from ~10 to ~300 μm, depending on the amplitude of the sonotrodes and the gas pressure. Increasing the amplitude of the sonotrodes and/or the gas pressure can reduce the mass median diameter of droplets. The cooling rates of droplets are remarkably high, approximately in the order of magnitude of 10^6 K/s, as estimated from the dendrite spacing measurements. Since the relative gas-droplet velocity is very low (<0.5 m/s), it is hypothesized that the agitating effect of the ultrasound leads to the rapid cooling by enhancing convective heat transfer. The as-atomized particles possess excellent sphericity and are free of contamination. Some challenging technical issues associated with the USWA process include, for example, how to keep the system resonating for very large temperature changes in the gas, and how to

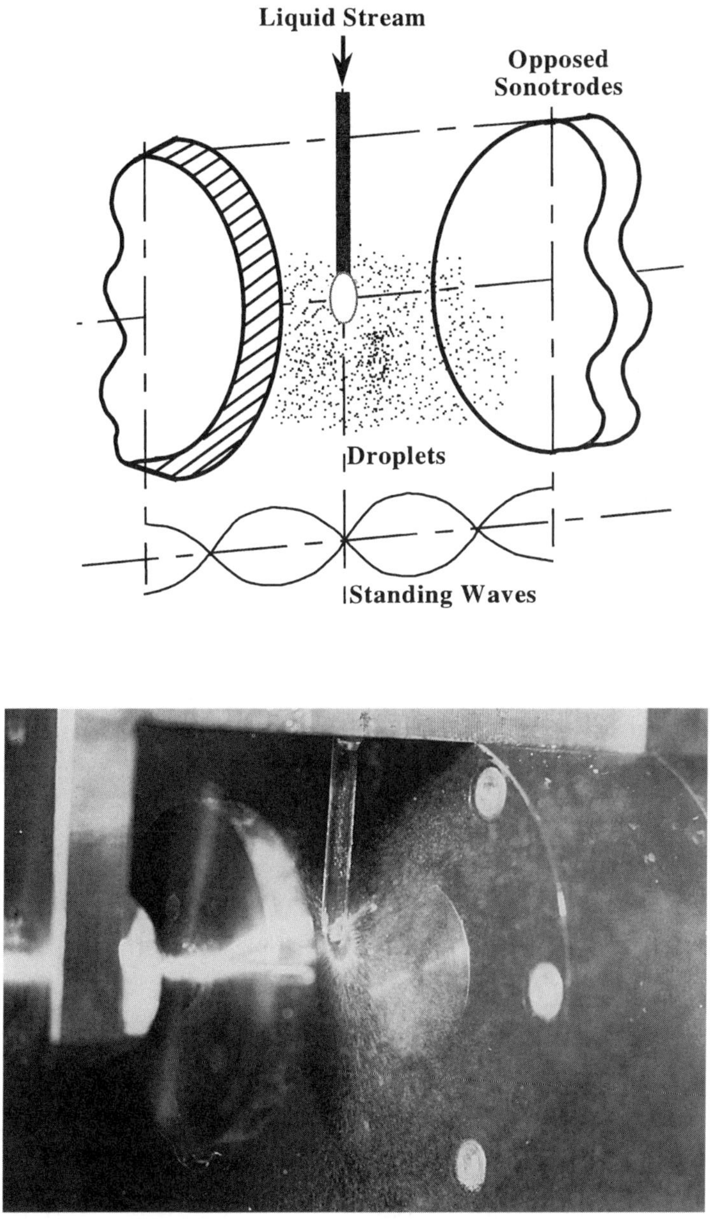

Figure 2.24. Ultrasonic standing wave atomization of melt. *Top:* schematic. *Bottom:* photograph. *(Courtesy of Prof. Dr-Ing. Klaus Bauckhage at University of Bremen, Germany.)*

avoid heating up and contamination of the ultrasonic vibrating surfaces. The USWA technique has been demonstrated to be capable of atomizing extremely viscous fluids (such as synthetic resins and highly concentrated suspensions) or fluids of high surface tension (such as molten metals/alloys) into narrow-sized, very fine, spherical droplets/powders.

2.2.16 Steam Atomization

Atomization of melts was first conducted for lead and tin *via* steam or air in the early 1920's. *Steam atomization* provides a choice of conditions between those of gas and water atomization. Low-pressure steam can produce relatively coarse droplets of less spherical shape (irregular, spheroidal). On average, relatively high quench rates, somewhat greater than 10^3 °C/s, can be achieved for finer droplets. Steam atomization has been successfully applied to most carbon steels and low alloy steels, stainless steels, Co-base alloys and superalloys, non-Al-Ti Ni-base alloys. Typical oxygen levels for untreated powders are 500 to 1500 ppm. Self-milling of such powders can remove 80%–90% of the surface oxides. To further lower the oxygen level, inexpensive chemical treatments can be used. Typically, cleaned stainless steel powders exhibit oxygen levels of 0.02%–0.04%, which are extremely clean according to ASTM standards.

2.2.17 Other Atomization Methods

In addition to the methods described above, there are other lesser known but interesting atomization techniques. In the eighteenth century, lead shot was made by pouring lead melt through a perforated basket down a shot tower, similar to the techniques used today for granulating precious metals. During World War II, a novel atomization process was invented for iron powder production in Germany, i.e., the *DPG spinning disc* method. In the DPG process, a partially water-atomized iron melt was directed to a rotating disc and further atomized by passing through slot orifices on the disc. Reynolds

Metals Company used a commercial centrifugal atomizer to produce elongated droplets quenched with air. This atomizer has a spinner consisting of a steel cylinder with numerous fine holes along the barrel. The atomization system, while generating large output and high yield, allows variation of the spinner speed and hole size, adjustment of the melt temperature, and control of the atmosphere if necessary.

A simple experimental apparatus[196] has been used to produce individual mercury and sodium amalgam droplets in the size range of 0.3–2.5 mm diameter. A high resistance to the flow of the liquids is provided by a capillary tube consisting of a 0.20 mm OD tungsten wire placed in a 0.25 mm capillary bore. Flow rates of 1 μl/s and lower can be obtained, enabling the formation of droplets as small as 1 mm diameter in a time interval of 20 s. The droplets are formed at a plastic tip and stripped off by a stream of argon when reaching the desired size. The variation in the droplet size can be controlled within +/-2% of the mean value.

In *pressure atomization*, a liquid metal is pressurized by an inert gas through a very small diameter nozzle at the bottom of a melt crucible to form droplets. The process is relatively gas efficient and energy efficient. Using a swirl nozzle, alloys of melting temperatures up to 1300 K have been atomized, generating a mass median size as small as 50 μm and a narrow size range (a standard deviation of 1.4) with good sphericity. The nozzles used may be made from stainless steel, graphite, or ceramics allowing sufficient nozzle lives. Gas pressures of about 2 MPa are feasible for the production of fine particles. To achieve a large throughput, pressures in excess of 10 MPa or multiple arrays of small nozzles may be used. The method can process low melting-temperature metals at a rate over 1 ton/hour and make particles of 100–1000 μm. The two-fluid prefilming version of pressure atomization technique[159] is even more energy efficient due to the internal-mixing effect, as discussed previously. Another variation involves the generation of droplets by forcing liquid metal through a porous ceramic filter. The pore size of the filter may range from 40 to 150 μm, generating aluminum droplets with a mean size of 500–900 μm. Backup pressure above the melt can speed up the process. The use of gas blast or vibrations can increase the cooling rates of droplets and result in smaller droplets by detaching droplets from the filter pores earlier. This technique can generate fairly spherical droplets and control droplet size by changing filter pore geometry and size. Oxides and foreign particles may be filtered out during the process. Moreover, the associated apparatus is simple.

As discussed in the next chapter, a liquid stream may spontaneously break up into droplets. Superimposition of vibrations as longitudinal oscillations (50–1000 Hz) on a stream through a nozzle orifice can promote the breakup and make it occur at a certain time, giving rise to almost identical droplet size and excellent uniformity (a standard deviation less than 1.05). This is the basic idea of the *vibrating-orifice atomization* technique. Since it makes one droplet at a time, the throughput is low and falls roughly with the square of orifice diameter or droplet diameter. The nozzle diameter and melt stream speed are important process parameters influencing droplet size. The melt density and surface tension are also of importance to the size distribution of droplets generated.

Electromagnetic forces may be used to disrupt a liquid metal. This requires passing a current through the liquid metal. Apparently, the current will cease once the breakup of the liquid metal begins. In *electromagnetic atomization*,[197] the current is passed from the liquid metal through a nozzle to an electrode. The gap between the nozzle and the electrode is placed in a strong magnet field. Intensive atomization then takes place due to very strong expulsive forces generated during pouring the liquid metal against the electrode. Expulsion velocities may exceed 100 m/s, depending on the strength of the magnetic field. The breakup of the liquid metal stream is controlled by the excitation of the Rayleigh instability. Very uniform tin droplets of 600–850 μm have been produced by pulsing the current every 27 ms and using a 0.5 mm diameter nozzle at a flow rate of 0.05–0.12 kg/min.[5] The atomization can be conducted in a vacuum, making the process cost lower than that using other gases such as argon. Another process, termed *magnetohydrodynamic atomization*, is conceptually similar to electromagnetic atomization in terms of the breakup mechanism. It is also similar to vibrating-orifice atomization because it involves vibrations. However, the vibrations are imposed on the metal stream by pulsing a current through the liquid metal in the tundish placed in a magnetic field. Impulses and waveform can be well controlled and altered to allow optimum performance and control over droplet size distribution. The atomized lead droplets are 2–4 mm in diameter with a narrow size distribution within 10%.[5]

Impaction atomization, referred to as *Granshot process*, has been used by Uddeholm in Sweden to generate large particles of iron and other alloys. In this process, a liquid metal stream is poured onto a flat, circular refractory brick in air. The liquid metal splashes and ejects off the brick surface into a thin sheet and breaks up *via* ligaments into droplets during flight. It is followed by water quenching of droplets/particles. The atomization regime of the Granshot process is the same as the sheet regime for a rotating cup or disc. This technique is a major industrial method for granulation of liquid metals at large throughput up to several tons per minute. It has been used for the atomization of steels, cast iron, ferroalloys and mattes in large quantities. Melt stream can be over 20 mm in diameter. The particle size ranges from 3 to 15 mm, and smaller ones are spherical while larger ones are flat with a thickness of 4–7 mm. Droplet size and shape can not be readily changed because they are mainly determined by physical and chemical properties of the liquid metal/alloy of interest. For some alloys, however, particle shape tends to be largely influenced by oxide film formation. In addition, the process requires much higher stream velocity and is limited to smaller scales for the production of powders, making the advantages of the method no longer apparent. Therefore, this technique does not appear to have established significant commercial applications in powder production.

Electrostatic atomization technique has not been developed suitably for atomizing liquid metals/alloys at sufficiently high metal flow rates. Both the throughput and capacity are low. In addition, it is difficult to raise a liquid metal to over 10 kV. Thus, despite its potential advantages such as narrow size range and low velocity spray, this method is difficult to establish industrial applications for powder production. Recently, attempts have been made to scale-up electrostatic atomization using linear arrays of Taylor cones.[198] The feasibility of increasing the liquid throughput rate in electrostatic atomizers has been evaluated, and conditions for high-quality powders, composed of dense, spheroidal, submicrometer, and nanocrystalline oxide particles, have been found. The method has been applied to the synthesis of yttria powders.

3

Fundamental Phenomena and Principles in Droplet Processes

Droplet properties have an important impact on the transport phenomena in associated droplet processes. A thorough understanding of fundamental phenomena, principles and mechanisms in droplet processes is therefore needed in order to enhance efficiency of droplet generation, and to control droplet properties. In this chapter, the mechanisms governing droplet generation and deformation in various droplet processes are reviewed in detail.

3.1.0 DROPLET FORMATION

Atomization generally refers to a process in which a bulk liquid is disintegrated into small drops or droplets by internal and/or external forces as a result of the interaction between the liquid (dispersed phase) and surrounding medium (continuous phase). The term *dispersed phase* represents the liquid to be atomized and the atomized drops/droplets, whereas the term *continuous phase* refers to the medium in which the atomization occurs or by which a liquid is atomized. The disintegration or breakup occurs when the disruptive forces exceed the liquid surface tension force. The consolidating effect of liquid surface tension tends to pull a liquid into a form that exhibits the minimum surface energy, while the stabilizing effect of liquid viscosity tends to oppose any change in liquid geometry. External forces, such as aerodynamic forces, centrifugal forces and electrostatic forces, acting on the liquid surface may distort the bulk liquid and promote the disruption.

When a bulk liquid is exposed to another continuous phase, for example, a liquid column, jet or sheet emerging from a nozzle into a quiescent or moving gaseous environment, or a liquid film or free surface in a blowing gas flow, the competition between the cohesive and disruptive forces will set up on the liquid surface, leading to oscillations and perturbations in the liquid. Under favorable conditions, the oscillations may be amplified to such an extent that the bulk liquid disintegrates into droplets. This initial breakup process is often referred to as *primary breakup*, *primary disintegration*, or *primary atomization*. A population of larger droplets produced in the primary atomization may be unstable if they exceed a critical droplet size and thus may undergo further disruption into smaller droplets. This process is usually termed *secondary breakup*, *secondary disintegration*, or *secondary atomization*. Therefore, the final droplet size distribution produced in an atomization process is determined by the liquid properties in both the primary and secondary disintegration.

3.1.1 Droplet Formation in Atomization of Normal Liquids

Atomization of normal liquids has a wide range of applications, as discussed previously. In many applications, the fundamental phenomena and principles during atomization are common or similar. According to the geometry feature of bulk liquids, droplet formation may be loosely classified into the following primary "idealized" modes:

(a) Liquid dripping

(b) Liquid column/jet breakup

(c) Liquid ligament breakup

(*d*) Liquid sheet/film breakup

(*e*) Liquid free-surface breakup

Among these modes, the liquid dripping is the simplest mode of droplet formation frequently observed in nature. The liquid jet breakup mode is of primary importance for the design of plain-orifice atomizers that are used extensively in diesel and rocket engines. The liquid ligament breakup mode may occur haphazardly during plain-jet air-blast atomization and rotary disk atomization in industrial gas turbine and spray drying applications, respectively. The liquid sheet/film breakup mode is directly relevant to the design of rotary atomizers for spray drying and to the performance of pressure-swirl and prefilming air-blast atomizers that are widely used in spray combustion such as oil burners and gas turbine combustors. It is well-established[199] that the droplet sizes produced by pressure atomization or air-blast atomization are proportional to the square root of the initial thickness of the sheets or the initial diameter of the ligaments from which the droplets are formed. The liquid free-surface breakup mode may be an idealized operating mechanism for the formation of waterfall mists and ocean sprays.[200]-[203] Most of the fundamental modes may produce monosize droplets with satellites. Although some idealized modes such as the liquid dripping, sheet and free-surface breakup may be observed in real world, the ideal orderly conditions for the fundamental modes seldom occur in most atomization processes, and liquid breakup tends to be chaotic, leading to a wide range of droplet size with a maximum to minimum size ratio in excess of 10. However, some atomization techniques, notably centrifugal atomization, may operate in the *Direct Droplet Formation* regime and thus may approach the idealized breakup mode as evidenced by the narrow droplet size distribution.

Various mechanisms have been proposed to account for these breakup modes, such as dripping mechanism, Rayleigh mechanism,[37] Weber theory,[38] Ohnesorge criteria,[40] Castleman hypothesis,[204] Taylor mechanisms and instability theory,[205][206] as well as mechanisms proposed by Fraser et al.[116][207][208] by Mansour and Chigier,[209] and by Faragó and Chigier,[210] among others.

Recent research efforts[210]-[212] have focused on establishing the most important theories and mechanisms governing the fundamental physical processes of breakup of liquid jets and sheets. Thermodynamic analyses and wave growth theories have been developed on the basis of experimental observations and measurements of wavelength, amplitude and frequency. Detailed studies on the physical processes that influence or can be made to influence the liquid breakup will establish the link between atomizer design and spray characteristics, and allow reasonably complete, qualitative and quantitative descriptions of the dominant physical processes. The ultimate objective of the theoretical analyses is to predict the temporal and spatial joint distribution functions of droplet sizes, three-dimensional (3-D) components of velocities, temperatures, and other droplet properties in sprays and their dependence on design features (geometry and configuration of atomizer and spray chamber) and operation conditions (mass flow rate ratio of liquid to gas, pressure and temperature). Extensive reviews have been made by Giffen and Muraszew,[213] Chigier,[214] and Lefebvre.[1]

Liquid Dripping. The most elementary mode of droplet formation is perhaps the liquid dripping from an orifice where a hanging or pendant drop forms quasi-statically under the action of gravity. This occurs typically when a liquid is slowly discharged from the end of a thin tube, for example, a dripping faucet or a burette (Fig. 3.1). Therefore, it is referred to as *dripping mechanism*. When the gravity force on the liquid exceeds the attaching surface tension force, the liquid will be pulled away from its attachment, leading to the formation of a droplet. Due to the low flow velocity of the liquid in the dripping mechanism, the gravitational and surface tension forces on the drop govern the formation process of the drop and determine the drop mass and size. For example, for the slow emission of a liquid with surface tension σ, and density ρ_L from a thin circular tube of diameter d_0, the size D of the spherical drop formed can be calculated from the following equation based on the force balance:

Eq. (1) $$D = \left(\frac{6 d_0 \sigma}{\rho_L g} \right)^{1/3}$$

For a liquid (for example, water), the size of a drop formed at a sharp-edged tube opening can be reduced from 3.6 mm to 784 μm if the opening diameter decreases from 1 mm to 10 μm. Apparently, the dripping mechanism is concerned with large drops and low liquid flow rates. Therefore, it is a functional mechanism in drop formation processes common in nature.

Figure 3.1. Liquid breakup *via* dripping mode.

Various equilibrium forms of suspended drops can be obtained for different values of the radius of curvature at the drop tip and for different values of surface tension and gravitational forces, as calculated by Tanasawa and Toyoda.[41] These researchers also presented photographs of various modes of dripping from a cylindrical tube as well as successive stages of dripping from a large plane surface. Small satellite droplets were observed following large drops. For very viscous liquids, a long continuous filament tails the main drop and subsequently breaks up into a large number of small droplets.

Recently, the size and shape of a liquid droplet at the molten tip of an arc electrode have been studied,[215] and an iterative method for the shape of static drops has been proposed.[216] Shapes, stabilities and oscillations of pendant droplets in an electric field have also been addressed in some investigations.[217][218] The pendant drop process has found applications in determining surface tensions of molten substances.[152] However, the liquid dripping process is not an effective means for those practical applications that necessitate high liquid flow rates and fine droplets (typically 1–300 μm). For such fine droplets, gravitational forces become negligible in the droplet formation mechanism.

Another notable example of the dripping mechanism is the slow formation of a drop by breakaway from a liquid film on the bottom surface of a flat, horizontal plate under the action of gravity. The size of the drop formed can be estimated using the following expression derived on the basis of a force balance between gravity and surface tension:[1]

Eq. (2) $$D = 3.3\left(\frac{\sigma}{\rho_L g}\right)^{0.5}$$

From a liquid film such as a water film, the diameter of a drop formed under the action of gravity is calculated to be 9 mm with the above equation. Similarly to the liquid dripping mode, the liquid film breakup mode governed by the dripping mechanism is also typified by large droplets and low liquid flow rates.

Liquid Column/Jet Breakup. When a liquid jet issues from a nozzle, oscillations and perturbations form on the jet surface as a result of the competition of cohesive and disruptive forces. Under certain conditions, for example, if the liquid flow rate exceeds a certain value, the equilibrium between the surface tension and gravitational forces can no longer be maintained, and the transition from dripping to laminar flow will occur. The oscillations may be amplified to such an extent that the liquid breaks up into droplets. The continuous length of a liquid jet prior to breakup and the droplet size are important parameters characterizing a liquid jet breakup process. These parameters reflect the growth rate of disturbances and the wave number of the most unstable disturbance, respectively.

The phenomenon of liquid jet breakup has been the subject of theoretical and experimental investigations for more than one century.[37]-[41][115][210][219]-[230] Reviews of liquid jet breakup mechanisms have been made by Tanasawa and Toyoda,[41] McCarthy and Molloy,[230] and Reitz and Bracco,[229] among others.

In *Rayleigh's theory* on jet instability,[37] a non-viscous liquid jet affected by surface tension forces under laminar flow conditions becomes unstable only to the axisymmetrical disturbances whose wavelength exceeds the perimeter of the jet, i.e., disturbances on a jet with wavelengths greater than its circumference grow. Further, one class of disturbances grow fastest and govern the jet breakup process. The exponential growth rate q_{max} and wavelength λ_{opt} of the fastest-growing disturbance may be expressed, respectively, as:

Eq. (3) $$q_{\max} = 0.97\left(\frac{\sigma}{\rho_L d_0^3}\right)^{0.5}$$

Eq. (4) $$\lambda_{opt} = 4.51 d_0$$

In the idealized mode, liquid jet breakup and droplet formation are fairly regular. The liquid jet running downwards collapses, forming droplets of uniform size at uniform spacing. After breakup, the liquid jet of length $4.51d_0$ converts into a spherical droplet so that:

Eq. (5) $$D = 1.89d_0$$

Namely, the average droplet diameter D is roughly twice the liquid jet diameter d_0. Another analysis based on the consideration of surface energy gave $\lambda_{opt} = \pi\, d_0$.[88]

Rayleigh's theory was based on the analysis of the breakup of a non-viscous jet with adequate surface tension and at low relative velocity. In more general cases, liquid jets may be viscous or turbulent, and surrounding air or gas may exert influences on the jets. Drag forces tend to tear the surface of a turbulent jet into fine ligaments that subsequently break up into droplets under the action of surface tension. In addition, the droplets formed usually coalesce, giving rise to small satellite droplets between large droplets. Hence, the droplet size predicted by Rayleigh's theory represents the upper bound of actual droplet sizes. Although the Rayleigh mechanism was derived from the analysis of the idealized breakup of non-viscous liquid jets under laminar flow conditions, it has been widely accepted in subsequent theory developments as first approximation.

To account for the effect of liquid viscosity on liquid jet breakup, Weber extended Rayleigh's theory to a more general theory for low-velocity jet breakup. In *Weber's theory*,[38] it is assumed that any disturbance causes rotationally symmetrical oscillations on a jet. Surface forces may either damp out or amplify the disturbance if the wavelength of the initial disturbance is less or greater than a minimum value λ_{min}, respectively. If the disturbance is amplified, the jet may eventually break up. The minimum wavelength is equal to the jet circumference for both non-viscous and viscous liquids, consistent with Rayleigh's theory. There exists a particular wavelength λ_{opt} that is most favorable for liquid disintegration:

Eq. (6)

$$\lambda_{opt} = \begin{cases} 4.44 d_0 & \text{non-viscous liquids} \\ \sqrt{2}\pi d_0 \left(1 + \dfrac{3\mu_L}{\sqrt{\rho_L \sigma d_0}}\right)^{0.5} & \text{viscous liquids} \end{cases}$$

where μ_L is the dynamic viscosity of liquid. The expression for non-viscous liquids is close to Rayleigh's analysis result. Interestingly, liquid viscosity has no effect on the minimum wavelength, but it increases the optimum wavelength. Haenlein's experiments[39] showed that the wavelength corresponding to maximum instability may range from $30d_0$ to $40d_0$ for high-viscosity liquids. Weber also found that air friction or relative air velocity shortens both the minimum wavelength and the optimum wavelength. For non-viscous liquids, for example, these values are decreased from $3.14d_0$ and $4.44d_0$ to $2.2d_0$ and $2.8d_0$, respectively, when relative air velocity is increased from zero to 15 m/s. In addition, air flow may induce wave formation when relative air velocity exceeds a certain minimum value. If all the liquid within one wavelength of the jet forms a droplet ideally, then the droplet diameter may be expressed as:

Eq. (7) $$D = \left(1.5\lambda_{opt} d_0^2\right)^{1/3}$$

This is essentially identical to Rayleigh's formulation (Eq. 5) if the wavelength of the fastest-growing disturbance is calculated using the same equation (Eq. 4). This expression is also virtually identical to the formulation for the maximum droplet diameter derived by Tyler[231] assuming that the volume of the droplet obtained by primary disintegration of a jet is equal to the volume of a cylinder of diameter d_0 and length equal to the wavelength of the most rapidly growing disturbance. These are perhaps the first plausible conjecture that fitted well with experimental data.

Weber's theory has been further extended by many investigators[39]–[41][204][220][227] to account for high-velocity jet breakup and droplet formation under the influence of ambient air. Various mechanisms of jet breakup[40]–[41][210][220][227][232] have been proposed and divided into breakup regimes to reflect the differences in the appearance of jets and to identify the dominant forces leading to jet breakup as operation conditions are changed.

For *steady injection of a liquid through a single nozzle with circular orifice into a quiescent gas* (air), the mechanisms of jet breakup are typically classified into four primary regimes (Fig. 3.2)[40][41][220][227] according to the relative importance of inertial, surface tension, viscous, and aerodynamic forces. The most commonly quoted criteria for the classification are perhaps those proposed by Ohnesorge.[40] Each regime is characterized by the magnitudes of the *Reynolds number* Re_L and a dimensionless number Z:

Eq. (8) $$\mathrm{Re}_L = \rho_L U_L d_0 / \mu_L$$

Eq. (9) $$Z = \mathrm{Oh} = \mathrm{We}_L^{0.5}\,\mathrm{Re}_L^{-1} = \frac{\mu_L}{(\rho_L \sigma d_0)^{0.5}}$$

with $$\mathrm{We}_L = \frac{U_L^2 \rho_L d_0}{\sigma}$$

where U_L is the liquid velocity at the nozzle exit, We_L is the *Weber number*, Z is known as *Ohnesorge number*, Oh, and sometimes also referred to as *Stability number* or *viscous group*. Ohnesorge number represents the ratio of an internal viscosity force to a surface tension force. Depending on jet velocity, nozzle diameter and properties of liquid and air, the disintegration of a round liquid jet in a quiescent gas may occur in the following four regimes,[39][40][204][220][227][229] as depicted in Fig. 3.2[40][41][220][227] and summarized in Table 3.1.

Table 3.1. Classification and Criteria of Breakup Regimes of Round Liquid Jets in Quiescent Air

Regime[40][220][227]	Predominant Breakup Mechanism[40][227]	Criteria*[227]
Rayleigh Jet Breakup (Varicose Breakup)	Surface Tension Force	$We_A < 0.4$ or $We_A < 1.2 + 3.41Oh^{0}$
First Wind-Induced Breakup (Sinuous Wave Breakup)	Surface Tension Force, Dynamic Pressure of Ambient Air	$1.2 + 3.41Oh^{0.9} < We_A < 13$
Second Wind-Induced Breakup (Wave-like Breakup with Air Friction)	Dynamic Pressure of Ambient Air	$13 < We_A < 40.3$
Atomization	Unknown, but Plausibly: Aerodynamic Interaction, Turbulence, Cavitation, Bursting Effect	$We_A > 40.3$ or $Oh \geq 100\,Re_L^{-0.92}$ [220]

*$We_A = U_L^2 \rho_A d_0/\sigma$ where ρ_A is the density of gas (air).

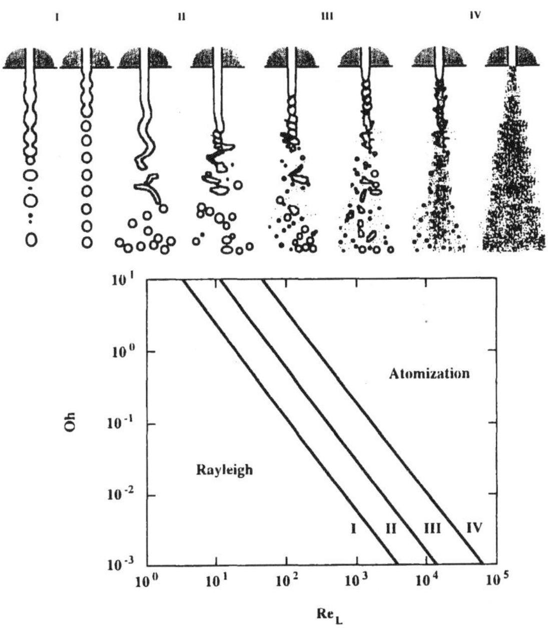

Figure 3.2. Breakup regimes of round liquid jets in quiescent air. I: Rayleigh Jet Breakup (Varicose Breakup); II: First Wind-Induced Breakup (Sinuous Wave Breakup); III: Second Wind-Induced Breakup (Wave-like Breakup with Air Friction); IV: Atomization.

I. Rayleigh Jet Breakup. This regime is sometimes also termed *varicose breakup* to describe the appearance of the liquid jet. Radially symmetric waves (dilational waves) are formed as a result of the interaction of primary disturbances in the liquid and surface tension forces. The breakup of the liquid jet is caused by the growth of axisymmetric oscillations of the jet surface, induced by surface tension. There is a balance between the inertial and surface tension forces.[229] The jet breakup length is linearly proportional to the jet velocity, typically many nozzle diameters downstream of the discharge nozzle exit. Liquid viscosity moves the most unstable waves to longer wavelengths but its stabilizing effect is secondary, as in other breakup regimes.[229] The breakup time is proportional to $d_0^{1.5}$ for non-viscous liquids and to d_0 for viscous liquids. Droplet diameters are larger than the jet diameter. With increasing jet velocity, the surface tension effect is augmented. The aerodynamic forces of surrounding air are no longer negligible and may attenuate the waves formed. A static pressure distribution may be produced across the jet, accelerating the breakup. Droplet diameters become approximately the same as the jet diameter. This regime is characterized by low Reynolds numbers and large droplets of nearly uniform size.

II. First Wind-Induced Breakup. With increasing jet velocity, the effect of aerodynamic forces becomes increasingly important and the relative influence of surface tension is lessened. The breakup of the jet is caused by jet oscillations with respect to the jet axis (*sinuous waves*). The magnitude of the oscillations increases with air resistance, and the wave growth is opposed by surface tension. The inertial effect of the surrounding air enhances the wave growth rate as the Weber number is increased.[229] Liquid breakup occurs a few nozzle diameters downstream of the discharge nozzle exit. Average droplet diameters are smaller than the jet diameter.

III. Second Wind-Induced Breakup. Further increasing jet velocity, the dynamic pressure of the surrounding air becomes predominant. The breakup of the jet is caused by the unstable growth of short-wavelength surface waves due to the relative motion between the jet and the surrounding air. The maximum growth rate occurs at progressively shorter wavelengths with increasing Weber number. At sufficiently large Weber numbers, the maximum growth rate and the corresponding wavelength become independent of the jet diameter.[229] Liquid breakup occurs several nozzle diameters downstream of the discharge nozzle exit. Droplet diameters are smaller than the jet diameter. This regime is characterized by a wide range of droplet sizes generated at intermediate Reynolds numbers.

IV. Atomization. At high Reynolds numbers, the jet breaks up completely within a short distance from the discharge nozzle exit in a chaotic and irregular manner, and the spray formed is conical. The aerodynamic interaction, i.e., the relative motion combined with air friction, causes irregularities in the originally smooth liquid jet surface, leading to unstable wave growth on the jet surface, and produces unstable ligaments. As the relative velocity increases, the sizes of the ligaments decrease and their lives shorten. Much smaller droplets are generated upon the disruption of the ligaments. Droplet sizes spread over a wide range. Average droplet diameters are much smaller than the jet diameter.

In practice, there is no sharp demarcation between these clearly identified regimes. Further, the breakup mechanism corresponding to regime IV has not been unambiguously described. This is unfortunate because plain-orifice atomizers operate normally under the conditions corresponding to regime IV, and thus this regime is of practical interest in high-pressure sprays such as those in diesel engines. Although various mechanisms and hypotheses have been suggested, no one single mechanism is responsible for all jet breakup processes, and generally a combination of factors is involved. However, it is now widely accepted that the action of the surrounding air (gas) is the primary cause of the atomization of high-velocity jets,[204] and jet turbulence is a significant contributing factor[233] because it ruffles the jet surface and makes it more susceptible to aerodynamic effects. Primary atomization is deemed to be caused mainly by the action of internal forces, such as turbulence, inertial effects, change in velocity profile (flow relaxation or bursting effect), and/or surface tension,[230] while secondary atomization always involves the action of aerodynamic forces in addition to these internal forces.[230] Other factors influencing jet breakup may include nozzle length to diameter ratio,[227][234] ambient pressure,[234] cavitation, and liquid pressure oscillation, etc.

It should be noted that ambient gas density ρ_A is used in the aerodynamic Weber number We_A for determining the criteria in Table 3.1. However, the classifications of the modes of jet breakup developed by Ohnesorge,[40] Miesse,[220] and Reitz[227] were based on the liquid Reynolds and Ohnesorge numbers that did not include the effect of the relative velocity between the gas and liquid in two-fluid atomization. Faragó and Chigier[210] indicated that the modes of water jet disintegration in an air stream are a function of two parameters, i.e., the liquid Reynolds number and the aerodynamic Weber number. For other fluids, the viscosity and density ratios may be considered as additional non-dimensional groups. The dimensional analysis of disintegration of liquid jets in quiescent air, in parallel gas flow, and of spherical droplets in a gas stream showed that, generally, at least four independent non-dimensional groups are needed to ensure similarity among different liquid breakup processes.

Chigier[211] made a comprehensive listing of the triggering mechanisms for the initiation of liquid surface disturbances. These include: *(a)* pulsations and disturbances in liquid and air supply lines to the atomizer, *(b)* sharp edges causing separated flows, *(c)* wall surface roughness, *(d)* boundary layer growth, *(e)* turbulence generation and cavitation inside the atomizer and at the orifice exit, and *(f)* interaction of liquid surface with surrounding air flows downstream of the atomizer exit. Once the disturbances have been triggered, further wave growth depends largely on the interaction of the liquid surface with surrounding air flows. For a given group of atomizers, the methods for interfering with wave growth can be found once wave growth characteristics have been identified. Aerodynamic, acoustic, and electromagnetic fields may be applied. Acoustic instabilities in combustors, particularly in rocket engines, interact with hydrodynamic instabilities on liquid jet surfaces, generating pulsation and formation of clusters in sprays. For more rapid and improved atomization, in-phase fields may be used to augment the growth rate. For damping wave growth and instabilities, out-of-phase fields may be used.

Tanasawa and Toyoda[41] divided the mechanisms of jet breakup into four regimes: *(a)* laminar flow, *(b)* transition, *(c)* turbulent flow, and *(d)* spray. They presented photographs of these modes that were taken using electric sparks with very short duration (ns). In the *laminar flow regime*, the cylindrical liquid column is transparent since the liquid surface is very smooth. Bead-like swelling and contraction accompanied by rotationally symmetrical oscillations of the liquid surface occur at the tip of the liquid column, and the beads leave the column in the form of oscillating droplets. In the *laminar-to-turbulent transition regime*, the tip of the laminar liquid column becomes unstable and begins to oscillate in longitudinal direction. The bead-like swelling and contraction with shorter and more random period occur near the discharge nozzle, and propagate downwards to the tip of the liquid column where a short length of liquid column sheds and breaks up into several droplets. In the *turbulent flow regime*, lateral vibrations start as early as the liquid issues from the discharge nozzle, and travel downwards to the tip of the liquid column with increasing amplitude, mingling with spiral motions. The liquid column becomes opaque due to the roughened/ruffled surface with small waviness. The wavy motions are induced by the air resistance. On the crests of the waves there exists a negative pressure due to the high speed air flow, whereas in their dents a positive pressure. Thus, the waviness due to the turbulence is amplified by the air resistance. With increasing Reynolds number, the ruffled liquid surface may be dragged by the surrounding air and drawn out into fine ligaments and/or narrow leaf-like membranes around the liquid column. These ligaments and membranes are quickly cut off by the rapid growth of dents in their surfaces and break up swiftly into several droplets. In the *spray regime*, the liquid column spreads into a multitude of fine membranes mixed with fine ligaments of herring-bone-like configuration due to the large air resistance during its penetration into the surrounding air. The ligaments and membranes then break up into droplets. The spray formed consists of thin core stream with larger droplets and surrounding stream with finer droplets. The former results from the core jet, and the latter from the surrounding, scattered, fine ligaments and membranes.

Based on their experiments, Tanasawa and Toyoda[41] proposed a new dimensionless number for the classification of jet breakup regimes. This number, called *Jet number*, is defined as:

Eq. (10)
$$\mathrm{Je} = \frac{U_L^2 \rho_G d_0}{\sigma}\left(\frac{\rho_L}{\rho_G}\right)^{0.45}$$

where ρ_G is the density of ambient gas/air. The Jet number is a combination of the aerodynamic Weber number and the liquid-gas density ratio. It is an index of the degree of disruption and can be used to classify the modes of column/jet breakup:

(a) Je < 0.1: Dripping

(b) Je ≈ 0.1–10: Jet breaks up into droplets due to longitudinal oscillations

(c) Je ≈ 10–500: Jet breaks up into droplets due to lateral oscillations with additional effect of air friction

(d) Je > 500: Jet breaks up into multitude of droplets due to membrane formation

Although the Jet number is used for the classification, it should be noted that the intrinsic characters of a liquid flow depend on the Reynolds number. By combining the Jet number with the Reynolds number and introducing a correction factor for viscosity, the modes and configurations of liquid jet breakup can be clearly defined and described.[41]

For *steady injection of a liquid through a single nozzle with circular orifice into a co-flowing gas* (air), the breakup of the liquid jet has been found to be highly unsteady, and unstable liquid structures disintegrate in a bursting manner.[210] At high air flow rates, the unstable liquid cylindrical jet deviates from the main stream direction by a flapping motion. The pressure difference between the leading and trailing edges of the deviating liquid cylinder transforms the round jet into a curling liquid sheet. Further aerodynamic influence generates a stretched membrane bounded by thicker rims that eventually bursts into ligaments and droplets of various sizes due to its highly unstable nature. There is evidence of strong interactions between the liquid jet and large-scale structures in turbulent gas flow that result in bursting and formation of clusters of droplets and generation of pulsations. Ligaments may also be directly stripped from waves and liquid surfaces by aerodynamic shear. From an analysis of numerous high-speed spark photographs, various modes of round liquid water jet disintegration in a coaxial air stream have been identified by Faragó and Chigier[210] over a liquid Reynolds number range of 200 to 20,000 and an aerodynamic Weber number range of 0.001 to 600. Accordingly, the mechanisms of jet breakup have been classified into the following three primary categories,[210] as illustrated in Figs. 3.3–3.6 and summarized in Table 3.2.

Table 3.2. Classification and Criteria of Breakup Regimes of Round Liquid Jets in Co-flowing Air as Compared to Those of Thin Liquid Sheets and Spherical Droplets in Air Stream[210]

Breakup Regime/Mode				Criteria*
	Round Jet	Thin Sheet	Spherical Drop	
Normal Pulsating	Axisymmetric Rayleigh-type	Sheet forms a round jet	No breakup	$We_A < 15$
	Non-axisymmetric Rayleigh-type		Membrane-type	$15 < We_A < 25$
	Membrane-type	Membrane-type	Membrane-type	$25 < We_A < 70$
	Fiber-type	Fiber-type $We_A > 50$	Fiber-type	$70 < We_A < 500$
Super-Pulsating				$150 < We_A < 500$ $Re_L/We_A^{0.5} < 100$

* $We_A = U_R^2 \rho_A d_0 / \sigma$ where U_R is the relative velocity between gas and liquid.

I. Rayleigh-Type Breakup. In this category, a liquid jet disintegrates into droplets without the formation of any liquid membranes or fiber ligaments. The mean droplet diameter is of the order of the jet diameter, and the maximum droplet diameter is about twice the jet diameter. This category can be further divided into two subgroups: *(i)* Axisymmetric Rayleigh-Type Breakup, and *(ii)* Non-Axisymmetric Rayleigh-Type Breakup.

(i) Axisymmetric Rayleigh-Type Breakup. At intermediate air flow rates, the emerging liquid jet is accelerated and sheathed by the high speed annular co-flowing air jet. The breakup length is reduced by nearly half compared to that in quiescent air. The disintegrating jet may begin to show a very slight tendency to form a helical axis (Fig. 3.3a). However, the jet, ligaments and droplets formed are still nearly axisymmetric around this slightly helical axis. Those jets with the corkscrew shape before disruption (Fig. 3.3a) are typical examples of the jets belonging to the first wind-induced regime where aerodynamic effects are significant. Nevertheless, these are still assigned to the axisymmetric Rayleigh-type breakup mode for convenience.

(ii) Non-Axisymmetric Rayleigh-Type Breakup. This mode is the boundary between the classical Rayleigh breakup and the membrane-type breakup modes. The mechanism is essentially a combination of the deformation of a round liquid jet by aerodynamic forces and the instability of the deformed jet. The liquid jet is first accelerated rapidly in the high speed air stream (Fig. 3.3b). The jet diameter is thus significantly reduced as it interacts with the surrounding air stream. The direction of the thinning capillary liquid jet is influenced by the interaction between the liquid jet and the turbulent structures of the surrounding air stream. The formation of

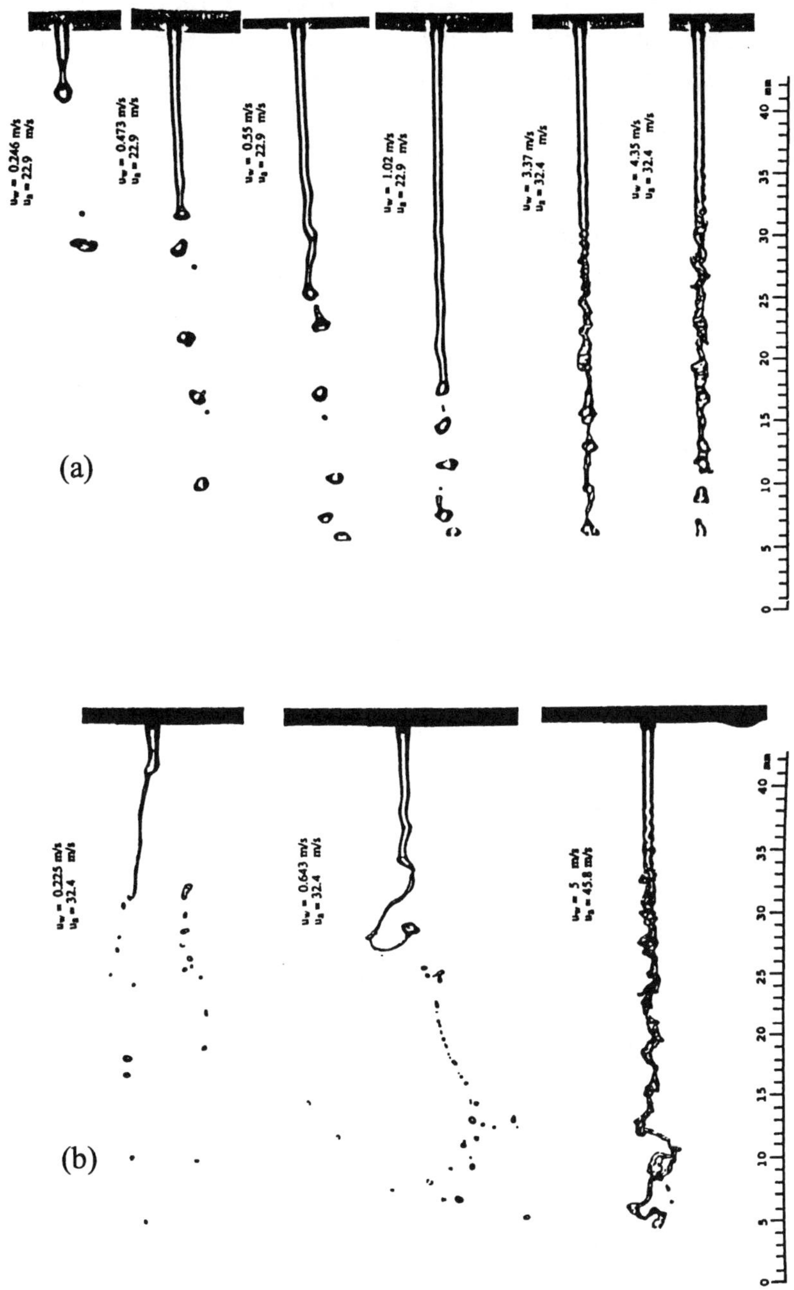

Figure 3.3. *(a)* Axisymmetric and *(b)* Non-axisymmetric Rayleigh-type breakup mode of round liquid jets in coaxial air flow. *(Reprinted with permission from Ref. 210.)*

waves on the round liquid jet may be caused by both the *Kelvin-Helmholtz instability*,[235]-[237] and the liquid-gas interaction. The aerodynamic forces thin the liquid jet, but the breakup itself may be due to the *Rayleigh capillary pinching mechanism*. The whole jet breaks up into droplets without the formation of intermediate membranes or fibers peeling off. This mode may also be the operating mode in the final stage of the membrane-type and fiber-type breakup modes.

II. Membrane-Type Ligament Breakup. At higher air flow rates, a round liquid jet develops into a thin liquid sheet (membrane) before breaking into droplets (Fig. 3.4). Aerodynamic forces exerted on the sheet create the Kelvin-Helmholtz instability due to air shear, leading to the formation of surface waves.[209][235] Liquid accumulates at the edges of the thin sheet, forming a liquid frame while waviness increases in amplitude. The diameter of the rim is smaller than the diameter of the emerging intact liquid jet. The frame then breaks up into droplets *via* the non-axisymmetric Rayleigh mechanism. The jet disintegration is completed within one or two wavelengths. The droplet diameter is considerably smaller than the jet diameter, with the mean droplet diameter being one order of magnitude smaller than the jet diameter and the maximum droplet diameter also smaller than the jet diameter. The morphology of the membrane-type disintegration of round liquid jets is very similar to that of thin liquid sheets, as described by Dombrowski and Johns,[238] and by Mansour and Chigier,[209] among others.

III. Fiber-Type Ligament Breakup. At even higher air flow rates, fibers are formed on the liquid jet and peel off the main liquid core (Fig. 3.5). The fibers then break up into droplets via the non-axisymmetric Rayleigh-type breakup mechanism. The liquid core is accelerated and develops waves. It eventually breaks up into droplets *via* the non-axisymmetric Rayleigh mechanism, or into ligaments. The ligaments may then break up into smaller droplets or ligaments by one of the three disintegration modes. The length of these ligaments is about two to five times the initial jet diameter, and new fibers are peeled off from the ligaments. The diameter of the newly formed fibers increases with increasing axial distance from the nozzle exit. Even further downstream, the ligaments form clouds of droplets and liquid fibers, leading to a periodic change from high to low density regions in the spray. Generally, the size of the droplets generated further downstream is larger than those formed closer to the nozzle exit. The completion of the jet disintegration requires several wavelengths. The droplet diameter is a few orders of magnitude smaller than the diameter of the emerging intact liquid jet. This mode resembles the short wavelength breakup mechanisms of jets in the second wind-induced and atomization regimes. The fiber-type breakup is the most important mode of liquid atomization for applications such as the atomization of liquid oxygen in the space shuttle main engine.

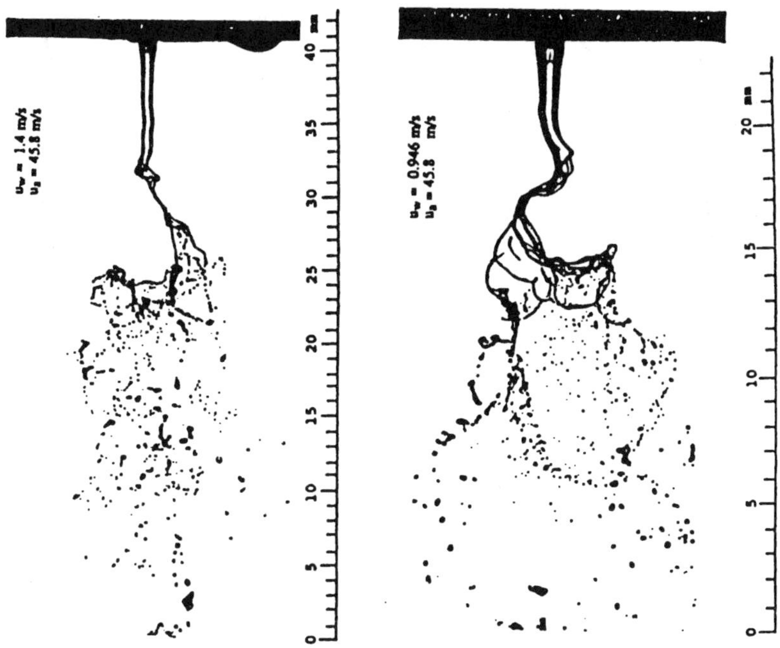

Figure 3.4. Membrane-type breakup mode of round liquid jets in coaxial air flow. *(Reprinted with permission from Ref. 210.)*

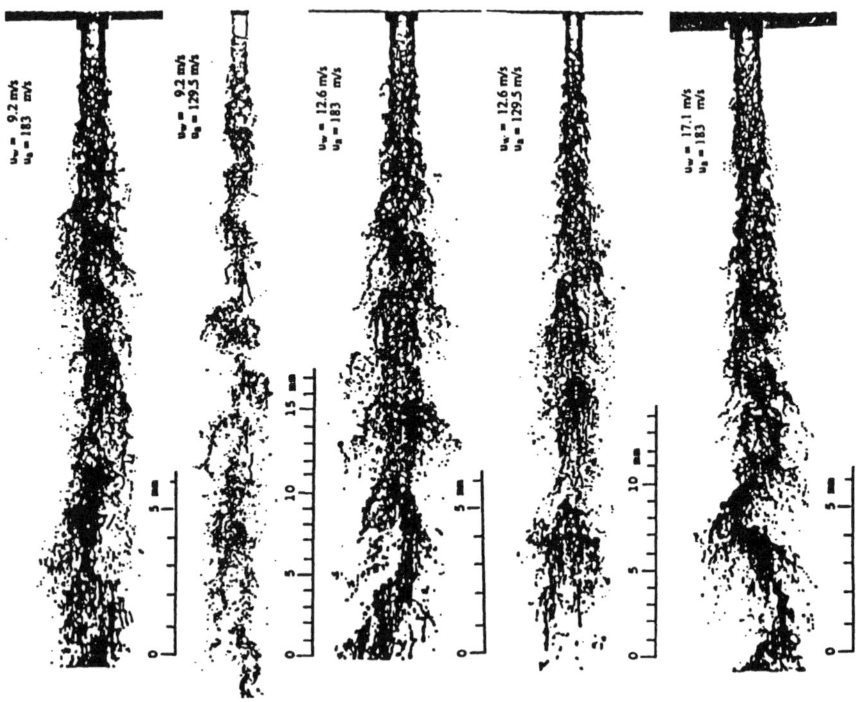

Figure 3.5. Fiber-type breakup mode of round liquid jets in coaxial air flow. *(Reprinted with permission from Ref. 210.)*

All these three modes can be divided into two groups: *(i)* Pulsating Jet Disruption as the normal sub-mode of atomization, and *(ii)* Super-Pulsating Jet Disruption that involves an extremely high periodic change between low and high density regions in the spray.

(i) Pulsating Jet Disruption. Different jet instabilities, such as capillary, helical, and Kelvin-Helmholtz instabilities, may dominate under different flow

conditions. All of these mechanisms involve the formation and development of waves and the disintegration of the wavy jet or ligaments after a critical wave form is attained. This produces a temporally periodic change of the liquid volume fraction and number density of droplets in the spray. It has been found that atomization is a pulsating, unsteady process, even if the emerging liquid jet and atomization gas are initially free of oscillation and vibration.

(ii) Super-Pulsating Jet Disruption. This mode (Fig. 3.6) occurs at low liquid flow rates and very high gas velocities. With reducing liquid Reynolds number and/or increasing aerodynamic Weber number, the amplitude of spray pulsation increases, while the breakup length of a round liquid jet in a co-flowing air stream decreases. Correspondingly, the axial distance, at which the wave formation and spray pulsation begin, also decreases.

Pulsation in a spray is generated by hydrodynamic instabilities and waves on liquid surfaces, even for continuous supply of liquid and air to the atomizer. Dense clusters of droplets are projected into spray chamber at frequencies very similar to those of the liquid surface waves. The clusters interact with small-scale turbulent structures of the air in the core of the spray, and with large-scale structures of the air in the shear and entrainment layers of outer regions of the spray. The phenomenon of cluster formation accounts for the observation of many flame surfaces rather than a single flame in spray combustion. Each flame surrounds a cluster of droplets, and ignition and combustion appear to occur in configurations of flames surrounding droplet clusters rather than individual droplets.

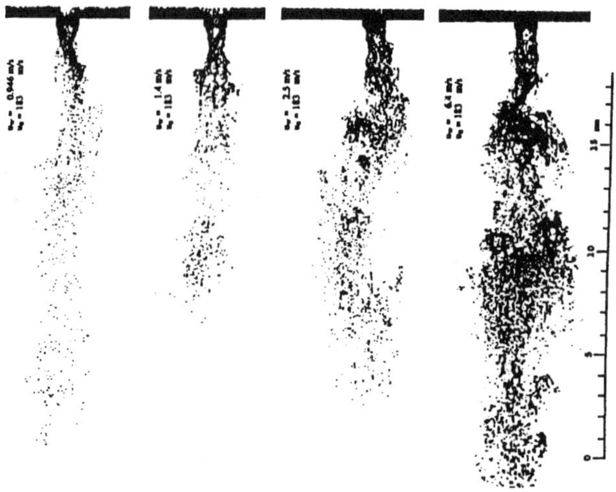

Figure 3.6. Super-pulsating disintegration mode of round liquid jets in coaxial air flow. *(Reprinted with permission from Ref. 210.)*

Based on the photographic observations, Faragó and Chigier[210] made a map of breakup regimes of round liquid jets in co-flowing air, as depicted in Fig. 3.7. In this map, the different jet disintegration modes are plotted as a function of the aerodynamic Weber number and the liquid Reynolds number at the nozzle exit. It can be seen that the axisymmetric Rayleigh-type disintegration is the dominant mode of jet breakup in the aerodynamic Weber number range of $We_A < 15$. The non-axisymmetric Rayleigh-type disintegration becomes the most prevailing mode in the range of $15 < We_A < 25$. At the transition from the axisymmetric to the non-axisymmetric Rayleigh-type disintegration mode, the corresponding aerodynamic Weber number increases slightly with increasing liquid Reynolds number. The membrane-type disintegration occurs in the range of $25 < We_A < 70$. This range seems to reduce with increasing Reynolds number. At higher Weber numbers, the fiber-type disintegration mode predominates, while the super-pulsating disruption mode takes place for $Re_L/We_A^{0.5} < 100$. For technical applications, the most important breakup mode is the fiber-type jet disintegration. In this mode, pulsating rates are high, and the largest portion of liquid is atomized in the secondary atomization region.

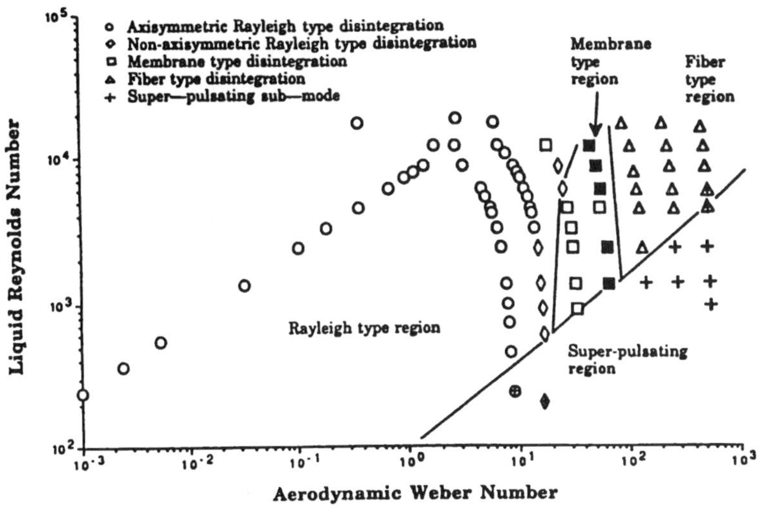

Figure 3.7. Breakup regimes of round liquid jets in co-flowing air. *(Reprinted with permission from Ref. 210.)*

Factors influencing jet breakup may include: *(a)* flow rates, velocities and turbulence of liquid jet and co-flowing gas, *(b)* nozzle design features, *(c)* physical properties and thermodynamic states of both liquid and gas, *(d)* transverse gas flow,[239] *(e)* dynamic change of surface tension,[115][240] *(f)* swirl,[241][242] *(g)* vaporization and gas compressibility,[243] *(h)* shock waves,[244] etc.

Chigier[211] found that turbulence in a liquid jet has important disturbing influences throughout the liquid flow. At the liquid surface, turbulent velocity fluctuations directly cause protuberances and roughness that result in direct stripping by surrounding air flow. Large eddy structures in the air flow penetrate into the liquid and leave distinct impacts of large scale vortical flow in the liquid. Both turbulent energy and turbulent scale play important roles in liquid breakup. Turbulence energy increases proportionally with increase in both liquid and air velocities. Turbulence intensities may be increased to above the levels associated with mean velocities by introducing bluff bodies, orifice plates, and transverse jets inside the atomizers that generate strongly separated and reverse flows. These augmented turbulence generators cause reduced droplet sizes and more uniform droplet size distributions.

In a supersonic gas flow with shock waves and expansion waves, the pressure distribution is highly irregular. When the surface of a liquid jet encounters such uneven pressure distribution, it may deform as a precursor to its total breakup process. Issac et al.[244] conducted extensive flow visualization studies on the atomization of a coaxial co-flowing Mach 1.5 jet using Schlieren/shadowgraphy, flash photography, and short duration (ns) laser imaging, and measured the resultant droplet size. In their studies, the operation conditions for the supersonic nozzle ranged from overexpanded to highly underexpanded mode. They observed and explained the formation of a bubble-like pattern as the liquid jet expands and interacts with the first set of waves. They identified three regions during the atomization based on photography evidence: *(a)* primary region enclosed by the first shock cell where the primary breakup occurred, *(b)* secondary region enclosing the bubble-like formation, and *(c)* third (subsonic) region further downstream. It was hypothesized that the atomization be initiated by the growth of 3-D structures on the liquid surface. Subsequent detachment of the 3-D structures into fine droplets was observed and deemed to be due to the intense shear at the liquid-gas interface. These observations provided experimental evidence supporting the boundary-layer stripping mechanism proposed by Taylor,[245] and Ranger and Nicholls,[246] i.e., the shearing action exerted by the high-speed gas stream on the liquid surface leads to the formation of a boundary layer and subsequent surface stripping.

Quantitative correlations for jet breakup length have been proposed by many investigators based on experiments.[38][41][220][247][248]

In *laminar flow region*, the breakup time t_b and length L corresponding to a small axisymmetric disturbance δ_0 are expressed as:[38]

Eq. (11) $$t_b = \frac{\ln(d_0 / 2\delta_0)}{q_{\max}}$$

Eq. (12) $$L = 1.03 d_0 \mathrm{We}_L^{0.5} \ln(d_0 / 2\delta_0)$$

and

Eq. (13) $$L = d_0 \mathrm{We}_L^{0.5} (1 + 3\mathrm{Oh}) \ln(d_0 / 2\delta_0)$$

for a liquid jet subjected only to inertial and surface tension forces, and a viscous liquid jet in the absence of air friction, respectively. The breakup length is proportional to $d_0^{1.5}$ for non-viscous liquids and to d_0 for viscous liquids. The variable group $\ln(d_0/2\delta_0)$ depends on experimental conditions such as nozzle geometry and liquid flow rate. Grant and Middleman[247] proposed an average value of 13.4 for glycol and glycerol-water solution, and a more general expression such as:

Eq. (14) $$\ln(d_0 / 2\delta_0) = 7.68 - 2.66\mathrm{Oh}$$

They also proposed an empirical correlation for laminar jet breakup length:

Eq. (15) $$L = 19.5 d_0 \mathrm{We}_L^{0.5} (1 + 3\mathrm{Oh})^{0.85}$$

To overcome the discrepancies between the predictions of Weber's correlations and subsequent experiments, Mahoney and Sterling[248] modified Weber's theory, and derived a universal equation for the breakup length of laminar Newtonian jets:

Eq. (16) $$L = d_0 \mathrm{We}_L^{0.5}(1+3\mathrm{Oh})\ln(d_0/2\delta_0)/f(\mathrm{Oh}, \mathrm{We}_L)$$

This correlation produces satisfactory agreement between the predictions and experiments.

To determine the upper critical point corresponding to the *transition from laminar to turbulent* jet flow, Grant and Middleman[247] derived the following empirical correlation:

Eq. (17) $$\mathrm{Re}_{crit} = 3.25\mathrm{Oh}^{-0.28}$$

Wu, Ruff and Faeth[249] studied the breakup of liquid jets with holography. Their measurements showed that the liquid volume fraction on the spray centerline starts to decrease from unit at L/d_0=150 for non-turbulent flows, whereas the decrease starts at about L/d_0=10 for fully developed turbulent flows. Their measurements of the primary breakup also showed that the classical linear wave growth theories were not effective, plausibly due to the non-linear nature of liquid breakup processes.

In *turbulent flow region*, Grant and Middleman[247] suggested the following empirical correlation:

Eq. (18) $$L = 8.51 d_0 \mathrm{We}_L^{0.32}$$

For turbulent liquid flows, the sizes of droplets generated are in the order of magnitude of the integral scale of the liquid turbulence. Droplet sizes are also related to the spectral distribution of the liquid turbulence. Miesse[220] analyzed his own experimental data and correlated the jet breakup length to d_0, We_L and Re_L. He indicated that the difference in atomizing characteristics of highly volatile liquids, (such as liquid nitrogen) and of more stable liquids (such as water) can be described completely by the liquid Weber and Reynolds numbers, and no other properties such as evaporation rate needs to be introduced. Miesse[220] further analyzed the breakup-length data of

Haenlein,[39] and found that for liquids of comparable viscosity, a better correlation may be obtained if the Reynolds number is defined with respect to the surrounding medium (air or gas). In this correlation, the breakup length is formulated as a function of d_0, We_L and Re_A where the density and viscosity of the air/gas are used in the aerodynamic Reynolds number. The experimental and analytical results[220] also showed that atomization may be brought about by increasing jet velocity, nozzle diameter, and/or liquid density, and/or by decreasing liquid surface tension. Under the experimental conditions, the breakup length increases with increasing jet velocity, nozzle diameter, and/or liquid viscosity, but decreases with increasing liquid density or surface tension. The wavelength of the most rapidly growing disturbance is essentially independent of spray chamber diameter. However, the maximum rate of growth decreases as the chamber diameter increases. Thus, the jet breakup length increases as the chamber diameter increases.

Hiroyasu et al.[234] and Arai et al.[250] indicated that the breakup length increases with increasing jet velocity up to a maximum value, and then turns to decline with further increase in jet velocity. A linear instability analysis and experimental confirmation showed that the Marangoni effect has a stabilizing effect on jet disintegration, but no effect on droplet size.[115] The influence of nozzle design on the stability and breakup of liquid jets has been reviewed by McCarthy and Molloy.[230]

Tanasawa and Toyoda[41] identified three regimes in the curve of the jet breakup length *vs.* the Reynolds number for *non-viscous liquids*: *(a)* laminar flow regime, *(b)* transition regime, and *(c)* turbulent flow regime. In the laminar flow regime, the breakup length increases with increasing Reynolds number and reaches its local maximum when the jet velocity attains a critical value. Beyond this point ($Re \approx 1400$), the jet becomes unstable and separation may occur. In the transition regime, the laminar and turbulent flows occur cyclically with random periods, shortening the jet breakup length. The jet surface exhibits irregular crests and dents accompanying lateral and spiral motions. In this regime, the jet breakup length decreases with increasing Reynolds number and reaches its minimum at Re ≈ 1800–2400. The transition begins at the local maximum jet breakup length and ends at the minimum jet breakup length. Further increasing the Reynolds number, the jet flow enters into the turbulent flow regime and the jet breakup length turns to increase with roughened/ruffled jet surface and small waviness. For *viscous liquids*, the transition begins at lower Reynolds numbers, and the range of transition is wider than that for non-viscous liquids. Atomization also occurs at lower Reynolds numbers (Re < 2400), even in the laminar flow regime because the air resistance is much larger than that in the turbulent flow regime of non-viscous liquids. In addition, there is no minimum jet breakup length for viscous liquids since the jet disintegration due to the interaction with the surrounding air is completed at far shorter jet lengths.

Shavit and Chigier[251] used fractals to quantify liquid surfaces during breakup and ligament formation. They measured fractal dimensions of gas-liquid interfaces in the breakup region of a disintegrating liquid jet that is surrounded by an annular gas jet, and both are injected through an air-blast coaxial atomizer. They found that the gas-liquid interfaces are categorized by a fractal set with dimensions and cutoffs related to the physical processes and operation conditions of atomization, such as axial locations and air and liquid velocities. The peak fractal dimension and the breakup point are approximately at the same axial location, and a good correlation exists between the peak value of the fractal dimensions and the average droplet size measured in the core of the spray.

In recent years, intensive research has been conducted by Chigier and co-workers[211] to obtain a better understanding of breakup processes of liquid jets and sheets. Surface waves, disturbances, and protuberances that are formed and grow along the length of liquid jets and sheets have been studied by high-speed microphotography and lasers. Wave amplitude growth rates, wavelengths, frequencies, and wave propagation speeds have been measured for different velocities of liquid and air. The experimental measurements showed that bursting and explosion phenomena characterize the final stages of liquid jet disintegration.

Liquid Sheet/Film Breakup. Liquid sheets encountered in practical atomization processes include flat sheets, conical sheets, and fan sheets.[209] A flat sheet may be generated by impinging two liquid streams against each other, or by impinging a liquid stream to the center of a rotating disk or cup. A conical sheet may be produced by deflecting a liquid flow in a pipe through an annular orifice, or by pressure-swirl, or prefilming air-blast atomizers, as discussed in the previous chapter. A fan sheet can be typically obtained from a fan-spray atomizer. The breakup of a liquid sheet is mainly governed by its initial velocity and physical properties of both the liquid and ambient gas. Pressure, aerodynamic drag, or centrifugal forces accelerate a sheet to a velocity that is required to expand the sheet against the surface tension force contracting it, as in the pressure-swirl, prefilming air-blast, or rotary atomization processes, respectively. Droplet size distribution is dictated by the orderliness of disintegration processes and the uniformity of generation of liquid threads/ ligaments.

Fraser et al.[116][207] defined three modes of sheet disintegration, as schematically depicted in Fig. 3.8.

(a) Rim Disintegration. Surface tension forces contract the free edge of a liquid sheet into a thick rim, that eventually breaks up following the mechanism governing the disintegration of a free jet. Following the breakup, the droplets generated continue to fly in the original flow direction, remaining attached to the receding surface by thin threads that also rapidly break up into rows of droplets. This mode typically generates large droplets and numerous small satellite droplets. It is the predominant disintegration mode of liquid sheets of high viscosity and high surface tension, such as liquid metals.

(b) Wave Disintegration. Instability of a thin liquid sheet resulting from the interaction with surrounding medium (air or gas) leads to rapidly growing surface waves. When the wave amplitude increases to a critical value, the areas of the sheet, that correspond to a half or full wavelength of the oscillation, are torn away before the

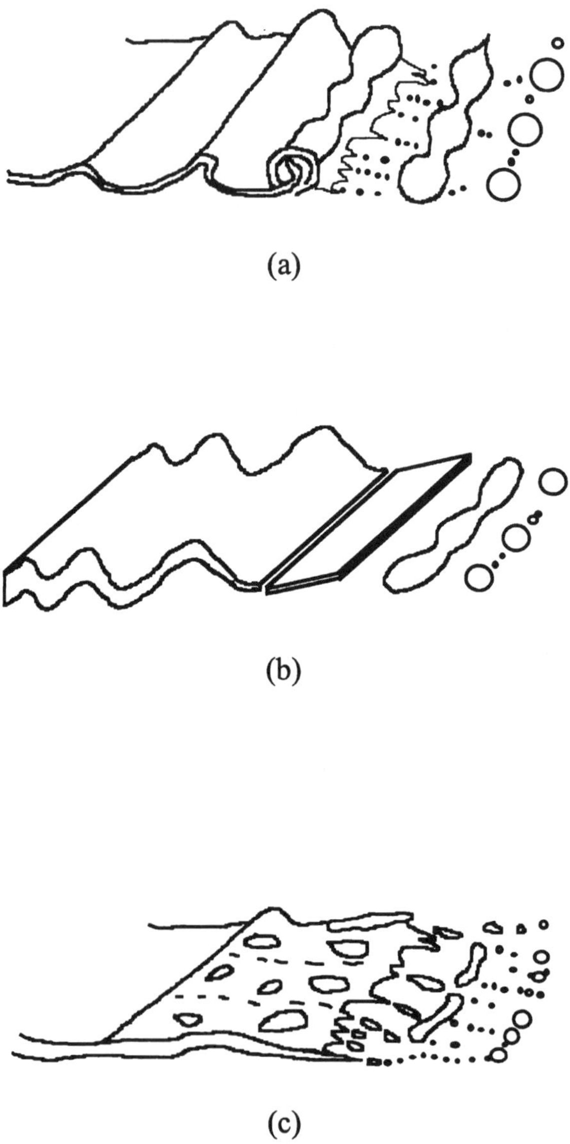

Figure 3.8. Liquid sheet/film breakup modes: Successive stages in the idealized breakup of *(a)* a sheet with a thick rim, *(b)* a wavy sheet, and *(c)* a perforated sheet.

leading edge is reached. Surface tension forces contract these torn areas rapidly into unstable ligaments that subsequently disintegrate into droplets by aerodynamic forces or liquid turbulence before a regular network of threads forms. The breakup of the ligaments follows the varicose instability mechanism. Wavy-sheet disintegration is highly irregular and consequently the droplet sizes vary significantly.

(c) Perforated-Sheet Disintegration. During expansion of a liquid sheet, holes may be generated in the sheet. These holes are delineated by thick rims formed from the liquid that is originally included inside the holes. The holes expand rapidly in size, until the rims of adjacent holes coalesce to form a web of strands/ligaments of irregular shapes. The ligaments subsequently break up into droplets of various sizes. Perforations have a similar evolution history since they occur in the sheet at the same distance from the origin (for example, orifice). The thread diameters are usually uniform and thus the droplet sizes tend to be constant.

Each of these three modes may occur in an atomizer generating liquid sheets, while in some atomizers two of the modes may occur simultaneously. Both the mean droplet size, and size distribution are determined to a large extent by the dominant mode. The extensive studies on the mechanisms of sheet breakup by Fraser, Eisenklam, Dombrowski, Hasson, Johns and co-workers[116][207][208][238] showed that a network of unstable liquid ligaments are principally caused by perforations in the liquid sheet, and perforations always result in larger droplets. The history of perforations determines the stage of growth at which ligaments break up. If the perforations are created by air friction, the ligaments break up very rapidly. Otherwise, if the perforations are caused by other means, for example, turbulence, the breakup of the ligaments occurs more slowly. Other possible causes for the perforations may be acoustical, electrical, dissolved air, suspended particles, particle impingement, etc. An increase in liquid viscosity reduces the growth rate of liquid instabilities and increases the wavelength of maximum growth rate.[252] Liquid sheets with high viscosity and high surface tension are most resistant to disintegration, while liquid density has a negligibly small effect on sheet disintegration.

Taylor[253] studied the formation of a liquid sheet by the impingement of two opposing jets and by the impingement of a single jet on a plate. The results showed that the thickness of the liquid sheet decreases as it spreads radially. Viscous shear forces at the plate surface decrease the liquid momentum continuously, causing additional reduction in the sheet thickness. For low Weber number jets ($We_{jet} < 800$), the momentum force of the liquid is balanced with the opposing surface tension force. For high Weber number jets ($800 < We_{jet} < 30000$), large-amplitude antisymmetric waves are generated, giving rise to earlier disintegration of the liquid sheet. The waves are induced by aerodynamic forces acting on the air-liquid interface. As the amplitude of the waves increases, strands are torn from the crests and holes are blown through the sheet. Eventually, droplets are shed from the expanding rims.

Taylor and Michael[254] analyzed the perforation process based on a balance of forces acting on a hole in a liquid sheet. The analysis showed that any hole with a radius greater than the thickness of the liquid sheet would grow, whereas smaller holes would close. Experiments also demonstrated the existence of a minimum initial hole size for growth to occur.

Taylor[205] also conducted mathematical analysis of the generation of ripples by wind blowing over a viscous fluid. Using a relationship between the growth of the amplitude of disturbance waves and the surface stress, Taylor derived a criterion for the instability of waves. In *Taylor's instability theory*, the disintegration of a liquid sheet/film is visualized as a process in which droplets are detached from the liquid surface with a wave of optimum amplitude. The diameter of the most frequent droplets is then formulated as a function of air velocity over the liquid surface, liquid density, surface tension and viscosity, as well as air density.

As indicated by Mansour and Chigier,[209] all previous experimental studies focused on *flat liquid sheets into quiescent air*. In many industrial applications, liquid sheets are injected into air streams or impinged by air streams. Thus, the influence of surrounding air flows on liquid sheet breakup needs to be ascertained. Mansour and Chigier[209] studied the development, stability, and disintegration of liquid sheets issuing from a two-dimensional (*2-D*) *air-blast atomizer* by still microphotography along with Phase-Doppler technique. In their experiments, a liquid (water) sheet emerges from the central slit of the 2-D atomizer with a high aspect ratio, and is impinged by two sheets of high-speed air on both sides at the nozzle exit. Based on their analyses of numerous photographs, Mansour and Chigier proposed the following breakup modes for flat liquid sheets into quiescent air, and air streams, respectively.

(A) Flat Liquid Sheets into Quiescent Air: Rim–Jet Disintegration. In *quiescent air* (without an air flow), a liquid sheet issuing from the 2-D nozzle will converge toward the axis of the sheet to form a round jet under the action of surface tension forces. The liquid jet subsequently breaks up into droplets.

At very low liquid flow rates (i.e., very low liquid velocities), a converging sheet is bounded by two thick rims at its two free edges. Two types of forces may act on the liquid: the surface tension forces pull the two rims together into a liquid jet, and the inertial force acts in the axial direction. The aerodynamic shear is negligible. As a result of convergence, the width of the sheet decreases linearly from the nozzle exit to the point of convergence. The thickness of the rims decreases with an increase in liquid flow rate. For small surface tension and negligible gravitational forces, liquid paths are in the axial direction. For dominant surface tension forces, liquid in the rims is contracted toward the axis. Liquid in the sheet between the rims is forced into a wedge and waves that are initiated at both rims propagate across the sheet, forming crisscross wave patterns. The amplitude of the waves is greatest near the rims and decays toward the axis. The waves from each rim interfere with each other, resulting in an increase in the amplitude if the waves are in phase. With increasing liquid flow rate, gravitational forces become important and the liquid jet formed is deflected downwards, particularly after the convergence. The waves due to surface tension appear farther downstream. The width of the sheet increases but this development is hindered by the action of surface tension forces. Around the edges of the sheet where the liquid velocity is lowest, the waves form immediately downstream of the nozzle. As the liquid flow rate further increases, the influence of gravitational forces lessens, and the convergence length increases linearly. Small distortions develop on the sheet surfaces, suggesting the transition to turbulence at Re = 1471. The surface disturbances become increasingly pronounced and the sheet loses its orderliness as the liquid Reynolds number is increased. The turbulent disturbances are dampened in the downstream region due to viscous dissipation of turbulent energy.

Point disturbances have also been observed at intermediate liquid flow rates. Vapor bubbles contained in the liquid and generated due to cavitation expand immediately after issuing from the nozzle, initiating point disturbances in the emerging liquid sheet. The disturbances propagate radially, forming surface ripples. The amplitude of the ripples increases with the liquid flow rate. As the liquid flow rate is increased beyond a certain value, the point disturbances can no longer be readily detected due to the ruffling of the surfaces by turbulence.

(B) Flat Liquid Sheets into Air Streams: Mechanical and *Aerodynamic Disintegration.* In *air streams* (with an air flow), a liquid sheet issuing from the 2-D nozzle will form a quasi-2-D expanding spray. The breakup modes are divided into two groups: *(1)* mechanical mode due to the action of liquid injection pressure, and *(2)* aerodynamic mode due to the action of air friction.

The *mechanical breakup mode* occurs around the rims of the sheet where the air-liquid relative velocity is low, forming relatively large droplets. At low relative velocities, aerodynamic forces are much smaller than surface tension and inertia forces. Thus, the breakup of the liquid rims is purely 'mechanical' and follows the Rayleigh mechanism for liquid column/jet breakup. For the same air pressure, the droplets detached from the rims become smaller as the liquid flow rate is increased.

The *aerodynamic breakup mode* occurs in the liquid sheet between the rims. In aerodynamic breakup, the perforation and wave breakup modes discussed above may occur simultaneously. At relatively low liquid flow rates, the liquid sheet between the rims exhibits small and ordered cell structures that may subsequently develop into perforations. These structures are bounded by large diameter ligaments that contain thin membranes. The ligaments are deemed to be the origin of large droplets, whereas the membranes break up into a multitude of small droplets. The disintegration of the membranes precedes the ligament breakup. When the liquid sheet extends to a certain axial location, the leading edge of the sheet will be anchored to the remaining liquid only through the cell frames. The wave front is blown into a balloon of variable thickness. Disruption may occur through the formation of holes. Where the perforations form in the sheet, the expansion of the holes by surface tension takes place very regularly. During the growth of the holes, thick rimes form around the holes. With increasing liquid flow rate, the structures become smaller and less ordered, and occur in a more chaotic manner.

The air flow is deemed to be responsible for the formation of large ordered and small chaotic cell structures. The aerodynamic shear occurs on the liquid surface that is in direct contact with the high-velocity air flow, and the degree of the shearing effect is a function of the liquid-air relative velocity. The aerodynamic shear causes small vibrations in the liquid sheet thickness that will develop into major disturbances. As the sheet thickness fluctuates, perforations develop in the sheet. The holes are eventually blown through the sheet so that disruption starts before the formation of main breakup region. The formation of ligaments or threads is a necessary stage in liquid sheet breakup.

Although a liquid sheet may leave the nozzle with some perturbations, the principal cause of the instabilities is the interaction of the sheet with the high-velocity air streams whereby rapidly growing waves are imposed on the sheet. Disintegration may occur when the amplitude of these waves reaches a critical value. Each full sinusoidal wave is initially distorted to yield two half-waves of very similar forms. The constant stretching of the half-waves increases the free surface energy of the already distorted liquid sheet. At the junction of the two half-waves, localized high tensile forces pull the half-waves in opposite directions, leading to thinning of the junction. When the local thickness of the sheet at the junction is reduced to below a critical value, perforations start to develop, eventually separating the half-waves and tearing the half-waves off the distorted sheet. These half-waves rapidly contract into unstable ligaments under the action of surface tension. The ligaments interact downstream with the air flow, and subsequently break up into droplets due to the aerodynamic shear.

The Phase-Doppler measurements showed that the spray two-dimensionality breaks down at a short distance from the atomizer in the downstream region. Boundary layer growth, and turbulent mixing result in the redistribution of droplets according to their size classes. The droplets detached from the central part of the sheet are subject to further breakup due to high local relative velocities, while the droplets detached from the rims are outside the main air flow field and have diameters close to those of the rims. The spray acquires the Gaussian velocity profiles in both the sheet-width (transverse) and sheet-thickness directions. The amplitude of oscillations in the liquid sheet is the primary parameter controlling the spray angle.

Faragó and Chigier[210] found that at similar aerodynamic Weber numbers, the disintegration modes of a thin liquid sheet in air streams are similar to those of a round liquid jet in a coaxial air stream (Table 3.2). At high aerodynamic Weber numbers, *Membrane-Type* or *Fiber-Type* breakup mode may set in.

Flat Sheets. Generally, the interface between a liquid sheet and air can be perturbed by aerodynamic, turbulent, inertial, surface tension, viscous, acoustic, or electrical forces. The stability of the sheet and the growth rate of unstable disturbances are determined by the relative magnitude of these forces. Theoretical and experimental studies[255][256] on disintegration mechanisms of flat sheets showed that the instability and wave formation at the interface between the continuous and discontinuous phases are the major factors leading to the breakup of a liquid sheet into droplets. At any given frequency, only two types of waves are possible. Either the two surfaces of the sheet are in phase to produce sinuous waves, or they oscillate out of phase to form dilational waves. For low ambient pressures and high liquid to gas density ratios, the growth rate of the sinuous waves is always greater than that of the dilational waves of the same frequency. For a liquid sheet with slightly disturbed surface and moving in air, surface tension forces tend to return the protuberance back, while the air tends to expand it further. This resembles the normal pattern of wind-induced instability, i.e., surface tension forces oppose any movement of the interface from its original position and tend to restore equilibrium, while aerodynamic forces augment any deviation from the interface and promote the instability. In addition to the temporal instability, spatial instability resulted from a sinusoidal disturbance with increasing successive maxima may occur even if a sheet is temporally stable.[257]

Short wavelengths of disturbance on thick sheets are stable unless wind velocities or Weber numbers are very high. The optimum wavelength for sheet disintegration corresponds to the maximum growth rate $\beta_{\max}$. For We >> 1, these are formulated as:[258]

Eq. (19) $$\lambda_{opt} = \frac{4\pi\sigma}{\rho_A U_R^2}$$

and

Eq. (20) $$\beta_{\max} = \frac{\rho_A U_R^2}{\sigma(\rho_L t_s)^{0.5}}$$

where U_R is the relative velocity between air (gas) and liquid sheet, and t_s is the thickness of liquid sheet.

Ligament diameter depends mainly on the film thickness, and thus thinner liquid films break down into smaller droplets. Rizk and Lefebvre[98] observed SMD $\propto t_s^{0.4}$, where SMD is the *Sauter mean diameter* of droplets. York et al.[255] and Dombrowski and

Johns[238] suggested that mean droplet diameter is proportional roughly to the square root of the film thickness. Therefore, it is critical to spread a liquid into a very thin sheet (film) in order to generate very fine droplets. Sheet thickness depends on flow conditions and air and liquid properties.[98] With increasing relative velocity, a liquid sheet disintegrates earlier, and ligaments are thinner and shorter, and generate smaller droplets. High values of liquid viscosity and flow rate result in thicker films, whereas variations in surface tension appear to have no effect on film thickness. However, sheets of low-surface-tension liquids disintegrate more readily and generate shorter ligaments.[98]

Weiss and Worsham[259] indicated that the most important factor governing mean droplet size in a spray is the relative velocity between air and liquid, and droplet size distribution depends on the range of excitable wavelengths on the surface of a liquid sheet. The shorter wavelength limit is due to viscous damping, whereas the longer wavelengths are limited by inertia effects.

Tokuoka et al.[260] conducted experimental studies on the disintegration of a liquid sheet by air impingement. Three types of disintegration patterns were identified for the breakup of a liquid sheet impinged by an air flow. The results showed that the amplitude of the waves on the sheet at relatively high liquid velocity and/or air impingement increases exponentially with time at the initial stage, and then reaches a critical value. Relatively large initial waves are agitated on the sheet by the air impingement. When the amplitudes of the waves are still small, the growth rate increases with increasing relative velocity. When the waves have grown to large amplitudes, the air flow may resist the growth. The waves are then deformed, colliding with ensuing waves, leading to disintegration into droplets.

Arai and Hashimoto[261] studied disintegration of a thin liquid sheet in a co-flowing air stream. For a constant sheet thickness, an empirical correlation was derived for the sheet breakup length as:

Eq. (21) $$L = 0.123\, t_s^{0.5}\, \mathrm{We}^{-0.5}\, \mathrm{Re}^{0.6}$$

where the Weber number and Reynolds number are defined as $We = t_s \rho_A U_R^2 / 2\sigma$ and $Re = t_s U_L \rho_L / \mu_L$. Clearly, the breakup length decreases with increasing relative velocity between liquid and air, while it increases with an increase in liquid sheet velocity or a decrease in liquid viscosity. Eroglu et al.[262] have also suggested the same correlation for the calculation of the breakup length of a low-velocity sheet surrounded by a high-velocity gas.

Recently, Knoll and Sojka[263] developed a semi-empirical correlation for the calculation of the Sauter mean diameter of the droplets after primary breakup of flat-sheets in twin-fluid atomization of high-viscosity liquids:

Eq. (22)
$$SMD = \frac{12\sigma}{\rho_L U_R^2 / \left[1 + 1 / \left(\alpha \frac{\dot{m}_G}{\dot{m}_L}\right)\right] + 4\frac{\sigma}{t_s}}$$

with
$$\alpha = \frac{1.62}{U_G^{1.3} \left(\frac{\dot{m}_G}{\dot{m}_L}\right)^{0.63} \mu_L^{0.3}}$$

where α is a parameter, U_G is the gas velocity, $\dot{m}_G$ and $\dot{m}_L$ are the mass flow rates of gas and liquid, respectively.

Wu, Ruff and Faeth[249] made an extensive review of previous theories and correlations for droplet size after primary breakup, and performed an experimental study of primary breakup in the near-nozzle region for various relative velocities and various liquid properties. Their experimental measurements revealed that the droplet size distribution after primary breakup and prior to any secondary breakup satisfies Simmons' universal root-normal distribution[264]. In this distribution, a straight line can be generated by plotting $(D/\mathrm{MMD})^{0.5}$ *vs.* cumulative volume of droplets on a normal-probability scale, where MMD is the *mass median diameter* of droplets. The slope of the straight line is specified by the ratio

MMD/SMD that may be 1.1, 1.2 or 1.5.[249] Thus, once the SMD is calculated, the entire droplet size distribution after primary breakup can be determined.

Conical Sheets. Conical sheets are deemed to be shorter than flat sheets because the radius of the curvature may have a destabilizing effect on fluctuations.[1] Fraser et al.[116] indicated that dilational waves in a sheet may be neglected because the degree of instability of these waves is always less than that of sinuous waves.

York et al.[255] analyzed pressure-swirl atomization and suggested the following breakup steps. First, waves form near the nozzle. The waves with the maximum-growth wavelength cause periodic thickening of the liquid conical sheet in the direction normal to the flow, and rings break away from the sheet. The volume of the liquid contained in a ring may be approximated by the volume of a ribbon whose thickness equals the sheet thickness at the breakup distance and whose width equals one wavelength. These rings or ligaments subsequently break up into droplets according to the Rayleigh mechanism. The mean droplet diameter produced by the pressure-swirl atomizer was then estimated as:

Eq. (23) $$D = 2.13\left(t_s \lambda_{opt}\right)^{0.5}$$

Fan Sheets. For fan sheets, wave disintegration mode may be assumed to be the working mechanism.[116] The most rapidly growing wave is detached at the leading edge in the form of a ribbon with a width equal to a half-wave-length, and a thickness equal to the sheet thickness at the breakup location. The ribbon immediately contracts into a ligament or filament, that subsequently breaks up into droplets following the Rayleigh mechanism. For a given fan spray atomizer and negligibly small liquid viscosity, the mean droplet diameter is estimated as:[116]

Eq. (24) $$D \propto \left(\frac{\dot{m}_L \sigma}{\rho_A^{0.5} \Delta P_L^{1.5}}\right)^{1/3}$$

where ΔP_L is the nozzle pressure differential or liquid injection pressure. This equation shows the influence of operation conditions (such as liquid injection pressure and flow rate) and properties of both liquid and ambient air on droplet size. As supported by subsequent experimental results, droplet size decreases with increasing ambient air density or pressure.

In practical fan sheet breakup processes, sheet thickness diminishes as the sheet expands away from the atomizer orifice, and liquid viscosity affects the breakup and the resultant droplet size. Dombrowski and Johns[238] considered these realistic factors and derived an analytical correlation for the mean droplet diameter on the basis of an analysis of the aerodynamic instability and disintegration of viscous sheets with particular reference to those generated by fan spray atomizers:

Eq. (25) $$D_m = \left[\frac{3\pi}{\sqrt{2}}\right]^{1/3} d_l \left[1 + \frac{3\mu_L}{(\rho_L \sigma d_l)^{1/2}}\right]^{1/6}$$

with

$$d_l = 0.9614\left[\frac{K^2\sigma^2}{\rho_G \rho_L U_R^4}\right]^{1/6}\left[1 + 2.6\mu_L\left(\frac{K\rho_G^4 U_R^7}{72\rho_L^2\sigma^5}\right)^{1/3}\right]^{1/5}$$

where D_m is the theoretical mean droplet diameter, d_l is the ligament diameter, and K is the fan spray parameter, $K = t_s L$. In this correlation, the variation of the mean droplet size with liquid viscosity obeys the following relation:

Eq. (26) $$\text{SMD} \propto \begin{cases} \mu_L^0 & \mu_L < 3\text{cP} \\ \mu_L^{1/6} & \text{for } 3 \le \mu_L \le 25\text{cP} \\ \mu_L^{7/20} & \mu_L > 25\text{cP} \end{cases}$$

The correlation derived by Dombrowski and Johns covers a large range of liquid viscosity and agrees favorably with experimental results. Crapper et al.[236] further applied second order and large amplitude theories to achieve better predictions. In addition, the effects of surface tension, and viscosity of a liquid sheet as well as the radial spreading and the resultant changes in the sheet thickness on the stability have been examined by Weihs.[257]

Recently, a study[265] has been conducted on the formation of droplets by the capillary wave instability of a *spherical liquid shell* with *pulsating cavity*.

Droplet Formation in Atomization Processes. On the basis of a number of experimental and theoretical studies on *plain-jet atomization*, some models can be deduced for the jet atomization mechanisms. For example, the breakup process of a plain jet may be hypothesized as follows: *(a)* Spreading of the liquid jet into a cone with an angle depending on liquid velocity; *(b)* Entrainment of air and initiation of small waves on the liquid surface; *(c)* Tearing of the liquid surface into fine ligaments; *(d)* Breakup of the fine ligaments into droplets; *(e)* Secondary breakup and coalescence/collision of droplets. The relative importance of these steps depends on the type of injector, relative velocity between air and liquid, liquid properties, liquid turbulence, air pressure and temperature, and air to liquid ratio. The spray formed is conical. The aerodynamic interaction causes irregularities in the originally smooth liquid jet surface, leading to unstable wave growth on the jet surface, and produces unstable ligaments. The aerodynamic force increases with an increase in the liquid velocity (or relative velocity), air density, and liquid-air contact surface area. As the relative velocity increases, the sizes of the ligaments decrease and their lives shorten. Small droplets are generated upon the disruption of the ligaments.

In *pressure-swirl atomization*, the complex atomization process may be conveniently subdivided into two main stages, as suggested by Lefebvre.[266] In the first stage, surface instabilities are generated as a result of the combined effects of hydrodynamic and aerodynamic forces. In the second stage, surface protuberances are converted into ligaments that subsequently break up into droplets. Lefebvre[266] stressed that the disintegration of a liquid sheet issuing from a nozzle is not caused solely by aerodynamic forces but must be the result, at least in part, of turbulence and/or other disruptive forces within the liquid itself. The disturbances within the liquid have a strong influence on sheet disintegration, particularly in the first stage. The aerodynamic forces play an important role in the disintegration of the liquid through its influence on the development of waves on the initially smooth liquid surface, and the production of the unstable ligaments. Therefore, an increase in the relative velocity between the liquid and surrounding air leads to a decrease in the size of the ligaments, and consequently a reduction in droplet size.

In *external-mixing air-assist atomization* and *plain-jet air-blast atomization* processes, the disruption of a liquid jet is achieved primarily due to the complex interaction between the liquid and air (gas), and relies on the large velocity difference between the two phases. The liquid breakup process during the aerodynamic atomization may be divided into three stages. First, instabilities and small waves develop on the liquid surface due to aerodynamic and shear forces. Second, the liquid jet is torn into ligaments or sheets due to the shear stresses caused by the pressure variations in the air flow around the small waves and bulges. Third, these ligaments or sheets break up into droplets *via* the Rayleigh mechanism or the mechanisms proposed by Fraser et al.[116][207] and by Mansour and Chigier.[209] The primary breakup may be followed by secondary breakup, depending on the two-phase flow conditions and the physical properties of the liquid and air. In general, the jet breakup modes proposed by Faragó and Chigier[210] may be considered as detailed and comprehensive mechanisms governing jet breakup in these twin-fluid atomization processes.

In *rotary* or *centrifugal atomization*, droplet formation may follow several distinct mechanisms, as proposed by many researchers[94][109]-[112] on the basis of experimental observations of rotating flat disk process (Fig. 3.9): *(a)* Direct Droplet Formation, *(b)* Direct Droplet and Ligament Formation, *(c)* Ligament Formation, and *(d)* Sheet/Film Formation. Liquid flow rate and rotational speed of disk are two critical factors determining the operating atomization mechanism. At low flow rates, droplet formation is a discontinuous one, and generally in the Direct Droplet regime. Under the action of centrifugal forces, the liquid spreads out across the rotating surface, and is thrown off from place to place at the periphery of the rotating disk or cup as discrete droplets of nearly uniform size. Each droplet may draw a fine ligament behind it. Eventually, the droplets separate from the ligaments, and the ligaments are converted into a series of fine droplets of nearly uniform size. With increasing flow rate, the droplet formation process may move into the Ligament regime. Ligaments are drawn along the entire periphery of the rotating disk or cup and the diameters of the ligaments are larger than those in the Direct Droplet regime. The ligaments then break up into droplets following the Rayleigh mechanism. Further increasing flow rate to a critical value at which the ligaments can no longer accommodate the flow of liquid, a thin continuous sheet or film forms, i.e., the droplet formation occurs in the Film regime. The sheet extends from the lip to an equilibrium length at which the contraction energy at the sheet edge due to surface tension is equal to the kinetic energy of the extending sheet. The sheet breakup then may follow the mechanisms proposed by Fraser et al.[116][207] A thick rim may form and disintegrate subsequently into ligaments and eventually droplets. Since the formation of the rim occurs without a controlling solid surface, the ligaments formed from the sheet are irregular in shape and size, resulting in a variation in the final droplet size. Generally, droplet size distribution becomes increasingly polydisperse as the droplet formation process moves from the Direct Droplet regime through the Ligament regime to the Sheet regime. It is advantageous to avoid the Sheet regime. One of the means to delay the transition from the Ligament regime to the Sheet regime is to serrate the outer edge of the cup or disk. Increasing the periphery speed of the rotating disk to a very high level, for example, more than 50 m/s, the liquid appearing at the edge of the disk can be immediately atomized by the surrounding air due to the high relative velocity.

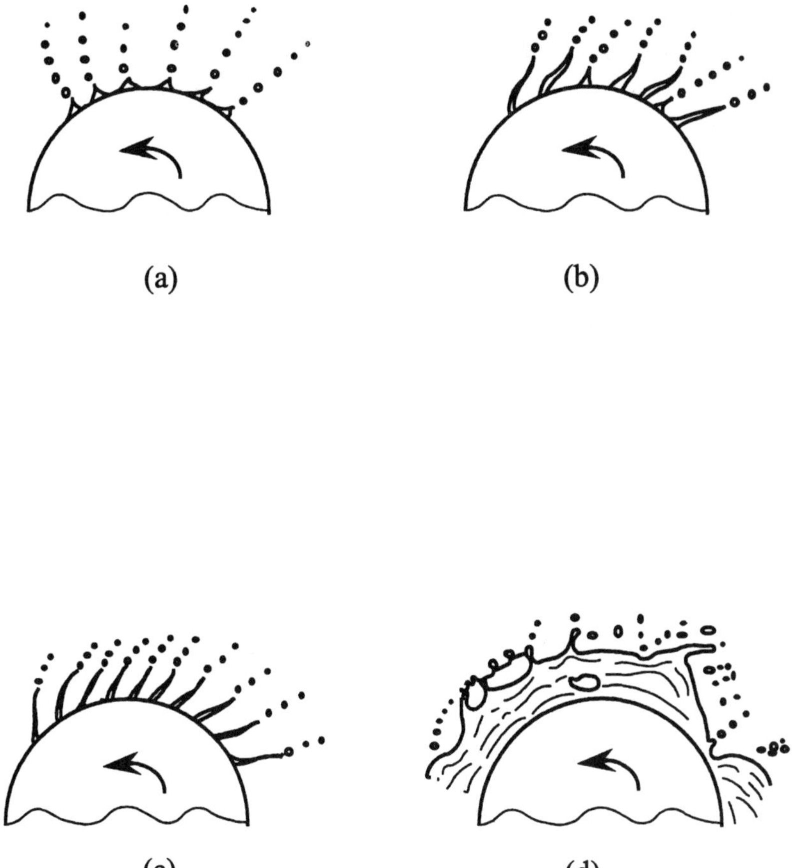

Figure 3.9. Regimes in rotary (centrifugal) atomization of liquids: *(a)* Direct Droplet Formation, *(b)* Direct Droplet and Ligament Formation, *(c)* Ligament Formation, and *(d)* Film or Sheet Formation.

For *ultrasonic atomization*, two rival theories have been developed: *(a)* capillary wave theory and *(b)* cavitation theory, as reviewed by Topp.[136] In the *capillary wave theory*, droplet formation may follow the wave disintegration mechanism. When a liquid is introduced onto a solid surface that is rapidly vibrated perpendicular to it, the liquid spreads in the form of a thin film in which capillary waves form. The gas interface of the liquid film in contact with the solid surface may become unstable under the influence of surface tension and density differences. This is the *Taylor instability* similar to that found in the rotating forces on a spinning disc. When the amplitude of the vibrating surface increases to the level at which the wave crests in the film become unstable and collapse, the liquid film breaks up and ejects away from the surface into a mist of droplets.[127] The phenomenon can be described by the Mathieu equation.[115] A solution of the Mathieu equation for inviscid liquids under very limited conditions of sheet thickness, amplitude and frequency gives a relation between the important parameters (liquid density, surface tension, excitation frequency of the solid) and the wavelength. Experimental observations revealed that the final droplet size is proportional to this wavelength.[127]

In the *cavitation theory*, the liquid-solid interface is subjected to violent pressure fluctuations when ultrasonic excitation takes place. This inevitably leads to cavitation. The theory postulates that it is the hydraulic shocks caused by the implosion of cavitation bubbles that initiate the liquid disintegration. It had been held for a long time that cavitation was the only mechanism responsible for the liquid disintegration.[115] In the early 1960's, it was suggested that the monosize droplets originated from the capillary wave disintegration, whereas cavitation caused the size dispersion of the spray formed. Later, an ingenious compromise was postulated and verified by Russian researchers that the capillary waves from whose crests atomization occurs were, in fact, initiated and driven by periodic shock waves generated by cavitation bubbles. It seems that atomization takes place from the crests of the capillary waves and from the portions of the liquid sheet disrupted by violent resonant bubbles and transient bubble activity. To obtain a near-monodisperse spray at high flow rates, the liquid film thickness must be maintained below the level at which bubbles can reach a resonant diameter. In practice, this is a matter of filming the liquid on the excited surface, and it appears that the hollow horn design may achieve this effect.[115]

3.1.2 Secondary Atomization

Secondary atomization (*secondary breakup*, or *secondary disintegration*) is a process in which a droplet splits up into smaller droplets as a result of the interaction between the droplet (*dispersed phase*) and surrounding medium (*continuous phase*). Aerodynamic forces acting on a droplet during its motion may cause deformation of the droplet and under certain conditions may overcome internal cohesive forces, disintegrating the droplet into smaller droplets. Since the final droplet size distribution produced in an atomization process will be determined by the liquid properties in both the primary and secondary disintegration, it is of importance to review various mechanisms developed for the secondary breakup of a single droplet under the action of aerodynamic forces.

The fragmentation of liquid droplets due to sudden exposure to a high-velocity gas stream has many important applications in the fields of aerodynamics and propulsion. For example, the phenomenon of supersonic rain erosion caused by the impingement of rain droplets at high relative velocities on exterior missile or aircraft surfaces can be greatly alleviated through proper aerodynamic design. This can be achieved by designing a body whose detached shock is sufficiently far removed to allow for droplet shattering in the region separating the shock from the body surface. In propulsion field, the rate of mixing and combustion of liquid fuel droplets can be greatly enhanced by virtue of the droplet fragmentation process. The droplet breakup time may become a rate-controlling mechanism in combustion systems.

Theoretical Considerations. A droplet generated in the primary atomization may be unstable and may further disintegrate into smaller droplets, depending on the competition between external aerodynamic forces and internal forces due to surface tension and viscosity at any point on the droplet surface. Therefore, analyses of droplet deformation and secondary breakup require a detailed knowledge of the distribution of the forces on the droplet. Further, the distribution of the forces varies with time as the droplet shape changes as a result of the deformation caused by these forces. When the internal forces at any point on the droplet surface are sufficient to balance the external aerodynamic forces, an equilibrium state is reached. As long as a change in external forces at any point on the droplet surface is compensated by a corresponding change in internal forces, a droplet can remain stable. However, if external forces are very large and any change in them can not be balanced by a corresponding change in internal forces, the external forces may deform the droplet and subsequently disintegrate it into smaller droplets. After the disruption, the higher surface tension pressure of a smaller droplet may be large enough to accommodate the variations in the external forces so that a stable stage may be attained and no further breakup can occur. Otherwise, further deformation and breakup may follow until the surface tension pressure of a further smaller droplet is large enough to maintain a force balance at all points on the droplet surface. When this equilibrium state is attained, the droplet is stable. This droplet size is termed *critical droplet size*. Conceptually, a stable droplet of the size equal to or less than the critical droplet size has an infinite breakup time, while the breakup time of any droplet larger than the critical droplet size increases with decreasing droplet size. Even for a droplet larger than the critical droplet size, no breakup may occur if the external aerodynamic forces on the droplet decrease due to the variation in flow conditions or the internal forces increase due to the variation in surface tension and/or viscosity during the breakup time to the extent that an equilibrium state is reached, even if initially the external forces are sufficient to produce droplet breakup. Therefore, a rigorous analysis of droplet breakup demands information of dynamic distribution of various forces and the transient effects of flow conditions and physical properties.

Basic Breakup Modes. Starting from Lenard's investigation of large free-falling drops in still air,[267] drop/droplet breakup has been a subject of extensive theoretical and experimental studies[268]-[285] for a century. Various experimental methods have been developed and used to study droplet breakup, including free fall in towers and stairwells, suspension in vertical wind tunnels keeping droplets stationary, and in shock tubes with supersonic velocities, etc. These theoretical and experimental studies revealed that droplet breakup under the action of aerodynamic forces may occur in various modes, depending on the flow pattern around the droplet, and the physical properties of the gas and liquid involved, i.e., density, viscosity, and interfacial tension.

According to Hinze,[270] droplet breakup may occur in *three basic modes*:

(1) A droplet is initially flattened to an oblate, lenticular ellipsoid and then may be converted into a torus, depending on the magnitude of the internal forces causing the deformation. The torus subsequently becomes stretched and splits into smaller droplets.

(2) A droplet is initially elongated to a long cylindrical thread or ligament of cigar shape, and then breaks up into smaller droplets.

(3) Some protuberances may be created on a droplet surface due to local deformations. Under favorable conditions, these bulges detach from the droplet and disintegrate into smaller droplets.

The first mode may occur when a droplet is subjected to aerodynamic pressures or viscous stresses in a parallel or rotating flow. A droplet may experience the second type of breakup when exposed to a plane hyperbolic or Couette flow. The third type of breakup may occur when a droplet is in irregular flow patterns. In addition, the *actual breakup modes* also depend on whether a droplet is subjected to steady acceleration, or suddenly exposed to a high-velocity gas stream.[270][275]

Subjected to steady acceleration, a droplet is flattened gradually. When a critical relative velocity is reached, the flattened droplet is blown out into a hollow bag anchored to a nearly circular rim which contains at least 70% of the mass of the original droplet. Surface tension force is sufficient to allow the bag shape to develop. The bag, with a concave surface to the gas flow, is stretched and swept off in the downstream direction. The rupture of the bag produces a cloud of very fine droplets presumably *via* a perforation mode, and the rim breaks up into relatively larger droplets, although all droplets are at least an order of magnitude smaller than the initial droplet size. This is referred to as *bag breakup* (Fig. 3.10).[286]

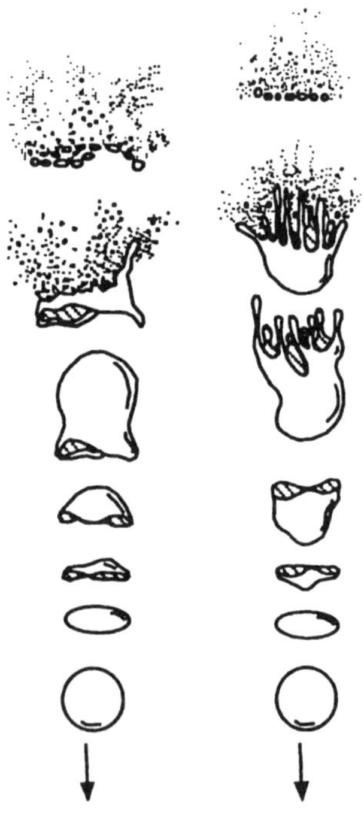

Figure 3.10. Some modes of droplet breakup. *Left*: bag breakup; *Right*: shear breakup.

Suddenly exposed to a high-velocity gas stream, a droplet is deformed into a saucer shape with a convex surface to the gas flow. The edges of the saucer shape are drawn out into thin sheets and then fine filaments are sheared from the outer part of the sheets, which subsequently disintegrate into smaller droplets and are swept rapidly downstream by the high-velocity gas. Unstable growth of short wavelength surface waves appears to be involved in the breakup process.[210] This is known as *shear breakup* (Fig. 3.10).[246]

In both these modes of droplet breakup, the initial stage of the process involves a flattening of the droplet into a disk shape, normal to the flow direction. In addition, there are two distinct size ranges of droplets with smaller droplets moving downstream of the larger droplets. In the bag breakup mode, a droplet breaks up *via* membrane formation with features similar to those seen in the co-flowing jet breakup cases. Structures present in the gas flow are responsible for the distorting and stretching of the liquid.[210] High-velocity droplet breakup mechanism is similar to the high-velocity jet breakup mechanism in the second wind-induced and possibly also the atomization regimes, which involve the unstable growth of short wavelength surface waves and the formation of fibers.[210] Generally, there seem to be similarities between the disintegration modes of round liquid jets in a coaxial air stream, and the breakup of droplets in an air stream. At similar aerodynamic Weber numbers, the disintegration modes are also similar (Table 3.2).

The experimental results of aerodynamic shattering of droplets behind shock waves revealed that the major variables affecting the high-speed breakup (Weber numbers much greater than 10) are the droplet diameter and the dynamic pressure of convective flow, with the liquid properties being less important. Ranger and Nicholls[246] extended previous experimental and analytical studies to a range of conditions typical of two-phase detonations, i.e., Mach numbers of 1.5–3.5 and droplet diameters of 750–4400 μm. Their shock-tube experiments revealed that droplet breakup can be temporally divided into two rather distinct stages.

The first stage, called *dynamic stage*, is the period during which a spherical droplet is flattened and deformed into a planetary ellipsoid with its major axis perpendicular to the flow direction as a result of the external pressure distribution. The eccentricity of the elliptical profile changes with time.

The second stage, termed *surface stripping stage*, is characterized by boundary-layer stripping from the deformed droplet due to the shearing action of the convective flow that rapidly reduces the droplet to clouds of micromist. The stripping stage is well developed after the droplet is contacted by the shock. At the high Mach number (3.5), after a droplet is collided by the shock, a well-defined wake forms behind it.

Interestingly, the shape of the wake is similar to that developed behind a hypersonic blunt body where the flow converges to form a narrow recompression neck region several body diameters downstream of the rear stagnation point due to strong lateral pressure gradients. The liquid material, that is continuously stripped off from the droplet surface, is accelerated almost instantaneously to the particle velocity behind the wave front and follows the streamline pattern of the wake, suggesting that the droplet is reduced to a fine micromist.

The experimental observations also revealed that the droplet breakup is a continuous process of disintegration that begins shortly after the initial contact between a shock and the droplet, and proceeds until the droplet is completely transformed into a cloud of mist. The collision between the shock and the droplet has little effect on the shattering phenomenon. The main function of the shock is to produce a high-speed convective flow. The droplet breakup is mainly caused by the interaction between the droplet and the convective flow established by the shock. The shearing action exerted by the high-speed gas stream on the droplet periphery causes the formation of a boundary layer in the liquid surface. The layer is established very rapidly after the droplet is intercepted by the shock, and stripped away from the equator continuously. This boundary-layer stripping mechanism accounts for the droplet breakup in the shear breakup regime.

Recently, many researchers investigated shape fluctuations, oscillations, deformation, and/or breakup of a droplet containing a polymer,[287] during elongational flow in blends of viscoelastic fluids,[288][289] in a gaseous medium,[290]-[295] in electric,[296]-[299] magnetic,[300] or acoustical fields,[301][302] in shock waves,[303] during interaction with other droplets,[304] and by surfactant.[305] Various deformation theories,[288][289] and droplet breakup mechanisms[306]-[309] have been proposed.

Breakup Criteria. Generally, droplet breakup in a flowing stream is governed by its surface tension and viscous forces, and dynamic pressure. For liquids of low viscosities, droplet breakup is primarily controlled by the aerodynamic force and surface tension force, and may begin when a critical condition, i.e., an equilibrium between these two forces, is attained:

Eq. (27)
$$\frac{\pi D^2}{4} C_D \frac{\rho_A U_R^2}{2} = \pi D \sigma$$

where C_D is the drag coefficient, and it is assumed that the relative velocity between the droplet and surrounding gas is high. From this equation, the critical Weber number We_{crit}, critical droplet size (maximum stable droplet size) D_{crit}, and critical relative velocity $U_{R\,crit}$ may be derived as follows:

Eq. (28)
$$We_{crit} = \frac{8}{C_D}$$

$$D_{crit} = \frac{8\sigma}{C_D \rho_A U_R^2}$$

$$U_{R\,crit} = \left(\frac{8\sigma}{C_D \rho_A D}\right)^{0.5}$$

Similar criterion has been obtained by Taylor.[205] Hinze[270] estimated We_{crit} to be 22 for a free-fall droplet, and 13 for a low-viscosity liquid droplet exposed suddenly to a high-velocity air stream. The latter value is comparable to those for water, methyl alcohol, mercury, and a low-viscosity silicone oil obtained by other investigators.[275]–[277]

For liquids of higher viscosities, the influence of liquid viscosity on droplet breakup needs to be considered. According to Hinze,[270] the critical Weber number may be modified to the following expression to account for the effect of liquid viscosity:

Eq. (29) $$We_{crit} = We_{crit,0}\left[1 + f(Oh)\right]$$

where $We_{crit,0}$ is the critical Weber number for zero viscosity. Empirical relationship has been proposed such as:

Eq. (30) $$We_{crit} = We_{crit,0} + 14Oh$$

In some practical processes, a high relative velocity may not exist and effects of turbulence on droplet breakup may become dominant. In such situations Kolmogorov,[280] and Hinze[270] hypothesized that the turbulent fluctuations are responsible for droplet breakup, and the dynamic pressure forces of the turbulent motion determine the maximum stable droplet size. Using Clay's data,[281] and assuming isotropic turbulence, an expression was derived for the critical Weber number:[270]

Eq. (31) $$We_{crit} = 1.18$$

Sleicher[278] has indicated that this expression is not valid for pipe flows. In pipe flows, droplet breakup is governed by surface tension forces, velocity fluctuations, pressure fluctuations, and steep velocity gradients. Sevik and Park[279] modified the hypothesis of Kolmogorov,[280] and Hinze,[270] and suggested that resonance may cause droplet breakup in turbulent flows if the characteristic turbulence frequency equals to the lowest or natural frequency mode of an entrained fluid particle. On the basis of this hypothesis, a formulation was derived for the critical Weber number in which Hinze's value of 1.18 is replaced by 1.04.

In *highly viscous flows*, the Reynolds numbers may be so small that dynamic forces may be no longer significant, and thus, droplet breakup may be dictated merely by viscous and surface tension forces. This may apply to the situation that, for example, some fluid globules are surrounded by a viscous fluid with a strong velocity gradient in the vicinity of the globules. A droplet is first elongated into a prolate ellipsoid by the viscous shear. When the viscous forces are sufficiently large compared to surface tension forces, droplet breakup occurs. This is often referred to as *Taylor mechanism*. This mechanism applies only when an undeformed or elongated droplet is small relative to local regions of a viscous flow, as indicated by Sevik and Park.[279] For large Reynolds numbers of external flows, the local regions may be very small in spatial dimensions compared to droplet sizes so that the dynamic pressure due to velocity gradient over the distance of the order of droplet size may become a dominant factor.

In practical situations, viscosities of both dispersed and continuous phases and their ratio may exert effects on droplet breakup. In particular, the viscosity of the dispersed phase may delay and impede breakup, as demonstrated by Meister and Scheele.[219] For a viscosity ratio of near unity, the perturbation wavelength has a minimum value corresponding to a maximum growth rate.[220][282] Either increasing or decreasing the value of the viscosity ratio reduces the tendency of droplet breakup.[220][270][282] Quantitative formulation encompassing the effect of the viscosity ratio has been proposed by Rumscheidt and Mason[283] such that:

Eq. (32) $$We_{crit} = \frac{1+(\mu_L/\mu_A)}{1+(19/16)(\mu_L/\mu_A)}$$

where μ_A is the viscosity of air (continuous phase). This may explain the facts that dispersion by pure viscous flow is restricted to emulsification processes, and it is very difficult to disintegrate liquids of large viscosity ratios. In shock experiments, the critical Weber number for breakup has been found to be $We_{crit} \approx 12$.[310]

Breakup Regimes. Hsiang and Faeth[285] made a comprehensive review of previous theoretical and experimental studies on secondary breakup of droplets. They extended the work of Hinze,[270] and Krzeczkowski,[311] and identified six regimes of droplet deformation and breakup, as depicted in Fig. 3.11,[270][285][311] and summarized in Table 3.3, where We_d and Oh_d are defined as $We_d = D\rho_G U_R^2 / \sigma$ and $Oh_d = \mu_L / (D\rho_L\sigma)^{1/2}$, respectively. This regime map can be used to determine if a droplet remains stable, or undergoes deformation or secondary breakup in addition to deformation. In this regime map, it can be seen that for a given Oh_d, a droplet may remain stable and not deform at a small We_d. With increasing We_d, a droplet may undergo nonoscillatory or oscillatory deformation. The transition from bag breakup,[286] through multimode breakup to shear breakup,[246] regime occurs with further increase in We_d. At very high We_d, catastrophic breakup[312] involving complex breakup mechanisms may take place. For high viscosity liquids, i.e., with increasing Oh_d, the breakup regime transitions are moved to higher Weber number values.

It is interesting to note that there are similarities in the breakup mechanisms, and even in the magnitudes of the Weber numbers between droplet breakup and jet breakup in various regimes. For example, the onset of the bag breakup occurs at the Weber number similar to that for jet breakup to enter the second wind-induced regime. This is not surprising, because the Weber number measures the relative importance of gas inertia forces to surface tension forces, while these forces are influential in distortion and breakup processes of both droplets and jets. For the same reason, the breakup mechanisms of droplets and jets in a high velocity gas should also be similar as long as the instability wavelengths of surface waves are much less than the diameters of the parent droplets or jets, and thus, the identity of the bulk liquid would become irrelevant.[210]

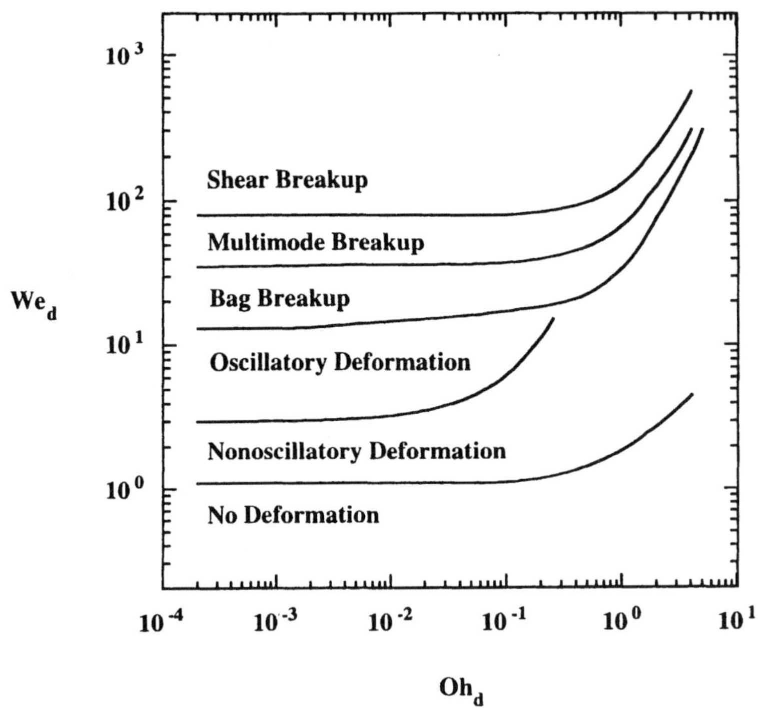

Figure 3.11. Regimes of droplet deformation and breakup.

Table 3.3 Values of Weber Number for Transitions from *No Deformation* Regime up to *Shear Breakup* Regime at $Oh_d < 0.1$

Transition to Regime	We_d [285]	We_d [311]	We_d [270]
Nonoscillatory Deformation	1.1	—	—
Oscillatory Deformation	3.0	—	—
Bag Breakup	13	10	13
Multimode Breakup	35	30	—
Shear Breakup	80	63	—

Hinze,[270] Hanson et al.[277] Gel'fand et al.[313] Krzeczkowski,[311] Reinecke and Waldman,[312] and Wierzba and Takayama[314] presented photographs of all the breakup regimes in the regime map. In high-pressure propulsion and combustion systems, acoustic and turbulent oscillations cause *oscillatory deformation* of droplets that may lead to secondary breakup. The breakup regime at the onset of secondary breakup has been termed *bag breakup*, also known as *umbrella* or *hat-type breakup.*[275] This regime involves deflection of a droplet into a thin disk normal to the flow direction, followed by deformation of the center of the disk into a thin bag (balloon-like) structure extending in the downstream direction. Both the thin center and thick rim subsequently disintegrate into droplets. The transition from bag breakup regime to shear breakup regime involves complex breakup processes, such as parachute breakup, chaotic breakup, bag-jet breakup, and transition breakup, etc.,[311] and therefore is referred to as *multimode breakup. Shear breakup* involves deflection of the periphery of the flattened droplet in the downstream direction and stripping of small droplets from the periphery. Shear breakup may occur when droplets are hit by shock waves.

Hwang and Reitz[315] studied the droplet breakup mechanisms in the *catastrophic breakup regime*. They found that the acceleration of a flattened droplet favors the development of the *Rayleigh-Taylor instability.*[206] The accelerating droplet breaks into large-scale fragments *via* the Rayleigh-Taylor instability. Ligaments and small droplets then form from the much shorter wavelength Kelvin-Helmholtz waves on the fragmented surfaces. Hwang and Reitz suggested that the droplet breakup mechanisms in the catastrophic breakup regime are similar to the high velocity jet breakup mechanisms in the second wind-induced regime, and possibly also the atomization regime, that involve the unstable growth of short wavelength surface waves.

The time t_b for a droplet to undergo deformation prior to secondary breakup is a function of Oh_d and a characteristic time $t*$:[285]

Eq. (33) $$t_b / t^* = 5/(1 - \mathrm{Oh}_d / 7), \qquad \mathrm{We}_d < 10^3$$

where $t^* = D(\rho_L/\rho_G)^{1/2}/U_R$. Similar formulations for t_b were derived and used by Ranger and Nicholls,[246] O'Rourke and Amsden,[310] and Reitz.[316] For a constant Oh_d, the breakup time is proportional to the droplet diameter and the square root of liquid-to-gas density ratio and inversely proportional to the relative velocity.[246] The distortion of a droplet prior to breakup has a significant effect on the droplet drag, and consequently on the droplet breakup process. During deformation, a droplet may flatten, leading to an increase in cross-stream diameter prior to breakup. The degree of droplet deformation at any given time t may be correlated to the maximum cross-stream droplet diameter $D_{c,\max}$ and a characteristic time t_c^*:[285]

Eq. (34) $$\frac{D_c - D_{ini}}{D_{c,max} - D_{ini}} = 0.615 \frac{t}{t_c^*}$$

where $t_c^* = t^*/(1 - \mathrm{Oh}_d / 7)$, D_c is the actual cross-stream droplet diameter, and D_{ini} is the initial droplet diameter. The maximum cross-stream droplet diameter is a function of We_d, as reported by Hsiang and Faeth[285] based on their measurement data:

Eq. (35) $$\frac{D_{c,\max}}{D_{ini}} = 1 + 0.19\mathrm{We}_d^{1/2}, \qquad \mathrm{Oh}_d < 0.1,\ \mathrm{We}_d < 10^2$$

Hsiang and Faeth[285] also presented measurement data for $D_{c,\max} / D_{ini}$ as a function of We_d in graphical form for Oh_d=1.5 and 3.1. Thus, the maximum cross-stream droplet diameter over a range of Oh_d may be determined by interpolating these data.

In the breakup regimes, a droplet may undergo secondary breakup when the breakup time is reached. The droplet size distribution after bag or multimode breakup may follow the Simmons' root-normal distribution pattern[264] with MMD/SMD equal to 1.1,

1.2 or 1.5.[285] A bimodal distribution pattern has been observed by other researchers[313] for different experimental conditions. For shear breakup, a bimodal behavior has also been found.[285] This is deemed to be caused by the core droplet that remains after the stripping of smaller droplets from its periphery ceases. In bag or multimode breakup, however, several large droplets form from the ring at the base of the bag, as compared to the single core droplet in the shear breakup process. A correlation for SMD has been derived by Hsiang and Faeth[285] based on their experimental measurements:

Eq. (36) $$\rho_g \mathrm{SMD} U_R^2 / \sigma = 6.2(\rho_L / \rho_G)^{1/4} [\mu_L / (\rho_L D_{ini} U_R)]^{1/2} \mathrm{We}_d$$

with a correlation coefficient of the fit being 0.91. Thus, the entire droplet size distribution after bag or multimode breakup may be determined once the SMD is calculated from Eq. (36).

Formulations for SMD of secondary droplets have also been derived by other researchers, for example, O'Rourke and Amsden,[310] and Reitz.[316] O'Rourke and Amsden[310] used the χ-square distribution[317] for determining size distribution of the secondary droplets. They speculated that a breakup process may result in a distribution of droplet sizes because many modes are excited by aerodynamic interactions with the surrounding gas. Each mode may produce droplets of different sizes. In addition, during the breakup process, there might be collisions and coalescences of the secondary droplets, giving rise to collisional broadening of the size distribution.

3.1.3 Droplet Formation in Atomization of Melts

Most commercial and near-commercial atomization processes for liquid metals/alloys involve two-fluid atomization or centrifugal atomization. As suggested by many experimental observations, two-fluid atomization of liquid metals is typically a three-stage process,[318][319] whereas centrifugal atomization may occur in three different regimes.[5][320] Many atomization modes and mechanisms for normal liquids may be adopted or directly employed to account for these atomization processes for liquid metals. For example, the Rayleigh mechanism may govern the disintegration of liquid ligaments formed during the breakup of a liquid jet in two-fluid atomization or generated in the Ligament regime of centrifugal atomization, whereas the mechanisms proposed by Fraser et al.,[116][207] and by Mansour and Chigier,[209] as well as the Taylor instability theory may be relevant to the disintegration of liquid sheets or films formed in two-fluid atomization, or generated in Sheet regime of centrifugal atomization. Simultaneous occurrence of two or more modes in a single process is possible, depending on the physical properties of melt (and gas) and atomizer geometry and configuration.

Droplet Formation in Gas Atomization. Experimental and modeling studies[160][161][169][318][319][321]–[325] have shown that *gas atomization of liquid metals* in spray forming and powder metallurgy processes may take place in two primary modes, i.e., liquid jet-ligament breakup and liquid film-sheet breakup.

Liquid Jet-Ligament Breakup (Fig. 3.12) may be an operating mechanism in gas atomization of melts with a free-fall atomizer. See et al.[318][319] made experimental and analytical studies on droplet formation during gas atomization of liquid metals with a free-fall atomizer using photographic techniques. These studies revealed that high velocity gas atomization of liquid metals is a three-stage process: *(1)* primary disintegration, *(2)* secondary disintegration, and *(3)* spheroidization and solidification. In the first stage, when a liquid metal stream issuing from a delivery nozzle enters into the influence region of discrete gas jets, it expands laterally and forms an expanding hollow cone above the geometric impingement point of the gas jets. The lateral spreading is caused by *(a)* the liquid being squeezed into the interstices of the dynamic pressure contours generated by the gas jets as they approach each other, and *(b)* an upward thrust generated by the reverse (recirculating) gas flow along the metal stream axis due to the negative pressure gradient in the upward direction. The magnitude of these effects depends upon factors such as metal specific gravity, metal stream diameter, nozzle design features (notably impingement angle), and gas jet velocity, among others. The periphery of the cone is presumably in the form of ligaments connected to the base of the cone. The cone spreads laterally until viscous and surface tension forces can no longer hold the liquid together as the cone falls under the action of aerodynamic forces and gravity into the region of higher gas velocities. The changes in dynamic pressure initiate and enhance the growth of disturbances on the liquid surfaces. As the cone thins out, ligaments and/or other irregular shapes are torn off or stripped from its periphery in a random manner, and subsequently break up into droplets following the Rayleigh mechanism. At high velocities, this tearing action may be due to the aerodynamic forces caused by the pressure variations in the gas flow around the waves and bulges on the liquid surfaces.[5] This primary disintegration occurs in time intervals much less than 5×10^{-4} s.[319] After the disintegration of each cone, another is formed which then undergoes disintegration. The cycle of the formation and disintegration of the metal cone repeats as the atomization proceeds. The initial expansion of the liquid metal stream is an important step in the primary disintegration because it determines the effective surface area of the liquid metal in contact with the gas. The exact manner in which droplets are formed depends on the dynamic instability of the surface disturbances, turbulence in the gas flow, and the variation in the gas velocity field around the metal surface. Although a fundamental understanding of the detailed atomization mechanism is yet to be established, it is widely accepted that in most cases the Rayleigh instabilities growing on the surfaces of the torn liquid ligaments lead to the formation of droplets.[4][147][238] Subjected to pressure forces and shear forces, a droplet formed by primary disintegration may undergo secondary disintegration if the dynamic pressure due to the gas jet velocity exceeds the restoring force due to the surface tension of the liquid metal. Solidification of the liquid metal/droplets may occur before, during and/or after the primary disintegration and/or the secondary disintegration, depending on process parameters and material properties. The final shape of a semi-solid or solid metal particle is determined by the relative magnitudes of the time needed for solidification to complete, and the time required for surface tension force to restore the metal droplet into a sphere.

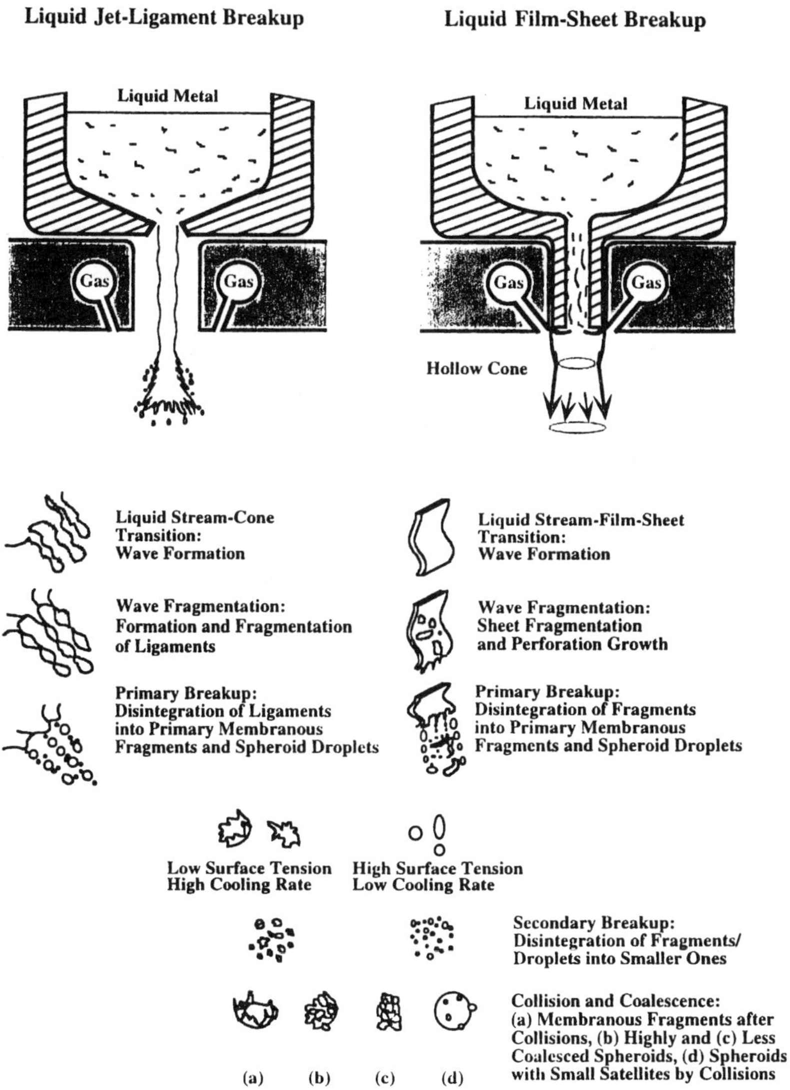

Figure 3.12. Schematic showing Liquid Jet-Ligament Breakup mode (*left*) and Liquid Film Sheet Breakup mode (*right*) in two-fluid atomization of melts.

Recently, detailed experimental observations using high-speed videography techniques[326] provided insightful information on the primary breakup stage in gas atomization of melts using a free-fall atomizer with two-level nozzle configuration. It is suggested that for the typical conditions prevailing in spray forming processes, the membrane-type breakup mode (or even fiber-type breakup mode at very high gas flow rates) proposed by Faragó and Chigier[210] may be the dominant operating mechanism(s) governing the liquid metal jet breakup. The interaction between the liquid metal jet and the gas jets from the top-level nozzles initiates small disturbances on the liquid surface. While the round liquid jet enters into the 'influence zone' of the gas jets from the second-level nozzles, it develops into a thin liquid membrane as a result of the distortion and thinning by the gas flow. Aerodynamic shear forces exerted on the membrane create the Kelvin-Helmholtz instability, leading to the formation of surface waves. Liquid accumulates at the edges of the thin membrane, forming a liquid frame while waviness increases in amplitude. The frame then breaks up into droplets *via* the non-axisymmetric Rayleigh mechanism. It is speculated that the 'liquid metal cone' observed in the experiments of See et al.[318][319] may be a temporal image of the Kelvin-Helmholtz waves because the jet breakup in a co-flowing gas is a pulsating process, as suggested by Faragó and Chigier.[210] However, in the work of Faragó and Chigier[210] no recirculating or transverse flow was present under the co-flowing condition. Therefore, the liquid jet-ligament breakup mechanism derived from the work of See et al.[318][319] might be different from the membrane-type breakup mode.

Liquid Film Sheet Breakup (Fig 3.12) is often encountered in high pressure gas atomization with a close-coupled atomizer.[324][325][327][328] This atomization mode is also a three-stage process, i.e., primary disintegration, secondary disintegration, and spheroidization/solidification.In the first stage, a liquid metal flows downward through a delivery tube initially. At the exit of the tube, it changes its flow direction to radially outward along the tube base plane, forming a thin liquid film, due to the recirculating gas flow in this region. The liquid film flowing radially then turns into a conical liquid sheet at the edge of the tube base plane, owing to the shear forces and pressure forces exerted by the gas flow parallel and perpendicular to the sheet surface, respectively. The pressure forces cause the sheet to align with the gas flow while the shear forces cause the sheet to accelerate downstream within the gas flow field. The conical liquid sheet forms a hollow cone beneath the delivery nozzle. The relatively high surface tension of metal melts produces a longitudinal streakiness in the spray due to the instability in the conical sheet near the nozzle. The sheet extending parallelly to the gas flow is highly unsteady and can vary from a length of essentially zero to a length extending a significant distance downstream into the gas flow field. While the sheet thins out, it fragments into droplets *via* the mechanisms proposed by Fraser et al.[116][207] and by Mansour and Chigier.[209] The droplets generated by the primary breakup accelerate, deform, and cool in the gas flow field. Secondary breakup of the droplets may occur, depending on their deformation and cooling/ solidification rates. Similarly to the situation in the liquid jet-ligament breakup mode, solidification of the liquid metal may occur at any stage during the atomization. Hence, the relative magnitudes of the time needed for solidification to complete and the time required for surface tension force to spheroidize a droplet determine its final shape.

In both atomization modes, as thin unstable ligaments, and/ or sheets disintegrate into round droplets, atomization gas may plausibly be trapped into the droplets under certain conditions. For alloys with alloying elements which readily react with atomization gas, for example, oxidize to form refractory oxides, solidification may be delayed and spheroidization is prevented so that rough flakes may form. For such alloys, the atmosphere in the spray chamber must be inert and protective to avoid the formation of any refractory and to foster spheroidal shape of droplets.

While gas atomization of liquid metals is generally viewed as a three-step process, the *Kohlswa ultrasonic gas atomization* is suggested to be a single-step process.[172] When 'particles' of an atomization gas at extremely high velocities strike a liquid metal stream, the liquid responds like a rigid solid but with low shear resistance. It is assumed that each gas particle may be able to knock out one metal droplet. However, the actual efficiency is low due to energy losses through reflection of gas pulses and heat generation, etc. To date, no mathematical model is available for quantifying the atomization mechanism in USGA.

Generally, gas atomization of melts produces a wide range of droplet size distributions. This is plausibly due to the uncontrolled disintegration of liquid stream into sheets and/or ligaments by gas jet(s) with a velocity distribution and the random breakup of the sheets and ligaments into droplets downstream of the jet impingement zone. For external-mixing atomizers, maximizing energy utilization of the high velocity atomization gas is of critical importance. At the high velocities used in gas atomization, the gas jets rapidly become turbulent and spread so that their velocities decay. A gas jet may maintain a centerline velocity approximately equal to its exit velocity for about six nozzle diameters downstream,[5] depending generally on Mach number.[175] If liquid breakup is completed within this high velocity zone, the fineness of atomization is optimized and a relatively narrow, mono-modal size distribution may be produced. If liquid breakup occurs downstream of the high velocity zone, a mono-modal, yet coarser, spray may be generated. Atomization spanning the high and low velocity zones may produce bimodal size distributions.[5] Therefore, using high gas velocities with optimized atomizer designs, such as HPGA and USGA, may be a way to increase the yield of fine droplets with a relatively narrow size distribution. Along with the rapid cooling potential, high velocity methods appear to be very attractive for many applications.

In the high-velocity atomization processes, the effect of the compressibility of atomization gas could be quite appreciable. Extending Taylor's analysis[245] for a gas flow over a liquid to a compressible gas flow, Bradley[329][330] developed an expression for the fastest growing wavelength that dominates the disruption at the interface. Supposing that surface tension nips off the crest of the wave into a filament whose diameter is some fraction of the fastest-growing wavelength, Bradley derived the following expression for the most frequently occurring droplet diameter:

Eq. (37) $$D = \frac{2\pi\varepsilon}{0.53k_{\max}}$$

where ε is a parameter taken as ¼, and $2\pi/k_{\max}$ is the fastest-growing wavelength. In Bradley's model, $k_{\max}$ is related to process parameters through a dimensionless parameter χ or χ_m:

Eq. (38) $$k_{\max} = \begin{cases} \dfrac{\chi \rho_G U_s^2}{\sigma} & 0 < \mathrm{Ma} < 1 \\ \dfrac{\rho_G U_s^2}{\chi_m \sigma} & 1 \leq \mathrm{Ma} < 10 \end{cases}$$

where U_s is the sonic velocity (speed of sound in gas), and Ma is the Mach number. The speed of sound in gas can be calculated using the following expression:

Eq. (39) $$U_s = (\gamma P_G / \rho_G)^{1/2} = (\gamma R T_G)^{1/2}$$

where P_G is the gas pressure, and the values of the gas constant R and the isotropic factor of gas γ can be found in Table 2.7 for various gases. The droplet diameter is then formulated as:

Eq. (40) $$D = \begin{cases} \dfrac{2.96\sigma}{\chi \rho_G U_s^2} & 0 < \mathrm{Ma} < 1 \\ \dfrac{2.96\sigma\chi_m}{\rho_G U_s^2} & 1 \leq \mathrm{Ma} < 10 \end{cases}$$

The dimensionless parameter χ or χ_m is a function of Mach number and can be determined from a universal curve for the atomization of various liquid metals, i.e., a plot of χ or χ_m *vs.* Mach number, as presented by Bradley in Ref. 329 for subsonic gas flow, and in

Ref. 330 for sonic and supersonic gas flow, respectively. Although the mathematical treatment of this model is sound and the derived equation for droplet diameter is simple for use, some important process parameters and material properties are not included, such as gas to liquid mass flow rate ratio, liquid viscosity, and nozzle geometry parameters. These parameters have been proven to affect the resultant droplet size in a significant manner. In addition, there seems to be no justification in assuming a constant value for the parameter ε. In fact, Rao and Mehrotra[331] have shown by analyzing atomization data for molten lead, tin, and lead-tin eutectic alloy that the parameter ε is a function of atomizing angle and nozzle diameter:

Eq. (41) $$\varepsilon = -0.12 + 0.57\vartheta + 0.38d_0 \qquad \text{for tin}$$

Eq. (42) $$\varepsilon = -0.14 + 0.66\vartheta + 0.48d_0 \qquad \text{for lead}$$

where ϑ is the atomizing angle, i.e., the angle between gas nozzle axis and liquid nozzle axis.

Another limitation of Bradley's model is that the liquid phase is assumed as an initially stationary flat layer of infinite depth over which a gas flows with uniform velocity. This is seldom the condition in practical atomization processes involving liquid metals. In Bradley's model, there is no indication of the origin of the spread in droplet size generated in gas atomization. Moreover, as pointed out by Mehrotra,[321] Bradley's model, and many previous models are pertinent to atomization process in which gas flows parallel to liquid surface. Thus, the models may not be able to adequately represent gas atomization processes of liquid metals involving ring-like atomizers. In such processes, gas streams typically impinge on a liquid stream at an angle. Thus, the velocity component perpendicular to the liquid metal surface is most likely to have a significant effect on the fluid dynamics of liquid metal breakup. The instability of liquid surfaces accelerated in a perpendicular direction has been analyzed by Taylor[206] and Lewis.[332]

Droplet Formation in Water Atomization. In *water atomization of melts*, liquid metal stream may be shattered by impact of water droplets, rather than by shear mechanism. When water droplets at high velocities strike the liquid metal stream, some liquid metal fragments are knocked out by the exploding steam packets originated from the water droplets and subsequently contract into spheroidal droplets under the effect of surface tension if spheroidization time is less than solidification time. It is assumed that each water droplet may be able to knock out one or more metal droplet. However, the actual number of metal droplets produced by each water droplet may vary, depending on operation conditions, material properties, and atomizer designs.

Droplet Formation in Centrifugal Atomization. The mechanisms of *centrifugal atomization* of liquid metals are quite similar to those for normal liquids. Three atomization modes have been identified in rotating electrode atomization process, i.e., *(1)* Direct Droplet Formation, *(2)* Ligament Disintegration, and *(3)* Film/Sheet Disintegration.[189][320] These are also applicable to the centrifugal atomization of melts in general. The *Direct Droplet Formation* mode is an operating mechanism at lower liquid flow rate. In this mode, liquid flows toward the periphery of a disk/cup or an electrode, and accumulates at the edge in the form similar to a small torus that deforms into many protuberances. From these protuberances, droplets are directly thrown off. The phenomenon is attributed to the *Taylor instability*,[205] which in this case is caused by centrifugal force, and opposed by surface tension force. Droplets formed in this mode are almost free of satellites. Under certain conditions, however, a bimodal droplet size distribution may be generated, if two droplets are thrown off from each protuberance, a large one from its head, and a small one from its tail, at equal numbers. With increasing liquid flow rate, the droplet formation process may transit into the *Ligament Disintegration* mode. In this mode, the protuberances develop to larger amplitudes than in the Direct Droplet mode before the elongated ligaments break up due to the *Rayleigh instability*. Droplet size distribution is typically bimodal. With increasing liquid flow rate, droplet size increases and the weight fractions of small and large droplets become similar. Further increasing liquid flow rate to a critical value at which the ligaments can no longer accommodate the flow of liquid, a thin continuous film or sheet forms, i.e., the droplet formation process transits into *Film/Sheet Disintegration* mode. In this mode, the sheet extends from the lip to an equilibrium length at which the contraction energy at the sheet edge due to surface tension is equal to the kinetic energy of the extending sheet. The sheet then breaks up *via* the mechanisms proposed by Fraser et al.[116][207]

Droplet Formation in Roller Atomization. In *roller atomization of melts*, the atomization mechanism was proposed to be similar in nature to the phenomenon of cavitation occurring in oil lubricated bearings.[188] During roller atomization, a liquid metal sheet forms in the gap between the rolls. Holes appear in the sheet, and seem to propagate as the sheet moves from the rolls. The sheet cavitates a small distance downstream of the nip between the rolls, forming perforations and threads. At low roll speeds (< 3000 rpm), the perforations formed propagate in the liquid sheet that issues from the gap between the rolls. At high roll speeds, the liquid may flow around the points of cavitation to form thin threads. The number and size of the perforations and threads are dependent on the number of cavities formed. It was speculated that droplets originate from the ligaments formed by impact of the rims produced from adjacent perforations (the perforated-sheet disintegration mechanism proposed by Fraser et al.).[116][207] The liquid threads break up into droplets by the propagation of liquid jet instabilities following the Rayleigh mechanism.

A mathematical analysis of the process is extremely difficult and requires to solve the Reynolds equation of lubrication theory and apply the solution to the cavitation boundary conditions. A two-dimensional analysis of the pressure distribution in the plane of the roll nip showed that the liquid pressure rises sharply to a large value near the nip, and drops equally sharply to a minimum just beyond the nip. Before large negative pressures are reached, the liquid may cavitate as a result of the expansion of entrained gases within the liquid.

3.2.0 DROPLET DEFORMATION ON A SURFACE

Droplet deformation during impact onto a surface is an interesting subject in many scientific and engineering fields. The icing and erosion/ablation of aircraft surfaces may be caused by the impact of droplets in clouds and/or rains on the surfaces during flight.[44][45] The erosion of turbine blades[46] operating in wet steam, the erosion of terrestrial surfaces during rain, and the erosion-corrosion of inhibitors in wet, CO_2-containing environment present in gas production and transportation facilities[333] are all related to droplet impact. The phenomena of droplet impinging and spreading on a solid surface are encountered in a wide range of processes and applications. These include, for example, spray combustion, spray cooling (quenching) of surfaces,[17] dispersed two-phase flow in once-through boilers, post-critical heat flux cooling, ink-jet printing,[47] metal welding/soldering, spray painting, lubrication, oil recovery from porous rocks,[48] and production of fine metal powders *via* impact atomization.[49] The spreading phenomenon is typically accompanied by simultaneous heat transfer and solidification of droplets on deposition surfaces in picoliter solder droplet dispensing for mounting of microelectronic components,[50] spray forming for near-net shape materials synthesis,[3] and thermal spray deposition for surface coating.[18][51][52]

Droplet impact on a surface is often accompanied by breakup of larger droplets into smaller ones due to droplet deformation, splashing and/or solidification on a cold surface, and/or vapor formation at liquid-solid interface on a hot surface. This phenomenon is typically observed in spray cooling of hot surfaces,[334] and in spray combustion where fuel droplets impinge on the hot wall of combustion chamber.[335] Impinging liquid droplets at a suitable impinging rate has been considered as a potential means for cooling integrated microelectronic equipment due to the high heat transfer rate, and reduced consumption of coolant as compared to liquid jet cooling.[336] In these applications, droplets may impact a dry or wet surface (a deep or shallow liquid pool). Bouncing of liquid may take place during impingement of droplets. Phase change such as solidification of liquid metals or evaporation of normal liquids may occur during spreading of droplets on a cold or heated surface. In most situations of practical interest, multiple droplets impact a surface, a physical phenomenon distinctly different from that in single droplet impact process.

The observations of deformation of a single droplet impinging on a surface started in 1876, as described extensively by Worthington.[337] Detailed studies of the deformation process were not possible until the development of high-speed photographic techniques and computer systems. To date, numerous theoretical, modeling, and experimental studies have been conducted to investigate the phenomena related to droplet deformation on a surface, such as spreading,[48][338]-[354] splashing/ejection and breakup,[355-360] wetting and contact angle,[361]-[366] heat transfer,[50][367]-[375] mass transfer,[376][377] solidification,[144][157][367][378]-[382] and shock wave formation[383] of droplets of normal liquids, melts, and polymer[384] on flat and non-flat surfaces, ceramic porous surface,[385] and liquid surface.[386] Extensive reviews of the previous studies have been made by Rein,[387] Liu et al.,[388][389] Dykhuizen,[390] and Dussan.[391]

3.2.1 Deformation of a Single Droplet on Flat and Non-Flat Surfaces

The deformation processes of a single droplet on flat and non-flat surfaces are shown in Fig. 3.13 and Fig. 3.14, respectively. In these figures, the impact Reynolds and Weber numbers are defined as $\mathrm{Re} = \rho_L u_0 D_0 / \mu_L$ and $\mathrm{We} = u_0^2 \rho_L D_0 / \sigma$, respectively, where u_0 and D_0 are the initial droplet velocity and diameter at impact, respectively. D/D_0 and H/D_0 are the dimensionless diameter and height of flattening droplet, respectively, and $t/(D_0/u_0)$ is the dimensionless time. The roughness or unevenness of a non-flat surface may be described by two parameters, i.e., roughness height and roughness spacing. Dimensionless roughness height and dimensionless roughness spacing may be conveniently defined as ε / D_0 and λ / D_0, respectively, where ε and λ are the amplitude and wavelength of the sinusoidal wave function for the surface shape,[389] respectively.

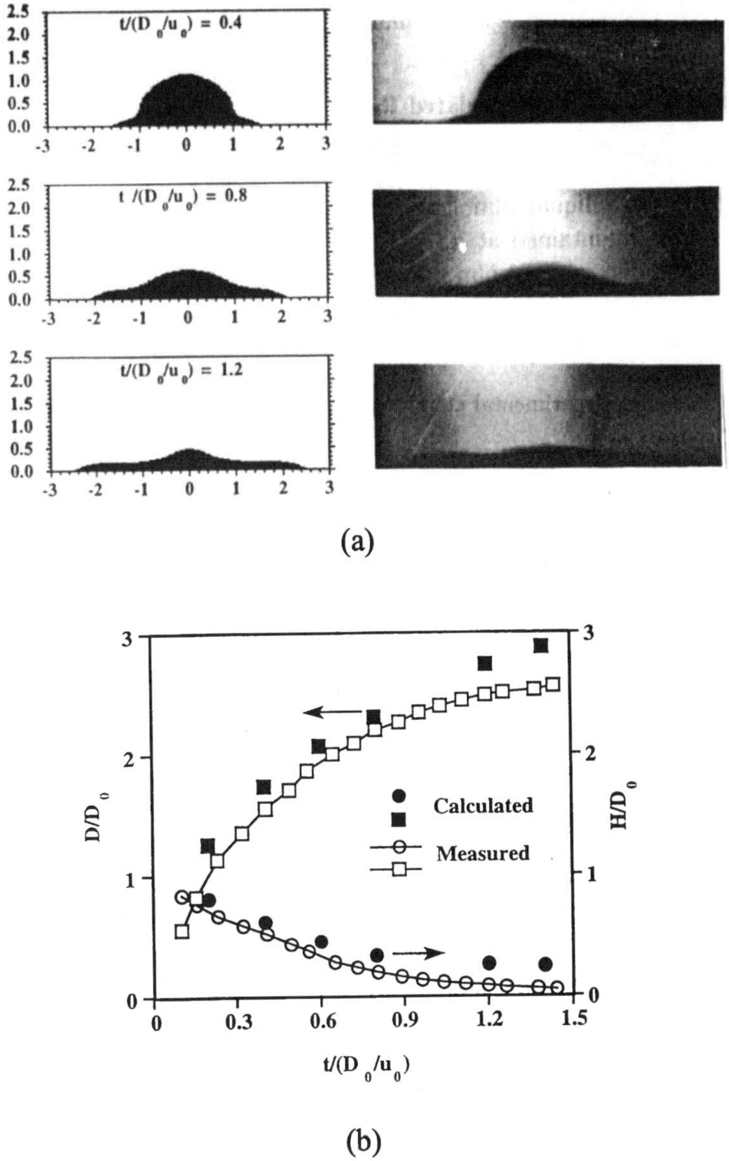

Figure 3.13. Deformation process of a single droplet impinging on a flat surface (Re = 1600, We = 26.7): *(a)* simulation (*left*), experiment (*right*), and *(b)* comparison between calculated and measured dimensionless diameter and height of a flattening droplet. *(Photograph: Courtesy of Prof. Dr. Jiro Senda at Doshisha University, Japan. Experimental data reprinted with permission from Ref. 334.)*

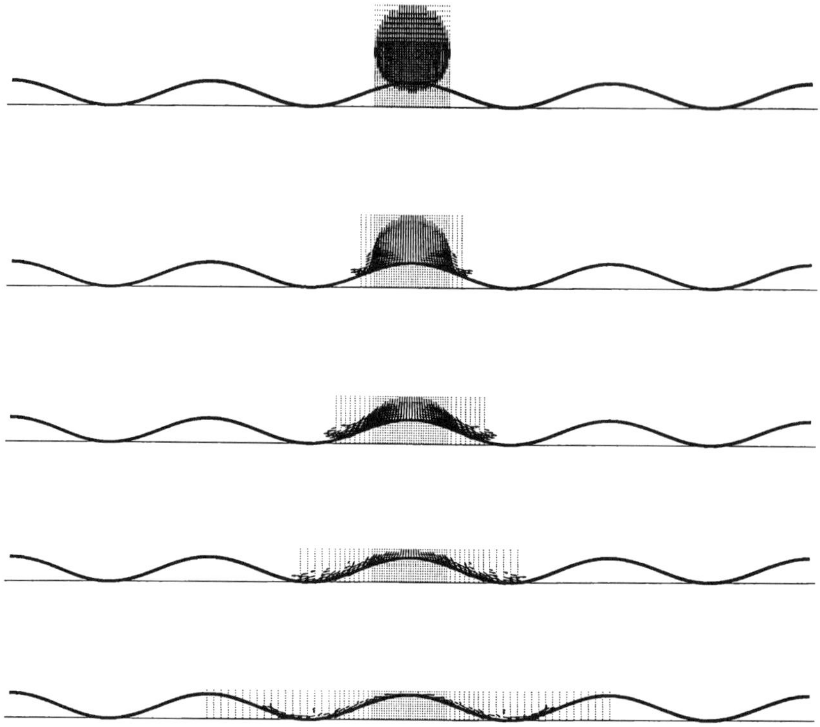

Figure 3.14. Deformation process of a single droplet impinging on a non-flat surface (Re = 38115, We = 8474, $\varepsilon/D_0 = 0.33$, $\lambda/D_0 = 2.8$). From top to bottom: $t/(D_0/u_0)$ = 0, 0.4, 0.8, 1.4, 2.3. *(Reprinted from Ref. 389, © 1995, with kind permission from Elsevier Science Ltd., The Boulevard, Langford Lane, Kidlington 0X5 1GB, UK.)*

During impingement on a *flat surface* (Fig. 3.13), a single droplet spreads uniformly in the radial direction from the impingement point as a thin liquid film, maintaining spherical shape at its top end. It eventually forms a thin, flat splat. The liquid that makes up the portion of the droplet that first impacts the surface ends up at the periphery of the splat. The spreading is fast initially, and the radial spreading velocity of the liquid may be as high as two to three times the impact velocity. The spreading rates decrease monotonically with time until vanish. The splat diameter approaches asymptotically its final steady state value (termed *final splat diameter* or *maximum spread diameter*) at some time (referred to as *deformation time*) depending on impact velocity and droplet properties (Fig. 3.15). For Reynolds numbers ranging from 300 to 30000, the dimensionless deformation time changes from about 1.4 to nearly 13, and the *dimensionless final splat diameter* (*maximum spread factor*, *flattening ratio*, or *degree of flattening*) varies from about 2.4 to nearly 11.3.

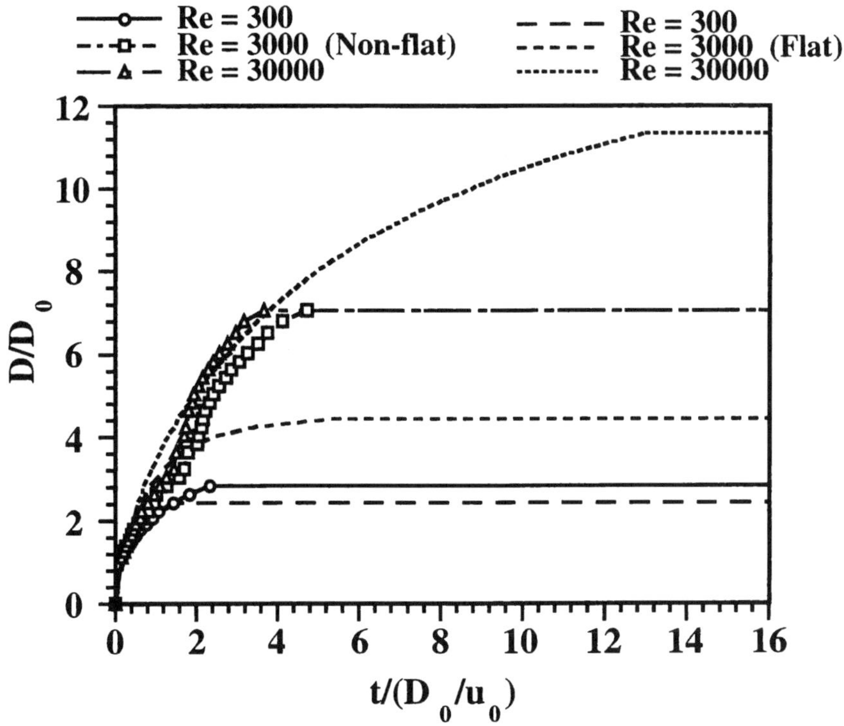

Figure 3.15. Spreading behavior of a single droplet impinging on flat ($\varepsilon/D_0 = 0$) and non-flat ($\varepsilon/D_0 = 0.33$, $\lambda/D_0 = 2.8$) surfaces at different Reynolds numbers (We = 2000) *(Reprinted from Ref. 389, © 1995, with kind permission from Elsevier Science Ltd., The Boulevard, Langford Lane, Kidlington 0X5 1GB, UK.)*

During impingement onto a *non-flat, waved surface* (Fig. 3.14), a single droplet spreads along the surface and eventually forms a thin, non-flat splat. Unlike the spreading behavior on a flat surface, the spreading rates on a non-flat surface do not decrease monotonically with time, but instead, exhibit a periodic repetition of a decrease-increase pattern (Fig. 3.15). The spreading of a droplet is fast when the liquid spreads from the top to the bottom of the sinusoidal wave (acceleration effect of surface). It slows down when the liquid spreads from the bottom to the top of the sinusoidal wave (deceleration effect of surface). If the wavelength of the surface is larger than the droplet diameter, the spreading process exhibits a periodic repetition of the acceleration-deceleration mode as the surface shape changes periodically, and ends when a violent breakup of liquid occurs (Figs. 3.14 and 3.16). If the wavelength of the surface is smaller than the droplet diameter, the surface hinders the droplet spreading (hindering effect), and the droplet breaks up and ejects from the surface in an irregular manner (Fig. 3.17). For Reynolds numbers ranging from 300 to 30000, the dimensionless deformation time changes from about 2.4 to nearly 3.6, and the dimensionless final splat diameter varies from about 2.8 to nearly 7.1.

It should be noted that the dynamic conditions of droplet impact processes discussed above cover a large range of the actual conditions in many industrial processes, such as spray forming, thermal spray, spray combustion, spray cooling, and aircraft flight. Under these conditions, the spreading behavior of droplets on a *flat surface* is essentially governed by inertia and viscous effects (Fig. 3.15). The surface tension effect on the deformation dynamics is not significant under these conditions (Fig. 3.18). However, with decreasing Weber number, surface tension force may become increasingly important to droplet deformation process. The effect of surface tension lies in that it makes the leading edge of the thin liquid film (splat) more rounded, and slows down the liquid velocity in the region of the leading edge.[335] This may cause some separation of the liquid edge from the surface.[371] It has been observed in experiments with high surface tension liquids[43] that splat edges indeed curl upwards, and splat rims are not in contact with the surface. This phenomenon is referred to as *recoil.*[366] A droplet recoils off the surface during its spreading when the liquid surface tension and viscosity overcome the inertial forces.[366] For We >> $Re^{1/2}$, however, the capillary effect becomes negligible in the droplet impact and spreading process. Generally, increasing impact velocity, droplet diameter and/or liquid density, or decreasing liquid viscosity, leads to an increase in the final splat diameter, while a large droplet diameter, a high liquid density, a low impact velocity, or a small liquid viscosity corresponds to a longer spreading time.

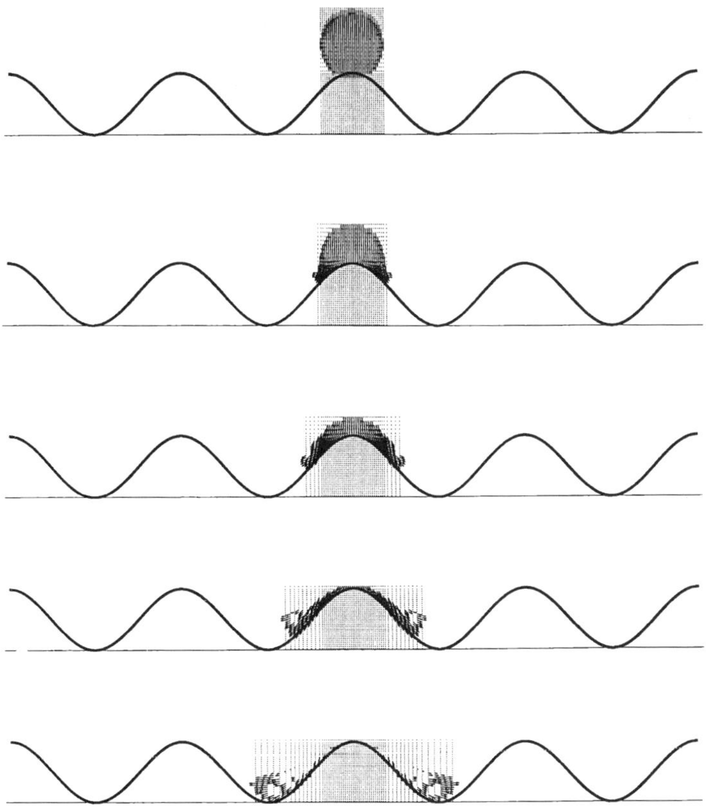

Figure 3.16. Deformation sequence of a single droplet impinging on a non-flat surface (Re = 38115, We = 8474, $\varepsilon/D_0 = 1$, $\lambda/D_0 = 2.8$). From top to bottom: $t/(D_0/u_0) = 0, 0.4, 0.8, 1.4, 2.3$ *(Reprinted from Ref. 389, © 1995, with kind permission from Elsevier Science Ltd., The Boulevard, Langford Lane, Kidlington 0X5 1GB, UK.)*

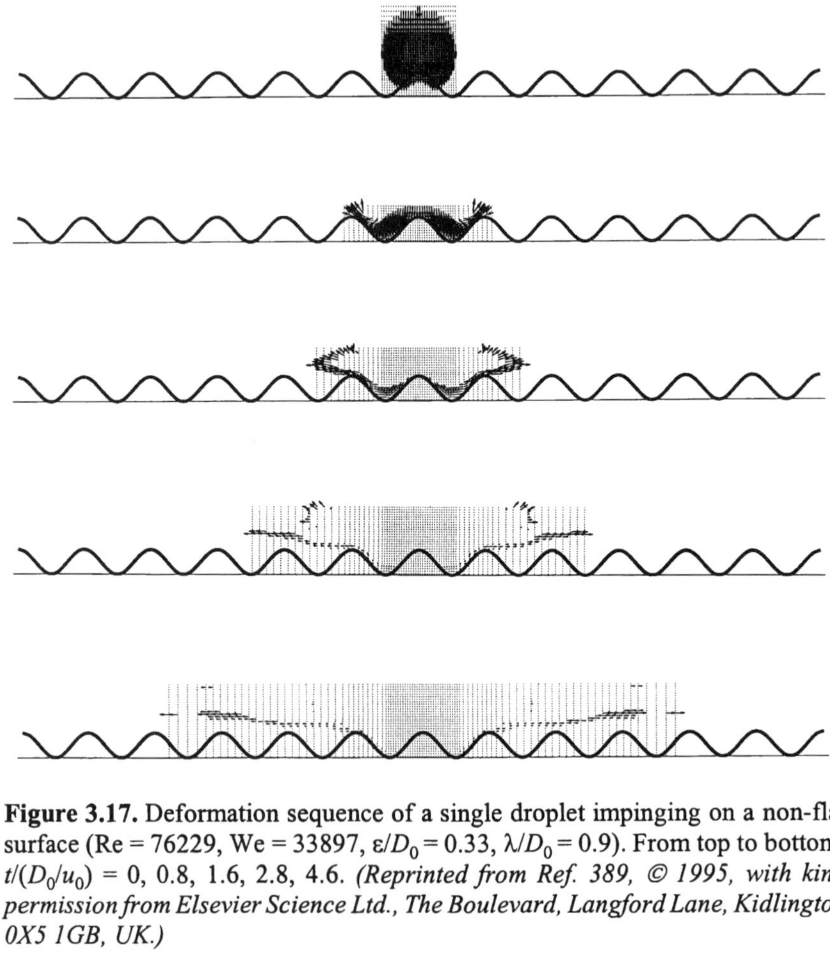

Figure 3.17. Deformation sequence of a single droplet impinging on a non-flat surface (Re = 76229, We = 33897, $\varepsilon/D_0 = 0.33$, $\lambda/D_0 = 0.9$). From top to bottom: $t/(D_0/u_0)$ = 0, 0.8, 1.6, 2.8, 4.6. *(Reprinted from Ref. 389, © 1995, with kind permission from Elsevier Science Ltd., The Boulevard, Langford Lane, Kidlington 0X5 1GB, UK.)*

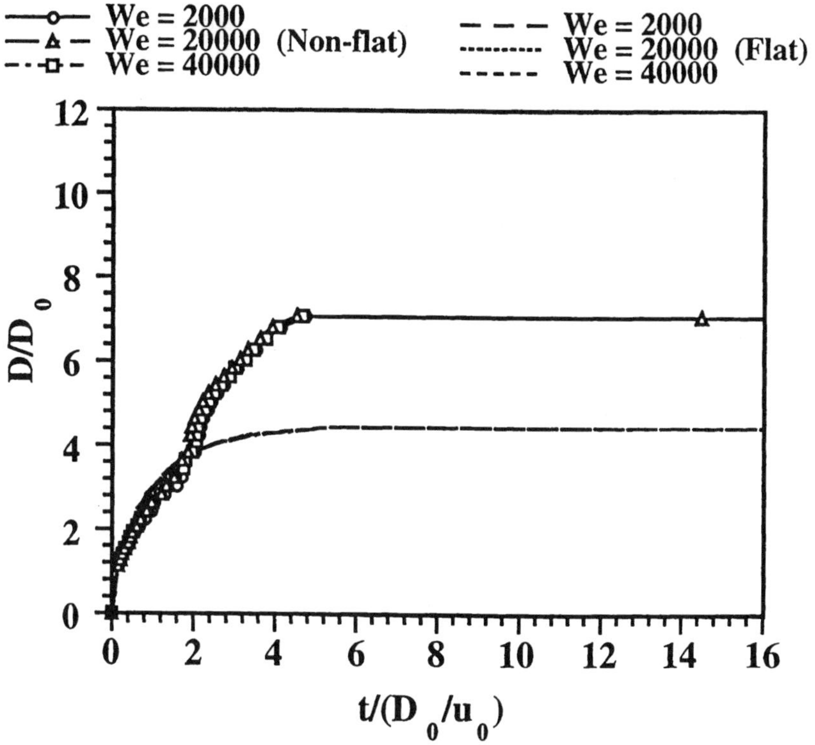

Figure 3.18. Spreading behavior of a single droplet impinging on flat ($\varepsilon/D_0 = 0$) and non-flat ($\varepsilon/D_0 = 0.33$, $\lambda/D_0 = 2.8$) surfaces at different Weber numbers (Re = 3000) *(Reprinted from Ref. 389, © 1995, with kind permission from Elsevier Science Ltd., The Boulevard, Langford Lane, Kidlington 0X5 1GB, UK.)*

The spreading behavior of droplets on a *non-flat surface* is not only dependent on inertia and viscous effects, but also significantly influenced by an additional normal stress introduced by the curved surface. This stress leads to the acceleration-deceleration effect, or the hindering effect depending on the dimensionless roughness spacing, and causes the breakup and ejection of liquid. Increasing impact velocity, droplet diameter, liquid density, and/or dimensionless roughness spacing (Fig. 3.19), or decreasing liquid viscosity, and/or dimensionless roughness height (Fig. 3.20), leads to an increase in the spreading rate. However, the final splat diameter, and the spreading time are determined by the time at which a violent breakup of liquid occurs, since the spreading process essentially terminates when a violent breakup of liquid occurs. A large droplet diameter, or a high liquid density leads to a larger final splat diameter, while a high impact velocity, a large dimensionless roughness height, a small dimensionless roughness spacing, or a small liquid viscosity corresponds to an earlier occurrence of a violent breakup of liquid. Decreasing the roughness height and increasing the roughness spacing of the impact surface may improve the extent of droplet spreading and reduce the breakup of liquid.

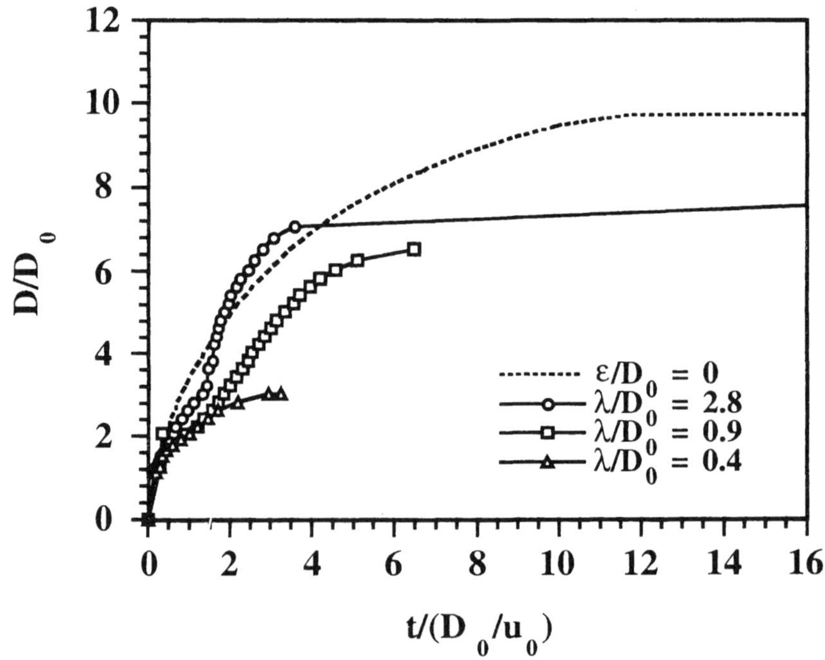

Figure 3.19. Spreading behavior of a single droplet impinging on non-flat surfaces of different roughness spacing (Re = 19060, We = 2119, $\varepsilon/D_0 = 0.33$). *(Reprinted from Ref. 389, © 1995, with kind permission from Elsevier Science Ltd., The Boulevard, Langford Lane, Kidlington 0X5 1GB, UK.)*

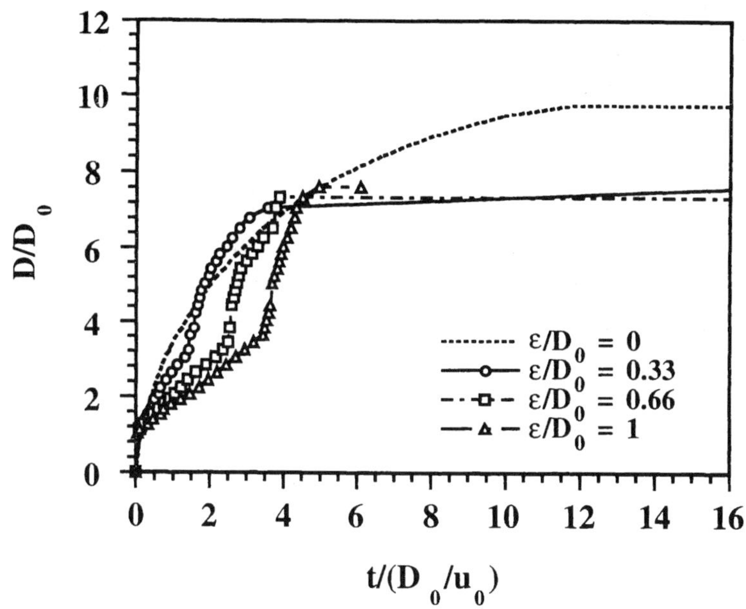

Figure 3.20. Spreading behavior of a single droplet impinging on non-flat surfaces of different roughness height (Re = 19060, We = 2119, $\lambda/D_0 = 2.8$) *(Reprinted from Ref. 389, © 1995, with kind permission from Elsevier Science Ltd., The Boulevard, Langford Lane, Kidlington 0X5 1GB, UK.)*

3.2.2 Splashing of a Single Droplet on Flat and Non-Flat Surfaces

Droplet splashing or breakup into secondary droplets during impingement on a surface is essentially governed by *(a)* impact surface conditions, *(b)* fluid dynamic conditions of droplets, and *(c)* heat transfer (surface temperature) and phase change (solidification or evaporation). A number of studies[46][392]-[395] on the impact of droplets on a surface revealed that droplets impacting on flat, smooth, dry surfaces do not splash, in other words, splashing occurs only if the surface is rough (non-flat) or wet (liquid film or shallow pool). This suggested that a good adhesion between a splat and the surface may be obtained, in terms of the even distribution of the liquid film and good contact, if the surface is flat, smooth and dry at the impact of droplet. There is evidence however that droplet splashing may occur even during normal impact on a smooth, dry substrate surface if the droplet impact velocity exceeds a threshold value,[396] and/or the droplet undergoes a phase change during the impact process.[390] Detailed sequences of droplet splashing on a solid particle and liquid droplet, in shallow and deep pools have been presented by Liu et al.,[18][389] and Harlow and Shannon,[397] respectively.

In addition to the impact surface conditions, the fluid dynamic conditions of a droplet may also determine if the droplet will splash during impingement. To distinguish pure spreading from splashing, many studies have been performed with water or other normal liquids on dry, wetted and liquid surfaces,[396][398][399] and with liquid metals (Pb and Sn) at different impact angles on non-flat, dry, metal surfaces of different roughness values under a variety of dynamic and geometry conditions.[400] These studies showed that the amount of splashing mass increases with increasing impact velocity, droplet size, and roughness height of the impact surface. A large surface roughness corresponds to an earlier and more violent break-up of impact droplets and bounce-off of secondary droplets. The influence of impact angle on the amount of splashing mass did not appear to be significant under the experimental conditions,[400] although the impact angle may affect the final splat geometry.[401][402] The dependence of the number of secondary droplets on surface tension seemed to be week as well.[393][396] The splash limits for normal and oblique impact of molten metal droplets on different surfaces have been identified on the basis of experimental observations.[400] Generally, the correlation between the Weber number and Ohnesorge number according to Walzel[398] may be used to delimit the regimes of droplet spreading and splashing on a dry, solid surface in a We-Oh map (Fig. 3.21):

Eq. (43) $$We = 7.9 \times 10^{10}\, Oh^{2.8}$$

If the fluid dynamic conditions of a droplet are within the spreading regime, i.e., below the threshold curve in the We-Oh map, liquid splashing and ejection from the impact surface can be eliminated. The corresponding threshold velocity can be written as:

Eq. (44a) $$u_{0s} = 2.8 \times 10^5 D_0^{-1.2} \rho_L^{-1.2} \sigma^{-0.2} \mu_L^{1.4}$$

The splashing criterion derived by Stow and Hadfield[396] showed that the threshold velocity may be estimated by:

Eq. (44b) $$u_{0s}^8 = \text{const} \times (\sigma / \rho_L)^2 \nu_L (u_0 / D_0)^3$$

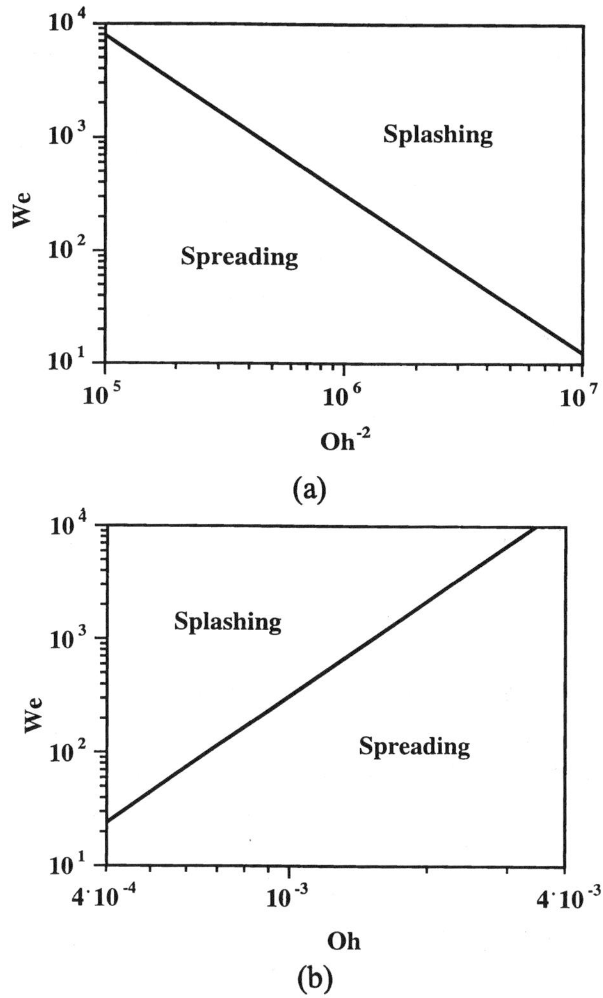

Figure 3.21. We-Oh map: regimes of droplet spreading and splashing on a dry, solid surface.

3.2.3 Droplet Deformation and Solidification on a Cold Surface

In many droplet impact processes, heat transfer and phase change (for example, solidification or evaporation) may take place simultaneously. In these circumstances, droplet temperature and surface temperature, in addition to initial droplet diameter, velocity, and properties, may significantly influence the final splat diameter and the deformation time because solidification may limit the extent of droplet spreading.[371] The actual spreading time may range from milliseconds down to microseconds, depending on droplet diameter, impact velocity, surface conditions, and liquid properties. For large droplets at low impact velocities, spreading and solidification may take place under similar time scales so that the onset of solidification may significantly interfere with the spreading process.[371] For small droplets at high impact velocities, spreading is essentially completed before the commencement of solidification. In the initial stage, droplet spreading is controlled by fluid dynamics, primarily by the rate at which the kinetic energy of the impacting droplet is consumed to overcome the viscous forces. After a droplet has spread to a certain length, perhaps $D/D_0 = 2$ to 3, heat transfer may take over, and the subsequent spreading is limited by the onset of solidification.

In case of *droplet solidification* during spreading, an increase in initial droplet velocity leads to a rapid increase in the final splat diameter and a decrease in the total solidification time (Fig. 3.22). Note that in this figure, the actual time step is the same for all three cases. However, with increasing droplet temperature and/or surface temperature, both the final splat diameter and the total solidification time increase.[52][388] Experimental studies on thermal spray droplet impact showed that the solidification time is typically two orders of magnitude longer than the flow time.[390] Even on a flat, dry surface, droplet splashing may occur if the surface temperature is below the melting temperature of the droplet.[46][52] This is due to the solidification of the lowest layer of the droplet first in contact with the cold surface as it flows radially, while the layers above the solid-liquid

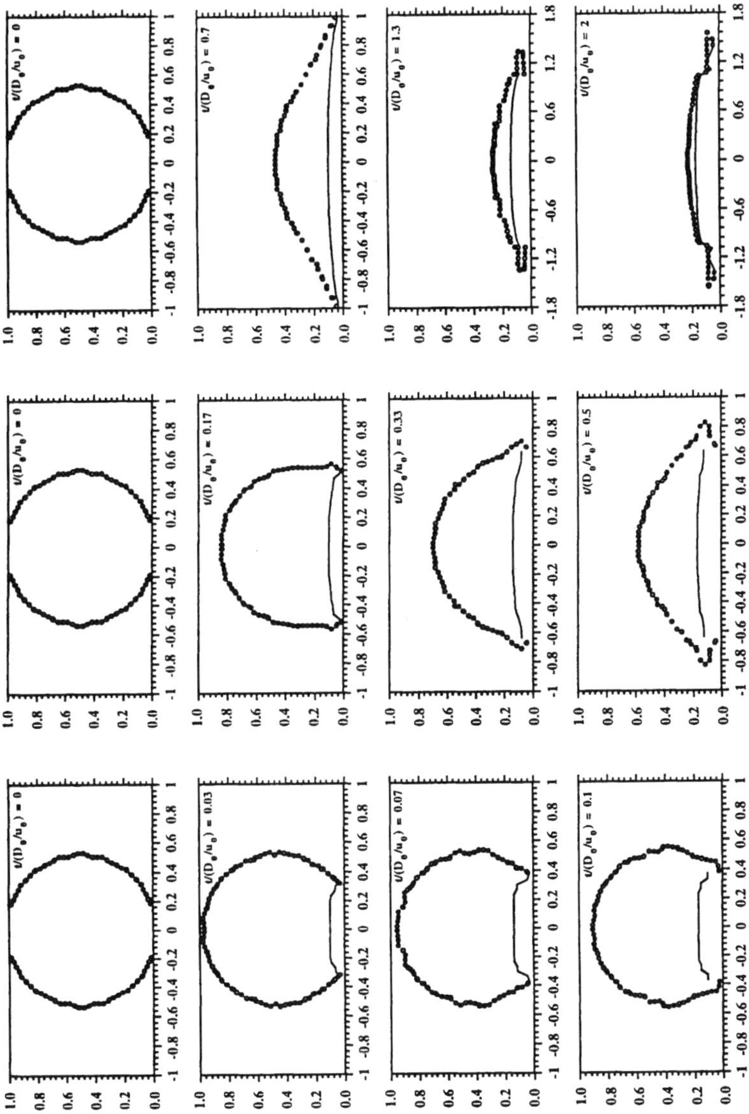

Figure 3.22. Spreading and solidification sequence of a single droplet impinging on a flat surface: Re = 867 (*left*), Re = 4335 (*center*), and Re = 17340 (*right*).

interface separates from the surface and ejects away as a result of the unique velocity distribution of the liquid at the leading edge.[46][52][144][395] A higher surface temperature, and/or droplet temperature may reduce droplet splashing due to a lower solidification rate.[46] However, there exists generally an optimum combination of deformation rates and solidification rates, or in other words, impact kinetic energy and thermal energy, in terms of the minimization of droplet splashing.[52][388] If the initial droplet velocity and/or the surface temperature is sufficiently high so that extensive deformation occurs prior to significant solidification, the separation does not take place or occurs at later time. If the initial droplet velocity and/or the surface temperature is sufficiently low so that significant solidification is completed prior to extensive deformation, the contact and eventual adhesion of the splat with the surface may be worse due to less flattening. At a lower surface temperature, the separation occurs earlier. Splashing and breakup always take place on a rough surface regardless of surface temperature.[390]

A number of researchers[390][402]-[406] have conducted experimental studies on the splat formation from a metal droplet impinging on a cold surface. Extensive photographic presentations of spreading, breakup, and solidification processes of liquid metal droplets on a cold surface have been given by Houben,[407] along with detailed descriptions of collision model, splat morphology, and influences of droplet size, impact velocity, temperature, and surface temperature. These studies suggested that droplet spreading, breakup and solidification may take place in many different modes, as depicted in Fig. 3.23. The final splat geometry may be either in a regularly-shaped disk pattern or in an irregularly-shaped platelet pattern, depending on droplet impact properties (size, velocity, angle, temperature, solid fraction), substrate surface properties (temperature, roughness, wetting), and their thermophysical properties (thermal conductivity, thermal capacity, density, surface tension, droplet viscosity, contact resistance). Star-shaped patterns (with corona) may result from splashing upon impact,[402] and may always occur on rough substrate surface regardless of substrate temperature.[390] On the other hand,

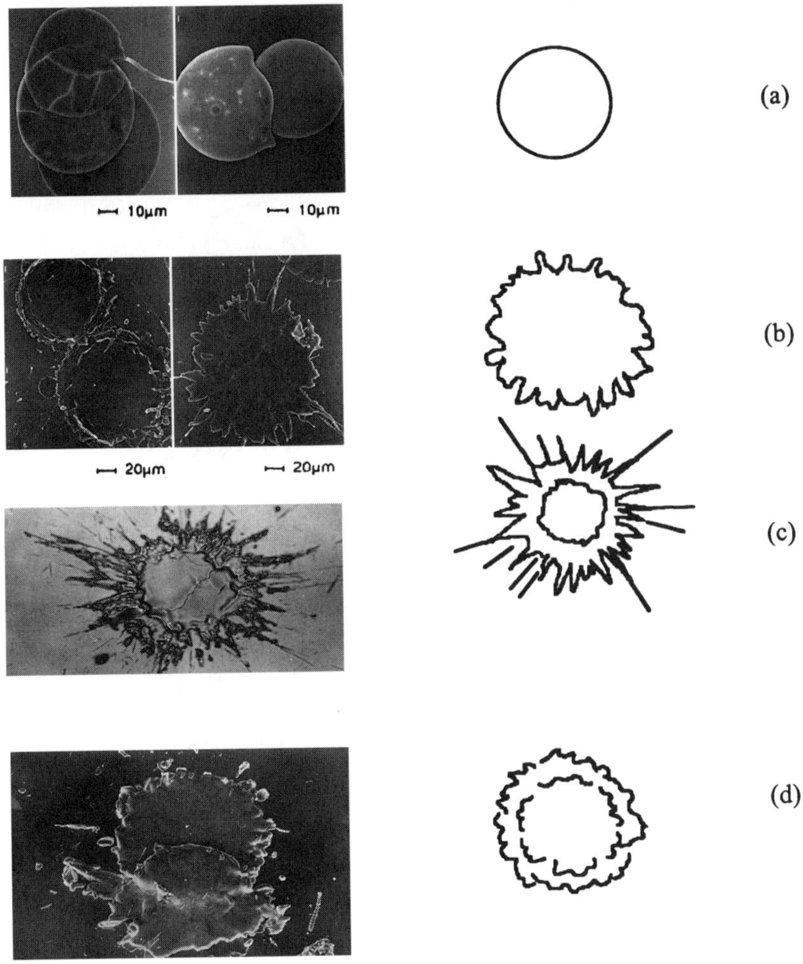

Figure 3.23. Schematic and micrograph of spreading, breakup and solidification modes of liquid metal droplets impinging on a cold surface. *(a)–(d)* "Pancake" pattern: regularly shaped disks with or without corona, *(e)–(h)* "Flower" pattern: irregularly shaped platelets with or without corona. *(Micrograph reprinted with permission from Ref. 403 and 407. Courtesy of Dr. J. M. Houben, Eindhoven University of Technology, Netherlands; and courtesy of Dr. Seiji Kuroda, National Research Institute for Metals, Japan.)*

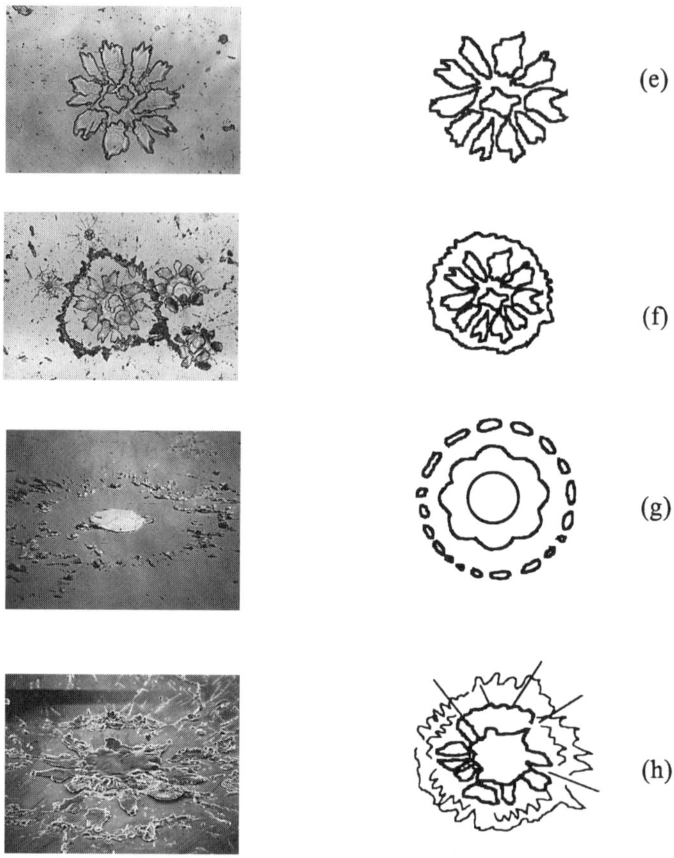

Figure 3.23. *(Cont'd.)*

decreasing substrate temperature may change the final splat shape from a regular disk to an irregular star. Sobolev et al.[408] identified a transition temperature that determines the splat morphology on a smooth substrate surface. At a substrate temperature below this transition temperature, droplet splashing occurs at the periphery leading to a star-shaped splat, while at a substrate temperature above this transition temperature, a regular disc-shaped splat forms. Generally, smaller droplet sizes, higher droplet velocities, higher droplet temperatures, colder and rougher substrate surfaces, and/or reduced wetting may lead to star-shaped patterns, while regularly-shaped patterns may result from larger droplet sizes, lower droplet velocities, lower droplet temperatures, and/or heated and smoother substrate.[46][390][402] Solidification, surface roughness, and wetting characteristics are among the most important factors influencing splashing and breakup of a droplet impinging on a cold surface.

It should be noted that the droplet spreading and solidification modes presented in Fig. 3.23 are more relevant to the conditions of high velocity and low thermal conditioning typical of thermal spray processes. For the conditions of low velocity and high thermal conditioning that characterize spray deposition processes, Matson et al.[409] presented an illustration of droplet impingement topology and classified droplet deposition and solidification behavior into seven modes (Fig. 3.24). The first and second modes involve rebound of a droplet off a fully solid and fully liquid substrate surface, respectively. The third mode is a thin pancake with lateral ejection of liquid. The fourth mode is a thick pancake or burst with surface oscillation or axial ejection of liquid. The fifth and sixth modes are spherical or semi-spherical pancake and bilaminate resulted from droplet spreading with subsequent surface recoil due to remelting, partial rebound, or bending. The seventh mode is a rigid pancake formed by surface motion or marginal spreading of a semi-solid droplet. These researchers suggested use of three additional dimensionless numbers, i.e., *Stefan number* (St), *Impact number* (I), and *Freezing number*

(F), in addition to the Reynolds and Weber numbers, to fully describe a droplet spreading and solidification process upon impact on a substrate. They introduced two new dimensionless numbers, defined as:

Eq. (45) $$\mathrm{I} = \frac{\pi \rho_L D_0^2 u_0^3}{12 A_c h (T_L - T_w)}$$

and

Eq. (46) $$\mathrm{F} = \left[\Delta H_m (1 - f_s) + c_{pL} \Delta T\right] \frac{\pi u_0 \rho_L D_0^2}{6 A_c h (T_L - T_w)}$$

where A_c is the contact area at the interface between a droplet and substrate, $A_c = \pi /8\ [1 + (D_s/D_0)^2]\ D_0^2$ formulated as a linear average of initial and final contact areas; h is the interface heat transfer coefficient; T_L and T_w are the initial droplet temperature and initial substrate temperature, respectively; f_s is the fraction solid of a droplet prior to impact; and ΔT is the difference between initial droplet temperature and melting (liquidus) temperature. The Impact number is seen to increase with increasing impact velocity and temperature difference across the interface. It is derived conceptually from expressions describing the kinetic energy transfer rate and the heat extraction rate through the interface. The Freezing number compares numerically the solidification time to the impact time. A small Freezing number results from a low velocity, gentle impact process, or poor interface heat contact. The most critical factors in using these new dimensionless numbers are to determine the heat transfer coefficient and the contact area between an impacting droplet and substrate. This requires the information of contact resistance, droplet deformation and final splat size.

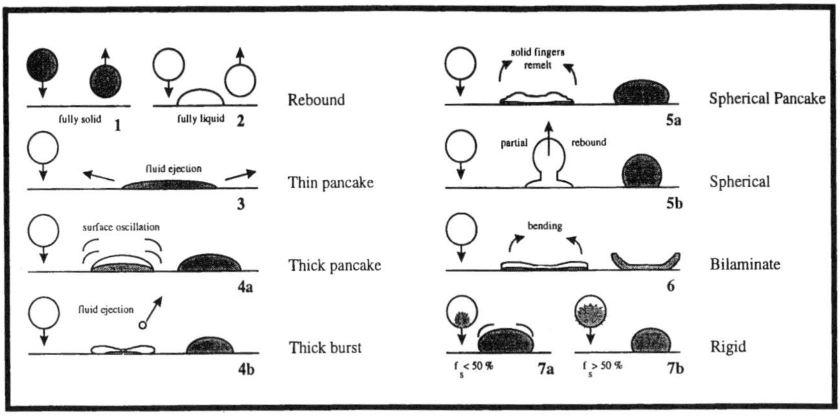

Figure 3.24. Classification of droplet impingement topology. *(Reprinted with permission from Ref. 409.)*

Under the conditions of high velocity and low thermal conditioning which characterize thermal spray (detonation coating) processes, evaluation of a correlation based on the Weber number is of reduced merit with a lower significance of surface-related phenomena relative to viscous effects. Hence, in designing a topology to investigate the thermal spray processes, it is appropriate to plot the Impact number as a function of the Reynolds number to identify domains of common deposition tendency. On the contrary, under the conditions of low velocity and high thermal conditioning which characterize spray deposition processes, viscous effects are, while still of importance, secondary effects in predicting impact behavior, as indicated by the analysis of the asymptotic behavior of the Reynolds number during spreading. Thus, in designing a topology to investigate spray deposition processes, it is more appropriate to plot the Freezing number as a function of the Weber number to identify domains of common deposition tendency, as illustrated in the *impingement map* (Fig. 3.25). At moderate impact velocities and low fraction solid, a combined approach may be more appropriate based on the formulation of the third new, yet dependent, dimensionless number, defined as a quotient of the Impact number and the Freezing number:

Eq. (47) $$\mathrm{N_Q} = u_0^2 / 2\left[\Delta H_m (1 - f_s) + c_{pL}\Delta T\right]$$

This number is conceptually an energy ratio, but independent of the interface heat extraction rate and thus the contact area. Since the interface heat transfer is assumed to control the solidification process of an impacting droplet, the choice of a dimensionless number should involve an evaluation of the influence exerted by this key factor. Therefore, the use of this newly defined dimensionless number is limited to an initial decision on which of the Impact number and the Freezing number is most appropriate for the application to a given material system at a know impact velocity.

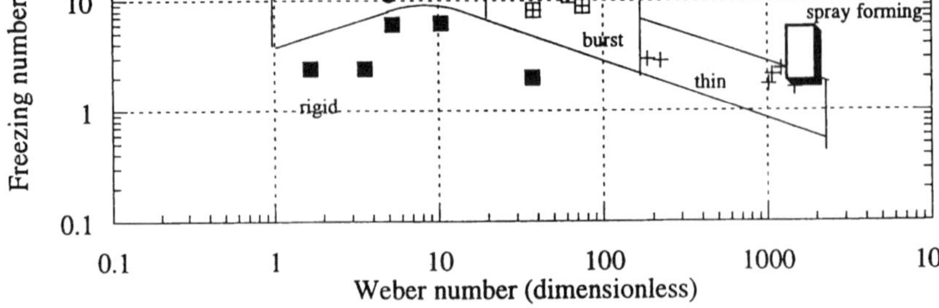

Figure 3.25. Impingement map *(Reprinted with permission from Ref. 409.)*

The occurrence of each deformation mode may be then correlated to a specific combination of the values of the Weber number and the Freezing number, as depicted in Fig. 3.25. The *thin pancake mode* is characterized by large values of the Weber number (200 < We < 1000), allowing for a large spreading, and extremely fast solidification with F ≈ 2. The liquid freezes as soon as it touches the substrate. Continuity and conservation of momentum require that the remaining liquid flow radially outward in the form of a thin sheet jet. This sheet freezes as soon as it contacts substrate, leading to the formation of a pancake with a low aspect ratio. In this case, solidification is so fast that droplet spreading may be largely retarded. Instabilities may cause the thin sheet jet to break up, leaving behind fine projections in a starburst pattern. The *thick pancake mode* represents the case where the heat extraction rate is sufficiently high to freeze the bottom part of a droplet while its upper part still moves and oscillates due to the conversion of kinetic energy to surface energy. This case corresponds to the conditions $1 < D_s/D_0 < 3$, We ≈ 80 and F ≈ 20. Bursting may occur at high velocities in the late stage of deformation when ejection is facilitated by a high impact kinetic energy. The *spherical pancake mode* is characterized by small values of the Weber number (We < 20), large F values (10 < F < 100), and a fully liquid droplet at impact. At a smaller F value (~10), a droplet may solidify into a semi-spherical pancake. Fingers stretching outward from the base of the semi-spherical pancake may lift up, fold inward, then remelt and finally retract into the bulk. At a larger F value (~100), the heat extraction process is very slow so that a droplet may take its equilibrium shape, i.e., spherical shape. For even larger F values (F > 100), a droplet does not freeze and is ejected upward as rebound. The *bilaminate mode* is characterized by a large We value (~300) and a large F value (~100). The base of a droplet freezes quickly and liquid floats on the top surface of the splat that subsequently solidifies at a slower rate controlled by the interface heat extraction rate. The upper part of the liquid region shrinks during solidification while the already solidified lower part is constrained due to the attachment to the substrate surface, leading to bending of the edges. The *rigid pancake mode* is encountered when a droplet solidifies after only minimum deformation with possible slight surface oscillations. This case corresponds to a small F value (F < 10), and a large fraction solid at impact. As the fraction solid increases, droplet deformation becomes more difficult as the mushy zone gets denser and stronger. During impact, the solid outer surface cracks open and the semi-solid interior flows out leaving behind a hollow shell (a half-moon shaped inclusion in the final splat). At a very large fraction solid, a droplet virtually does not deform, giving rise to a low heat extraction rate. Overall, these deformation modes may be concentrated in the impingement map and the relationship between each regime is coherent with the transition from one regime to the next following the anticipated trajectory. For example, a spherical pancake becomes a thick pancake by increasing the Weber number; A bilaminate mode becomes a thin pancake when reducing F. The transition is, however, not strictly bounded due to the many approximations made in defining the quantities such as contact area, heat transfer coefficient, and thermal physical properties. The impingement map may be useful for a wide range of engineering design applications, particularly to the investigation of porosity and oxidation level in sprayed preforms. The map presents a graphical characterization of desired operation regimes and provides guidelines for manipulating process parameters to get desired spray conditions, controlling the morphology of sprayed products, and avoiding regimes which promote poor product quality.

In addition to the deformation modes described above, there are some other modes of practical interest, for example, the *cyclic spreading*, *recoiling*, and *oscillation mode* typical of solder jetting process.[50] This mode leads to a final splat shape of multiple surface ripples at low substrate temperatures and hemispherical shape without surface ripples at high substrate temperatures.

Hofmeister et al.[410] employed two high-speed thermal imaging systems to record spatial and temporal temperature distributions at the splat-substrate interface, and to observe droplet spreading during impact and solidification on a quartz plate. They observed that a droplet superheated at impact solidified from the periphery of the splat, and the solidification proceeded radially inward to the center parallel to the splat interface with only slight undercooling. Undercooled droplets solidified immediately on impact with no additional undercooling. These experimental results are counterintuitive to our knowledge of transient nucleation behavior and shed some light on the physics of nucleation and solidification in droplet impact.

Sobolev et al.[408] used analytical models to investigate the mechanisms governing coating-substrate adhesion in thermal spray. They indicated that one of the most important mechanisms is the mechanical interlocking formed mainly due to the roughness of the substrate surface and the high pressure developed during droplet impact and solidification on the surface. Possible deformation, partial/complete melting, and subsequent solidification of the substrate surface, as well as splashing and rebounding of impinging droplets all influence the adhesion mechanisms. The roughness of the substrate surface was shown to be of critical importance for obtaining good adhesion.

3.2.4 Droplet Deformation and Evaporation on a Hot Surface

In case of *droplet evaporation* during impingement, droplet breakup behavior relates well with some specific surface temperatures, such as the maximum evaporation point and the Leidenfrost point.[335][411] The concomitant heat transfer process may be affected by many parameters, such as droplet size, contact angle, impact velocity, impact angle, liquid temperature, initial substrate surface temperature, surface roughness, and their thermophysical properties, as well as surfactant and gravity.[17] In most spray cooling systems, heat transfer during impingement and deformation of droplets on a surface may be categorized into the commonly known boiling curve regimes, i.e., nucleation, transition, and film boiling, in addition to thin film evaporation.[412][413]

Using different criteria, Shi et al.[336] suggested that heat transfer and surface cooling may be classified into two primary regimes: *(1)* cooling with wetting contact and *(2)* cooling with non-wetting contact, depending on substrate surface temperature. Liquid-solid wetting contact without bubble nucleation may be achieved in the initial stage when a liquid droplet impinges on a heated surface where heat flow is quasi-steady and conduction-dominant.[336][414] Following this initial transient stage, the heat transfer between the droplet and the surface will enter into nucleation, transition, and film boiling stages, depending primarily on the substrate surface temperature. In the *wetting contact cooling regime*, heat transfer rate is high due to the continuous or semi-continuous direct contact, followed by the nucleation boiling or transition boiling. In the *non-wetting contact cooling regime*, heat transfer rate is much lower due to the resistance of vapor generated between the liquid and hot surface, followed by the film boiling. At lower substrate surface temperatures, increasing droplet impact velocity may enhance the heat transfer rate due to the increased contact area, while it turns to lower the heat transfer rate at higher substrate surface temperatures due to rebounding of liquid from the surface. Heat transfer efficiency may be improved by lowering droplet temperature and impact angle. The initial substrate surface temperature has been found to be the most important parameter that influences the impinging heat transfer characteristics[336] because it determines to a large extent in which heat transfer regime the actual cooling process takes place. Numerical modeling[414] has been conducted to calculate the Nusselt number for droplet evaporation on a heated solid surface in the low superheat regime, including the effects of droplet contact angle, solid conductivity, and droplet concentration on heat transfer.

Chandra and Avedisian[411] studied the collision dynamics of a liquid (n-heptane) droplet on a polished, solid, stainless steel surface, and on a liquid film created by deposition of a preceding droplet using a flash photographic method. They presented a comprehensive series of clear images of droplet shape, morphology, and structure during the deformation process. In their experiments, the impact Weber number was fixed at 43, with an initial droplet diameter of 1.5 mm, and an ambient pressure of 0.101 MPa. Surface temperature ranged from 24 °C to above the Leidenfrost temperature of the liquid. The evolution of wetted area and spreading rate of a droplet both on a stainless steel surface and over a thin liquid film were found to be independent of surface temperature only during the early stages of impact due to negligible surface tension, and viscous effects under the experimental conditions. The influence of surface temperature on the overall deformation and breakup behavior of a droplet is illustrated in Fig. 3.26, where the transition from nucleation boiling to film boiling as the surface temperature is raised above the Leidenfrost temperature (200 °C) of n-heptane is also shown. At low surface temperatures, a single bubble was observed within the droplet at the point of impact during deformation (Fig. 3.26*a*). With increasing surface temperature, the population of bubbles within the droplet increased as a result of progressive activation of nucleation sites on the surface. At surface temperatures near to the boiling point (98.4 °C) of the liquid, a spoke-like cellular structure formed in the center of the circular liquid film of the flattened droplet due to the coalescence of a ring of bubbles that formed within the droplet during spreading (Figs. 3.26*b* and *c*). At higher surface temperatures, but below the Leidenfrost point, numerous bubbles appeared within the droplet (Figs. 3.26*d* and *e*). However, the overall droplet shape in the early stages of impact was unaffected by the occurrence of these bubbles. At even higher surface temperatures above the Leidenfrost point, no bubbles formed and droplet deformation/breakup entered into the film boiling regime (Fig. 3.26*f*). This was attributed to the fact that there is no longer solid-liquid contact and thus no opportunity for heterogeneous nucleation to occur on the heated surface.

In a subsequent study, Chandra and Avedisian[385] investigated droplet impingement on a ceramic porous surface. It was found that the spreading rate of a droplet on the porous surface at 22 °C is lower than that on the stainless steel surface, and for a given temperature the maximum spread diameter is smaller on the ceramic surface than on the stainless steel surface. No transition to film boiling was observed on the porous surface at 200 °C, unlike that seen on the stainless steel surface. The evolution of wetted area and spreading rate of a droplet on both the porous surface and the stainless steel surface was found to be independent of surface temperature during the early stage of impact. This tendency was attributed to the negligible surface tension, and viscous effects.

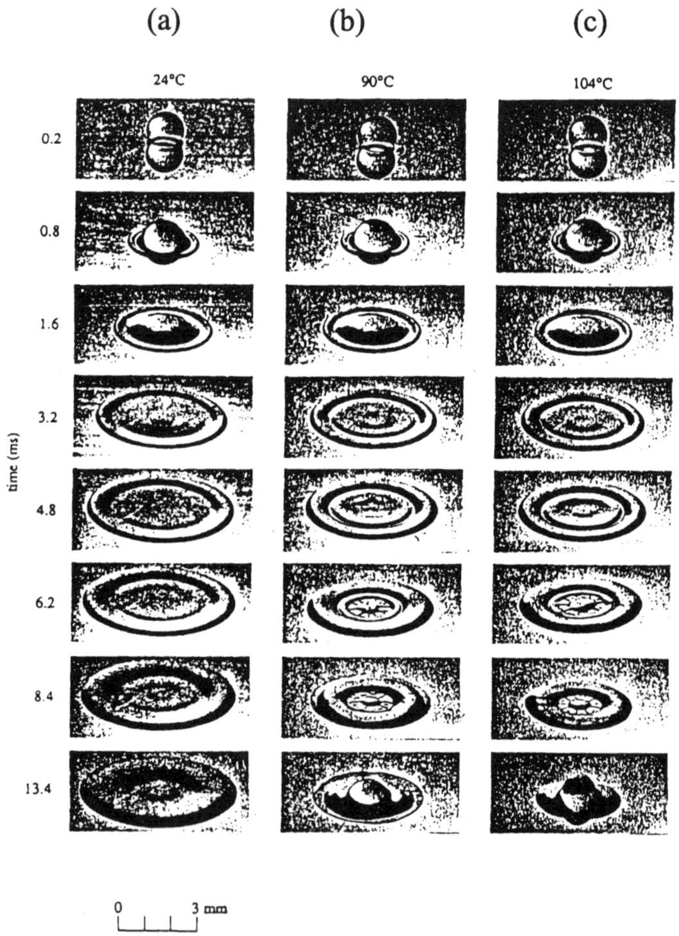

Figure 3.26. Deformation and breakup of droplets of n-heptane on a heated stainless steel surface (The angle of inclination of the camera to the horizontal is 30°): *(a)* T_w = 24 °C, *(b)* T_w = 90 °C, *(c)* T_w = 104 °C, *(d)* T_w = 160 °C, *(e)* T_w = 195 °C, *(f)* T_w = 205 °C. *(Reprinted with permission from Ref. 411.)*

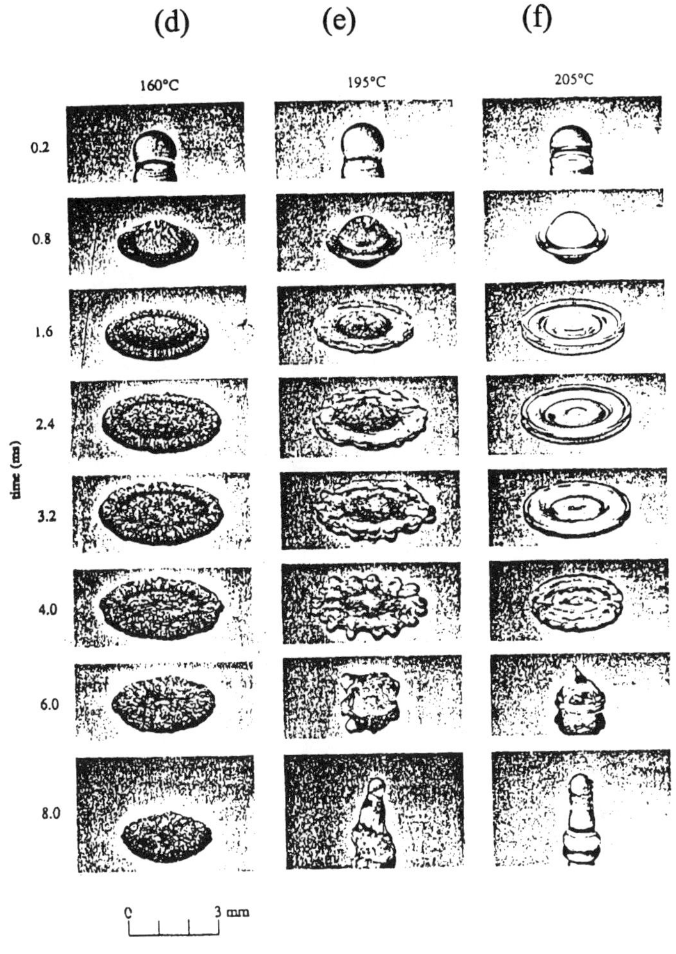

Figure 3.26. *(Cont'd.)*

On the basis of experimental observations (Fig. 3.27), Senda et al.[335][415] proposed six modes for water droplet deformation, and breakup during impingement on a hot surface coupled with heat transfer and evaporation (Fig. 3.28). Each mode occurs under a specific combination of surface temperatures and impact conditions, as described below.

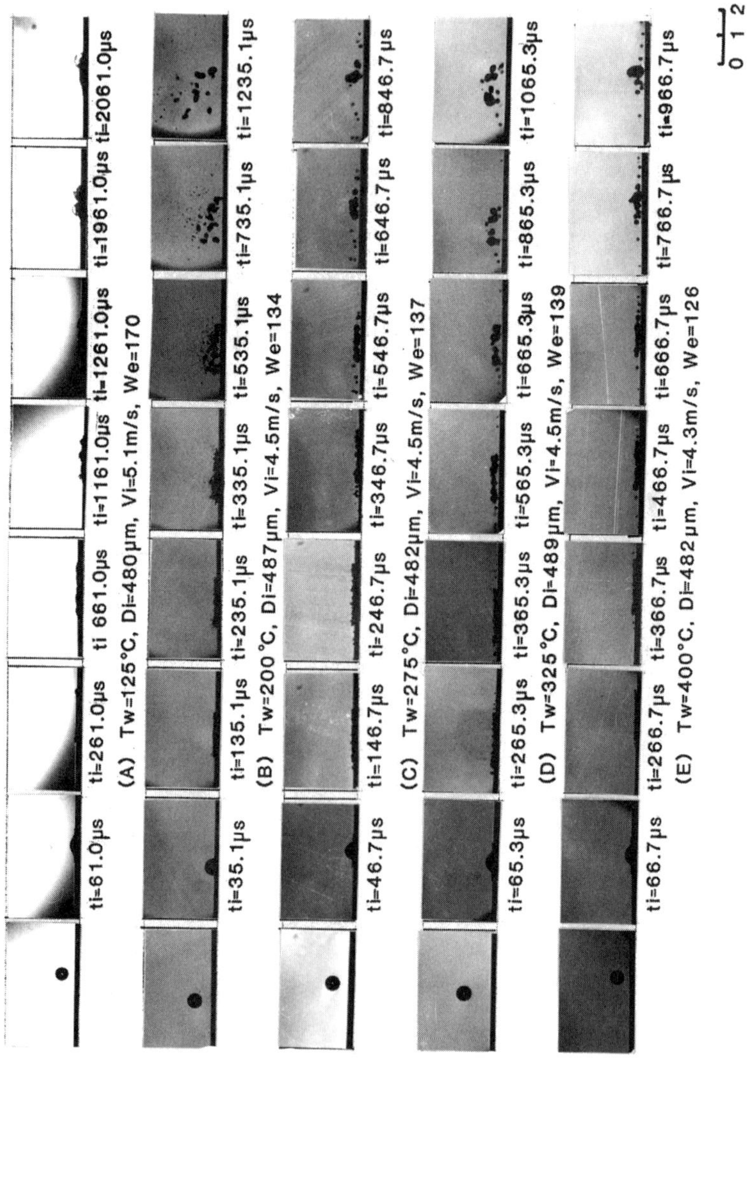

Figure 3.27. Spreading, breakup and evaporation of normal liquid droplets impinging on a hot surface. *(Photograph: Courtesy of Prof. Dr. Jiro Senda at Doshisha University, Japan.)*

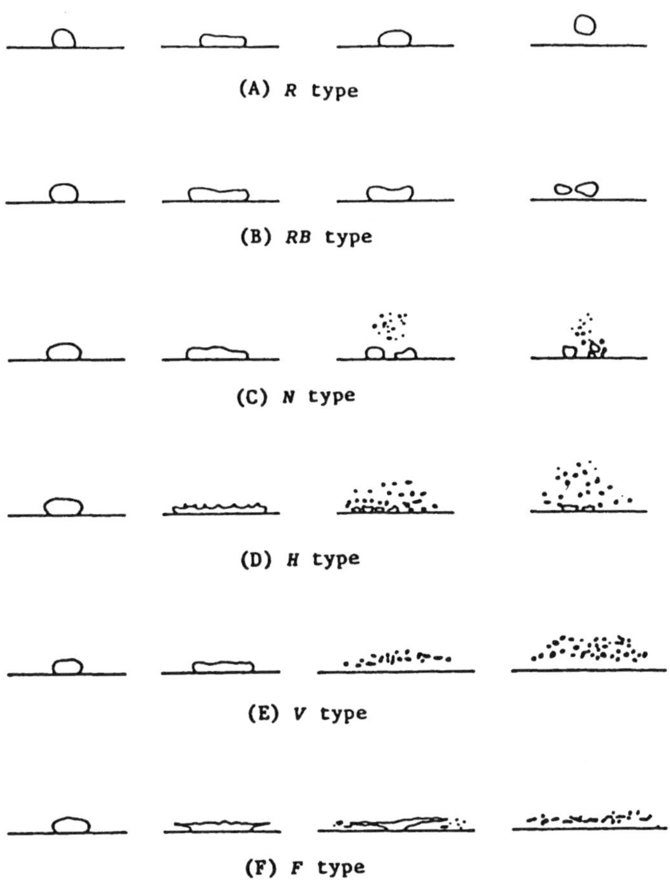

Figure 3.28. Schematic of spreading, breakup and evaporation modes of normal liquid droplets impinging on a hot surface. *(Reprinted with permission from Ref. 415.)*

Type R. This mode corresponds to low impinging velocities at any surface temperatures considered (200–400 °C). A droplet spreads as a radial film after impinging on a hot surface. Then, it shrinks and *rebounds* from the surface without breaking up. Hence, the mode is called *R* type.

Type RB. With increasing droplet diameter and/or impinging velocity, the shrinking process of the radial film may become unstable, leading to disturbances on the film surface as the film leaves the substrate surface. As the disturbances become larger, the radial film may *break* up during shrinking or during *rebounding*. The mode is therefore referred to as *RB* type.

Type N. With further increasing droplet diameter and/or impinging velocity, the breakup mode may change from the *RB* type to a completely different type depending on substrate surface temperature. At a surface temperature of 200 °C, the radial film breaks up due to the formation of vapor that blows through the center of the film before any shrinking takes place. Small droplets are blown upwards by the vapor, while the large pieces of the broken radial film remain on the substrate surface.

Type H. Further increasing droplet diameter and/or impinging velocity from the level in type *N*, the number of spots where the vapor blows through the radial film increases as compared to that in type *N*. Small droplets blown by the vapor are distributed even over the radial film, but the blowing-through of the vapor is weaker than in the type *N*. The pieces of the broken radial film that remain on the substrate surface are similar to those in type *N*.

Type V. At surface temperatures of 300 °C and 400 °C, the radial film does not shrink. The film breaks up and nearly all the small droplets resulted from the breakup are blown upwards by the vapor. The substrate surface becomes dry within a short time period.

Type F. As the droplet diameter and/or impinging velocity further increases from the level in type *V*, the small droplets resulted from the breakup disperse in the radial direction, leading to transition into type *F*. In this mode, the radial film leaves the substrate surface in whole because the vapor beneath the film blows through in the radial direction. Thereafter, as the diameter of the film further increases, it is torn and broken up by the vapor into small droplets while the vapor flows beneath it. The small droplets are dragged by the vapor and disperse in the radial direction at velocities faster than that of the radial film. It is interesting to note that droplet dispersion is vertically long-shaped in type *N*, while it tends to toward the radial direction in types *H* and *V*, and finally becomes completely in radial direction in type *F*.

Senda et al.[335][415] also derived equations describing the thickness and diameter of the radial film formed on a hot surface as a function of the Weber number, and correlated the mean diameter of droplets resulted from the breakup of the radial film with the thickness of the radial film and the Weber number.

Generally, the occurrence of a specific mode is determined by droplet impact properties (size, velocity, temperature), surface properties (temperature, roughness, wetting), and their thermophysical properties (thermal conductivity, thermal capacity, density, surface tension, droplet viscosity). It appeared that the surface temperature and the impact Weber number are the most critical factors governing both the droplet breakup behavior and ensuing heat transfer.[335][412][415]

The droplet impact regime maps developed by Bernardin et al.[412][413] revealed that the droplet lifetime increases systematically with decreasing surface temperature, and the impact Weber number has a strong influence on droplet spreading characteristics and integrity. In the film boiling regime, strong surface tension forces in a droplet of low Weber number (We=20) preserve the liquid continuum, causing rebound of the entire droplet following impact. In all boiling regimes, an increase in the impact Weber number decreases the droplet spreading time, and increases the instability responsible for liquid breakup, leading to a more violent breakup. Different Weber numbers correspond to distinct temperature dependent heat transfer regimes, possessing different droplet impact characteristics and heat transfer modes. However, temperatures corresponding to the critical heat flux and the Leidenfrost point seemed to be little sensitive to both the droplet impact Weber number, and impact frequency under their experimental conditions.[412]

Andreani and Yadigaraoglu[309] indicated that if the impact Weber number is smaller than 30, practically no breakup occurs;

If the impact Weber number is larger than a critical value of about 50–80, a droplet will form a thin liquid film on the impact surface, and soon disintegrate into smaller droplets. For intermediate Weber numbers, a certain fraction of droplets will break up after moving away from the hot surface, ejecting only a few droplets. The impact angle exhibits a strong influence on the critical Weber number. For compressible flow during droplet impact on a flat substrate, there is evidence that the liquid directed toward the surface may rebound from the substrate after shock wave detaches from the substrate, creating a mist.[390]

As reviewed by Bernardin et al.,[413] previous experiments of sessile droplet film boiling heat transfer revealed that surface features may promote liquid-solid contact in film boiling. A rough surface may require a thicker vapor layer between a droplet and the surface to sustain film boiling, and thus, possess a higher Leidenfrost point (LFP) temperature. The LFP temperature for water droplets on porous ceramic surfaces was found to increase with increasing porosity, possibly due to the increased vapor absorption in the surface pores, so that a larger surface superheat may be required to maintain a sufficiently thick vapor layer sustaining film boiling. A few investigations of the effects of surface roughness on impinging droplet heat transfer showed that surface roughness may promote droplet breakup,[395] and enhance liquid-solid contact in dispersed droplet flow, thus increasing film boiling heat transfer.

Bernardin et al.[413] further studied the effects of surface roughness on water droplet impact history, and heat transfer regimes over broad ranges of Weber number and surface temperature using still and high-speed photographic techniques, and a heat transfer measurement module. The droplet Weber number ranged from 20, 60, to 220, with substrate surface temperatures of 100–280 °C. Three different surface finishes, i.e., polished, particle blasted and rough sanded, were examined that correspond to arithmetic average surface roughness values of 97, 970, and 2960 nm. The experimental measurements showed that similarly to most boiling systems, the heat transfer during droplet impingement on rough surfaces may be categorized into the commonly known boiling curve regimes, i.e., nucleation, transition, and film boiling, in addition to thin film evaporation. However, the heat transfer processes on rough surfaces involve liquid-surface interactions far more complex than those on smooth surfaces, and the interactions are influenced to varying degrees by surface roughness. The LFP temperature is especially sensitive to surface finish, while the temperature corresponding to critical heat flux (CHF) is fairly independent of surface roughness. In addition, the influence of the surface roughness on the LFP temperature for an impinging droplet is quite different from that for a sessile droplet. For an impinging droplet, protruding features on rough surfaces tend to rupture the liquid film produced by the droplet impact, greatly reducing the pressure beneath the droplet and thus yielding a lower LFP temperature as compared to that on a smooth surface. For a sessile droplet, surface features penetrate the vapor layer beneath the droplet, yielding a higher LFP temperature on rougher surfaces. It was speculated that surface features influence the boiling regimes of droplets in two major ways. First, it leads to violent breakup of the spreading liquid film at high temperatures corresponding to the film boiling and the upper portion of the transition boiling regime. Second, it increases nucleation site density at lower temperatures corresponding to the nucleation boiling and the lower portion of the transition boiling regime. Enhanced nucleation at lower temperatures is largely responsible for the decrease of droplet lifetime on rougher surfaces. The droplet impact regime maps constructed by these researchers for each of the three surface roughness values encompass the parametric effects of the Weber number and surface temperature, and illustrate the complex liquid-solid interactions that occur during the lifetime of the impacting droplet within each regime (nucleation, transition and film boiling). These maps revealed some key trends in droplet impact processes and may serve as an effective reference for understanding droplet impact behavior and for analytical or numerical modeling of droplet heat transfer on surfaces.

Experiments of Chandra et al.[17] showed that ridges on a surface with a height of the same magnitude as the thickness of the vapor film under an impinging droplet cause vapor bubble nucleation in the droplet, and significantly reduce the droplet evaporation time. Addition of a surfactant (sodium dodecyl sulfate) to water droplets largely reduces the lifetime of the droplets in the state of evaporation or nucleation boiling, and enhances nucleation boiling heat flux by up to 300%. For surface temperatures below those required to initiate nucleation boiling, the primary effect of the surfactant is to reduce liquid-solid contact angle, increasing the surface area wetted by a droplet and decreasing the temperature at which a droplet starts to wet the surface. A decrease in the liquid-solid contact angle also reduces the droplet thickness on the surface, enhancing heat conduction through the droplet and thus increasing the droplet evaporation rate.[363] At higher surface temperatures, the surfactant promotes vapor bubble nucleation and foaming in the liquid, greatly enhancing heat transfer and lowering the vapor bubble nucleation temperature. Addition of the surfactant did not affect the droplet shape during the initial stage of impact, but did increase the maximum spread diameter and reduce the recoil height. An increase in the surfactant concentration reduces the Leidenfrost temperature, above which a droplet is levitated above the surface on a thin film of its own vapor. Changing the orientation of the surface with respect to gravity has no effect on heat transfer. In addition, nucleation boiling is not affected by a reduction in gravity. However, in low gravity, droplets can not remain in stable film boiling and the pressure of vapor under the droplets pushes them away from the surface. Vapor bubbles nucleating on the surface separate and rise into the liquid even in the absence of buoyancy forces.

3.2.5 Interaction, Spreading and Splashing of Multiple Droplets on a Surface

Interactions of *multiple droplets* on a surface during impact are of practical interest in many applications. Multiple droplets colliding with a non-flat surface (Fig. 3.29) may cause breakup and bounce of liquid, and formation of voids within liquid due to the entrainment of air/gas.[389] The ejection and bounce of liquid, as well as the void formation may be largely depressed by rapid solidification in surface impact of melt droplets. Experimental, theoretical, and analytical studies by Yarin and Weiss[357] showed that at low impact velocities, a fully liquid droplet in a droplet train striking onto a preceding, flattening droplet on a substrate surface results in spreading in the form of a thin liquid film and propagation of capillary waves (Fig. 3.30), with the impact energy totally absorbed by the waves; At high impact velocities, a cloud of small secondary droplets emerge and splashing of liquid sets in. The pattern of the capillary waves is predicted to be self-similar. The splashing threshold corresponds to the onset of a velocity discontinuity propagating over the liquid layer on the substrate surface. This discontinuity shows several aspects of a shock. In an incompressible liquid, such a discontinuity can only occur in the presence of a sink at its front. The sink results in a circular crown-like sheet virtually normal to the substrate surface, and propagating with the discontinuity radially from the center of impact. The crown is unstable owing to the formation of cusps at the free rims at its top edge under the action of surface tension while the free rims propagate with a velocity independent of their local curvature. At the cusp sites, thin liquid jets are formed, which are unstable, again owing to the action of surface tension, i.e., *capillary instability* of the Rayleigh type, leading to splashing. The thin liquid jets break up eventually and form very small secondary droplets. In the splashing phenomenon, the role of gravity is negligible owing to the small scale involved, and only inertia and surface tension are of importance.

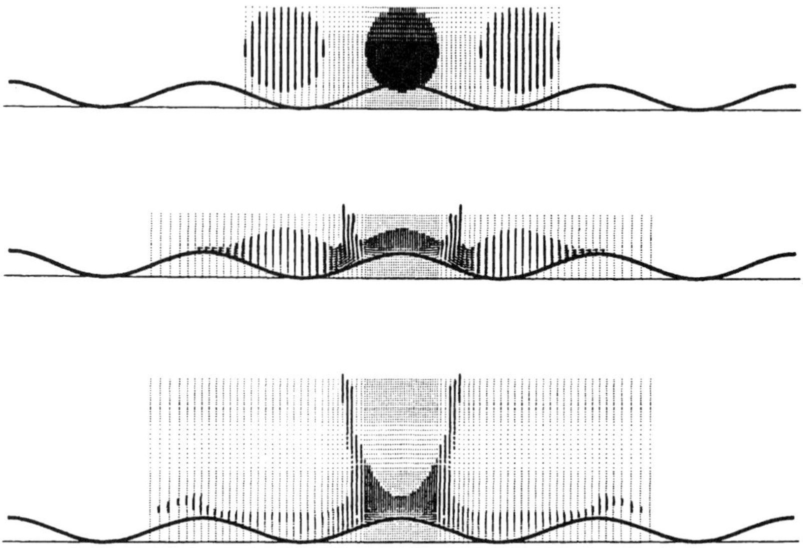

Figure 3.29. Interaction of multiple droplets impinging on a non-flat surface (Re = 76229, We = 33897, ε/D_0 = 0.33, λ/D_0 = 2.8). From top to bottom: $t/(D_0/u_0)$ = 0, 0.8, 1.6. *(Reprinted from Ref. 389, © 1995, with kind permission from Elsevier Science Ltd., The Boulevard, Langford Lane, Kidlington 0X5 1GB, UK.)*

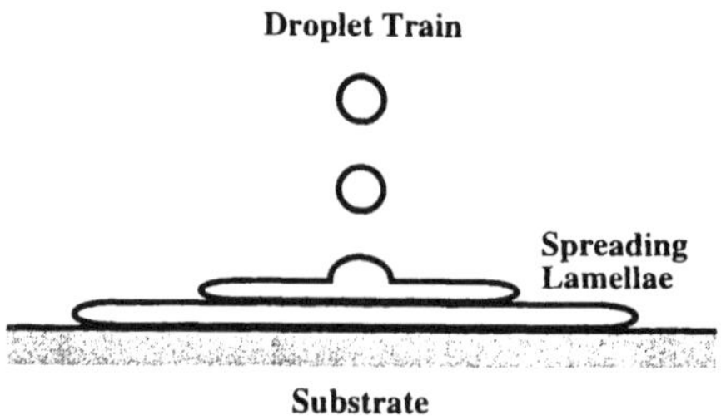

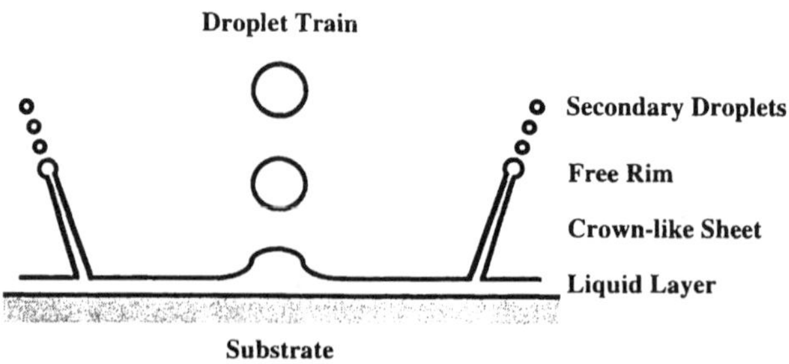

Figure 3.30. Schematic showing the impact phenomena of multiple droplets on a substrate surface: spreading pattern at low impact velocities (*top*) and splashing mechanism at high impact velocities (*bottom*).

Yarin and Weiss[357] discussed the underlying mechanism governing the spreading and splashing during impingement of a droplet train on a substrate surface, and indicated that the phenomena may be represented by a relatively simple quasi one-dimensional (1-D) model. They derived the governing equations for the surface tension dominated flow in the thin liquid film on the substrate surface, discussed the propagation of capillary waves over such a film, and found the self-similar regimes. They also derived the splashing conditions, explained the physical nature of splashing in terms of a kinematic discontinuity formation in an incompressible liquid, and addressed the structure of this discontinuity. The splashing threshold was described by a correlation between the *Capillary number*, Ca, and a non-dimensional viscous length λ_v such that:

Eq. (48) $$\mathrm{Ca} = c\lambda_v^{-3/4}$$

where:

$$\mathrm{Ca} = \mathrm{We}/\mathrm{Re} = u_0\mu_L/\sigma$$

$$\lambda_v = (v_L/f)^{1/2}\sigma/(\rho_L v_L^2)$$

$$c = u_0(\rho_L/\sigma)^{1/4} v_L^{-1/8} f^{-3/8} \approx 17\text{–}18$$

f = frequency of droplet train

The coefficient c, termed dimensionless impact velocity, was found to depend only slightly on the surface roughness as long as the roughness is much smaller than droplet diameter. Plotting Ca versus λ_v in logarithmic scales generates a straight line (Fig. 3.31). In the regime above the line, splashing occurs, whereas in the regime below the line, only spreading takes place during surface impact of a droplet train. The splashing threshold at $c \approx 17\text{–}18$ corresponds to a developed crown instability, strong enough to produce a cloud of secondary droplets, while crown formation begins slightly below the splashing threshold. Compressibility effect is of importance at the very early stage of splashing, leading to droplet jetting over the substrate surface. However, it is impossible to correlate the onset of splashing with the Mach number of the liquid based on its sonic speed, showing that the early events related to compressibility occur almost instantaneously in the time scale of the whole splashing process.

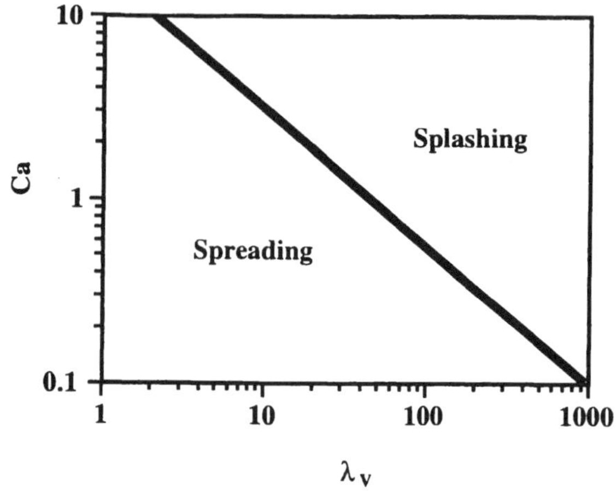

Figure 3.31. Splashing threshold for surface impact of a droplet train.

Yarin and Weiss[357] also determined the number and size of secondary droplets, as well as the total ejected mass during splashing. Their experimental observations by means of a computer-aided charge-coupled-device camera and video printer showed that the dependence of the critical impact velocity, at which splashing initiates, on the physical properties (density, viscosity, and surface tension) and the frequency of the droplet train is universal, and the threshold velocity may be estimated by:

Eq. (49) $$u_{0s} = 18(\sigma / \rho_L)^{1/4} \nu_L^{1/8} f^{3/8}$$

or

Eq. (50) $$u_{0s}^8 = \text{const} \times (\sigma / \rho_L)^2 \nu_L f^3$$

Comparing this splashing criterion to that derived by Stow and Hadfield[396] for single droplet impact (Eq. 44b), the scaling behavior may be reproduced if substituting f by u_0/D_0 in the above equation.

3.2.6 Sessile Droplet Deformation on a Surface

In addition to the phenomena discussed above, many other droplet deformation processes on a surface are driven by surface tension, gravity force, and/or droplet-surface chemical interaction, characterized by very low or zero impact velocities. The isothermal spreading of a sessile droplet on a reactive wetting substrate, for example, is driven by the liquid-substrate chemical reaction and limited by solute diffusion to the triple line.[416] Under such conditions, the spreading rate is independent of time and droplet volume, but proportional to the contact angle and linearly dependent on droplet solute concentration. Very slow droplet spreading processes can be dramatically affected by impurities, especially if they are volatile.[417] Such processes are also sensitive to any external perturbation like roughness or contamination. Inhomogeneities of concentration can give rise to surface tension gradients that may induce liquid flows or non-wetting fluid spreading, a phenomenon termed *Marangoni effects.*[418] Marangoni effects may be encountered in droplet deformation processes coupling with temperature/composition changes or chemical reactions. A typical example is solder droplet spreading on a metallic surface, a reactive form of the wetting problem. A metallic component may diffuse in the liquid toward the surface, where it is consumed by a reaction that forms a solid intermetallic phase. The resultant spatial variation in the composition of the droplet may cause composition gradients along the free surface of the droplet. Along with any thermal gradients, Marangoni effects may modify the transport characteristics of the spreading droplet. To account for the Marangoni effects, Braun et al.[419] extended the lubrication theory for the spreading of thin droplets in the presence of gravity and thermocapillarity to include mass transport and solutocapillarity. They used an approximate solute profile in the droplet to derive coupled evolution equations for the free surface shape and concentration field. The study showed that the reactive effects have relatively important impact on the flow patterns and spreading rates at the early stages of droplet deformation, and will phase out at the end of the spreading.

Carles et al.[417] examined the influence of surface and atmospheric contamination on the spreading dynamics of silicone oil droplets on glass surfaces. The results showed that an acceleration surface tension gradient develops under both contamination conditions. However, the spreading dynamics and profiles are different in the two cases. The atmospheric contamination gives rise to smooth surface tension gradients distributed over the whole droplet, and the profile is of a fairly general type, resembling those obtained in many cases where a flat film is made from a droplet spreading over a smooth surface. In this case, a droplet spreads and thins while losing its spherical cap shape. During the early stages, fringes are not visible due probably to too large thickness and too high slopes. Then, a bump develops near the contact line while the central part of the droplet becomes flat. This profile remains analogous to each other while it spreads and thins during the late stages of deformation. The radius increases almost linearly with time. In contrast, the surface contamination produces localized surface tension gradients near the contact line, and the profile is more comparable to those obtained when a liquid droplet spreads over a liquid layer. In this case, the profile does not retain the shape of a spherical cap, but instead shows an inflexion point near the edge, while the radius increases roughly as the square root of time.

Surface tension and gravity driven droplet spreading on a surface, has been studied with nonvolatile droplets containing surface-active agents at relatively low velocities.[354] At the velocities on the order of a few centimeters per second, a droplet moves spontaneously on a surface. For small droplets, this self-supported motion may be interpreted in terms of capillary theory, which provides insightful information about the deformation mechanism on the surface. For large droplets, gravity effect intervenes in. A remarkable change in the droplet profile has been observed and different spreading regimes have been identified.[354]

3.2.7 Spreading and Splashing of Droplets into Shallow and Deep Pools

In addition to the phenomena discussed above, other issues of practical importance include the phenomena of droplet impingement into shallow and deep pools. Splash and breakup of liquid almost always occur when a droplet at high velocity impinges onto *a pool of the same fluid*, regardless of pool depth if surface tension and/or viscosity are negligibly small.[397] However, the splashing behavior for finite ratios of droplet diameter to pool depth is distinctly different from that for very large, or infinite ratios of droplet diameter to pool depth. Even a shallow pool with a very thin film of liquid is sufficient to interact appreciably with the lateral jet of a flattening droplet to produce an upward motion, i.e., splash. For a shallow pool, the region of impact tends to empty and a liquid sheet is ejected from the periphery of the region of impact. In addition, secondary droplets tend to form at the edge of the splash sheet. For a deep pool, a crater forms in the region of impact, followed by the ejection of liquid from the center of the crater if impact kinetic energy is high enough. A splashing droplet may also rupture into two parts during penetration into a deep pool at high impact kinetic energy. The droplet material is deposited onto the bottom and the sides of a crater, the bulk of the material lying near the bottom. As the cavity sides collapse, the bottom simultaneously tends to rise back towards the original surface level. The two processes compete. For a low droplet impact velocity, the crater is shallow and the droplet rebounds to the surface before the sides collapse. For a high droplet impact velocity, the crater is much deeper and the sides collapse earlier, trapping much of the original droplet material well below the surface. The material deposited on the sides of the crater is carried up in the rebound splash, isolating two different parts of the original droplet fluid from each other. If the pool fluid is less dense and immiscible with droplet fluid, then each part of the droplet may coalesce and drift slowly downwards. As the droplet impact velocity approaches the order of 5000 m/s, the effects of compressibility become progressively more important. Upon impact, there is almost no splash. Shocks are generated in both the droplet and the pool, but their lateral motion in the pool is relatively slow so that very little expansion occurs. Once the shock reaches the top of the droplet, there is a strong reexpansion that violently expulses the material back out in the direction from which it enters. The outward moving velocity of the droplet top is very close to the initial impact velocity. The droplet material vaporizes and subsequently behaves much like a gas.

For a liquid droplet falling at low velocities onto *a liquid pool of a different fluid*, Marangoni flows induced by the surface tension difference between the droplet and the pool may dominate the interaction, and the Marangoni effects may control both the direction and intensity of the pool circulation.[420] The surface tension force may generate the strongest flow as compared to buoyancy force and stirring force induced by the falling droplet. The relative difference in the surface tension between the droplet and the pool determines the flow pattern and direction. If the droplet has a larger surface tension than the pool, the surface flow is directed inward, i.e., the Marangoni forces cause the pool surface to be drawn toward the droplet, and a deep flow loop may be created while the droplet is injected deep into the pool. If the droplet has a smaller surface tension, the Marangoni forces cause the pool surface to be drawn away from the droplet while the droplet spreads out onto the pool surface. If the droplet has the same surface tension as the pool, the droplet may penetrate into the pool.

Al-Roub et al.[421] identified three basic modes of liquid breakup during droplet impingement onto a liquid film: *(1)* rim breakup, *(2)* cluster breakup, and *(3)* column breakup. The rim breakup mode involves the breakup and ejection of one or a few small droplets at the outer edge of the film, while the cluster breakup mode involves the breakup of liquid into clusters of many small droplets at the outer edge of the film. In the column breakup mode, liquid breaks up into one or a few droplets from a column of liquid at the center of the spreading droplet as a result of the surface waves reflecting back to their source. The diameter and number of the secondary droplets have been formulated as a function of impact conditions (such as the Ohnesorge number and Reynolds number) and surface conditions.[356][422]

To determine if a droplet experiences spreading or splashing when it impinges onto a liquid film on a solid surface, the correlation between the Weber number and Ohnesorge number derived by Walzel[398] may be used:

Eq. (51) $$\mathrm{We} = 2.5 \times 10^3 \mathrm{Oh}^{0.4}$$

A We-Oh map can also be plotted on the basis of this correlation, similarly to that for the droplet spreading and splashing on a dry, solid surface (Fig. 3.21), but with smaller slope. The We-Oh relation delimits the regimes of droplet spreading and splashing onto a liquid film on a solid surface in the We-Oh map. If the fluid dynamic conditions of a droplet are within the spreading regime, i.e., below the threshold curve in the We-Oh map, liquid splashing and ejection from the impact surface can be avoided. The corresponding threshold velocity can be formulated as:

Eq. (52) $$u_{0s} = 50 D_0^{-0.6} \rho_L^{-0.6} \sigma^{0.4} \mu_L^{0.2}$$

Comparing Eq. (51) to Eq. (43) or Eq. (52) to Eq. (44a), it is clear that for the same liquid properties and droplet diameter at impact, splashing takes place at lower impact velocities on a liquid film than on a dry surface.

4

Empirical and Analytical Correlations of Droplet Properties

Atomization, or generally speaking droplet generation, is an extremely complex process that cannot yet be precisely predicted theoretically. The lack of general theoretical treatment of droplet processes has led to the development of numerous empirical correlations for droplet properties as a function of process parameters and material properties. In this chapter, empirical and analytical correlations for the prediction of droplet properties, such as droplet size distribution and droplet deformation characteristics will be summarized from experimental observations and theoretical analyses in available literature.

4.1.0 CONCEPT AND DEFINITIONS OF DROPLET SIZE DISTRIBUTION

Droplet generation is typically a random process. Droplet sizes are usually characterized by a wide spectrum. In many applications, mean droplet size and size distribution are the important factors to be controlled closely to achieve the desired rates of heat and mass transfer, or to meet specific application requirements because various dynamics and transport phenomena in droplet processes strongly depend on droplet size.

In most spray combustion systems, for example, droplet size frequency and spatial distribution govern fundamental flame characteristics such as heat transfer, flame length and stability, and emissions. A decrease in mean droplet size results in a higher heat release rate, easier lightup, a wider burning range, and a lower exhaust concentration of pollutant emissions. A uniform distribution of droplet sizes in spray may not always be desirable, since very small droplets are needed to provide high initial fuel evaporation rates for rapid ignition.

In spray deposition processes for near-net shape manufacturing, a small mean droplet size and narrow size distribution may lead to a refined microstructure and improved mechanical properties of as-sprayed preforms. However, in some other applications, such as agricultural aviation for crop spray,[88] small droplets of chemicals may not be desired under certain meteorological conditions because they can drift too far downwind due to their low settling velocities which might damage nearby crops, or present hazards to susceptible animals and human beings. On the other hand, excessively large droplets settle directly on the ground, resulting in nonuniform deposit of chemicals. Hence, both small and large droplets may cause inaccurate application of chemicals to the target area. Therefore, in such applications, the droplet size distribution needs to be controlled properly, and an atomization technique that can produce monodisperse sprays may be one of the solutions to eliminate the problems.

However, a monodisperse spray seldom exists in reality and is difficult to produce, although it can be defined mathematically as a spray consisting of droplets of the same size. Hence, a monodisperse spray usually refers to a spray in which droplets are very narrowly distributed. However, it is merely a relative term since a droplet size distribution that is sufficiently narrow in one application may be considered disperse in another. Thus, the choice of the exact criterion is sometimes arbitrary. In certain applications, a monodisperse spray is defined as a spray in which 95% of droplets are smaller than 1.2 times the average droplet size, and 5% of droplets are smaller than 0.8 times the average size.[88]

Droplet size distribution can be represented both graphically and mathematically. As a histogram of droplet size, both droplet number increment $\Delta N_i/\Delta D_i$, and volume increment $\Delta V_i/\Delta D_i$ can be plotted *vs.* droplet size D (Fig. 4.1), where i is the size range considered, N_i is the number of droplets in size range i, D_i is the middle diameter of size range i, and ΔN_i and ΔV_i are the number and volume increment within ΔD_i, respectively. The histogram based on volume is skewed to the right due to the weighting effect of larger droplets:

Eq. (1) $$\Delta V_i = \Delta N_i \frac{\pi}{6} D_i^3$$

As ΔD is made smaller, a histogram becomes a frequency distribution curve (Fig. 4.1) that may be used to characterize droplet size distribution if samples are sufficiently large. In addition to the frequency plot, a cumulative distribution plot has also been used to represent droplet size distribution. In this graphical representation (Fig. 4.2), a percentage of the total number, total surface area, total volume, or total mass of droplets below a given size is plotted *vs.* droplet size. Therefore, it is essentially a plot of the integral of the frequency curve.

Mathematical representation of droplet size distribution has been developed to describe entire droplet size distribution based on limited samples of droplet size measurements. This can overcome some drawbacks associated with the graphical representation and make the comparison and correlation of experimental results easier. A number of mathematical functions and empirical equations[423]-[427] for droplet size distributions have been proposed on the basis of

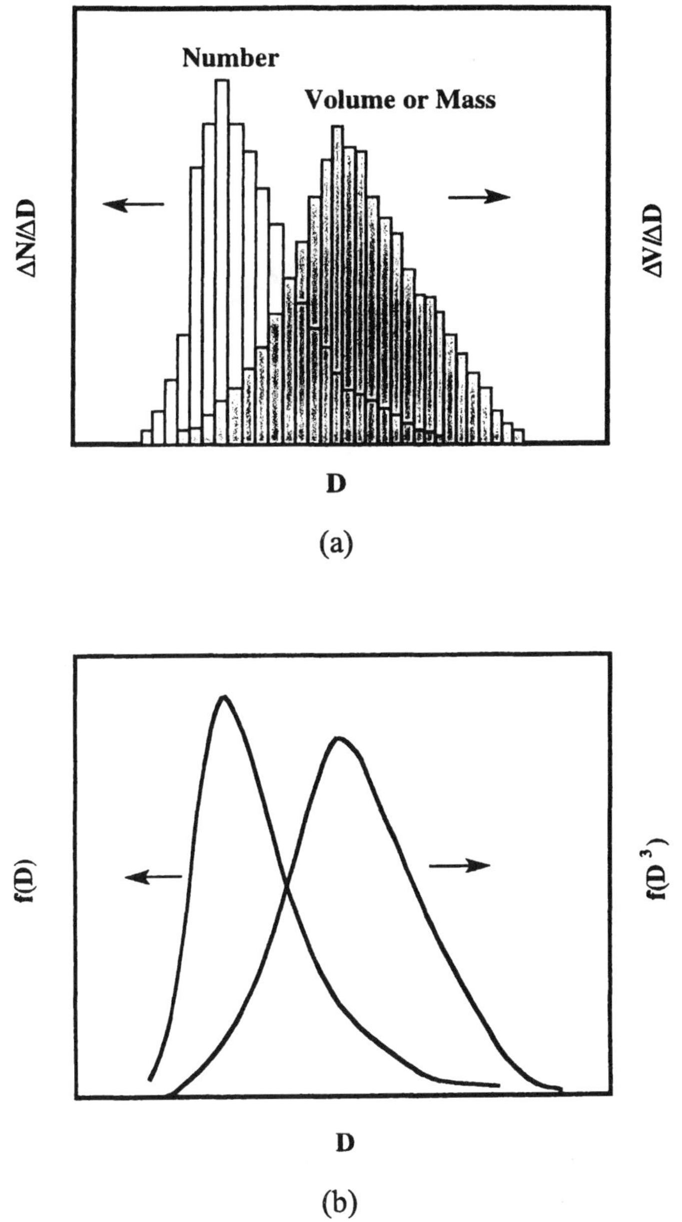

Figure 4.1. *(a)* Histograms and *(b)* frequency distribution curves of droplet sizes based on number and volume.

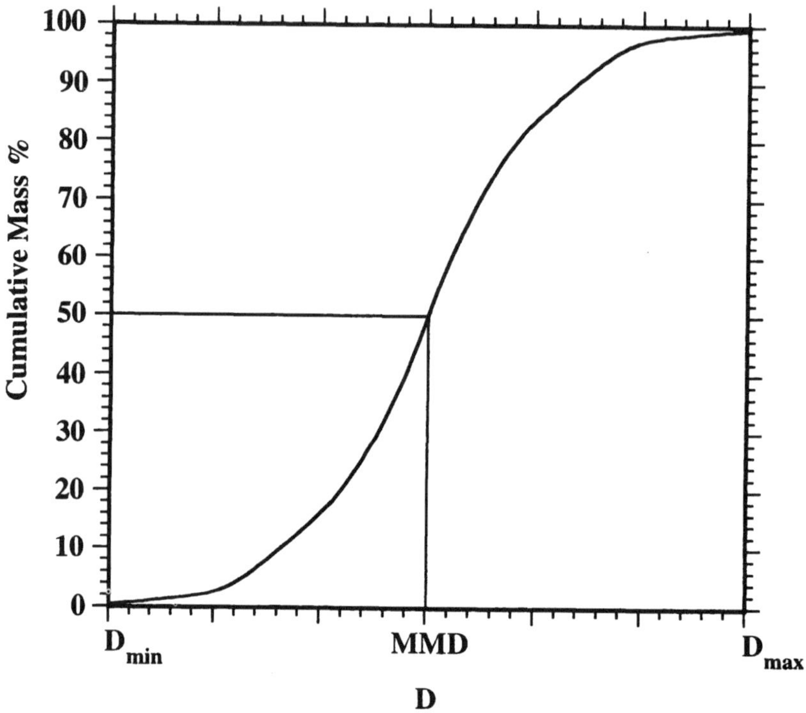

Figure 4.2. Cumulative droplet size distribution curve based on mass.

probability analyses, or empirical considerations due to the lack of fundamental theory describing droplet size distributions. Mugele and Evans[423] summarized various statistical formulas for mean droplet diameters and variances. The functions that are most frequently used in analyses and correlations of measured droplet size data include normal, log-normal, root-normal, Nukiyama-Tanasawa, Rosin-Rammler, and upper-limit distributions. However, it should be indicated that no single distribution function can characterize all experimental measurement data of droplet sizes, and none of these functions is universally superior to any other for representing droplet size distributions. Therefore, it is common practice to test all these distribution functions in order to find the best fit to a set of experimental data unless the precise mechanism of atomization in the droplet generation device has been understood and related to one or more distribution functions. For the selection of a suitable distribution function, the consistency with the physical phenomena involved and mathematical/computational simplicity/ease are some factors to be considered.

Many droplet size distributions in random droplet generation processes follow *Gaussian*, or *normal distribution* pattern. In the normal distribution, a number distribution function $f(D)$ may be used to determine the number of droplets of diameter D:

Eq. (2) $$f(D)=\frac{dN}{dD}=\frac{1}{\sqrt{2\pi}s_n}\exp\left[-\frac{1}{2s_n^2}\left(D-\bar{D}\right)^2\right]$$

where s_n is the *standard deviation*, a measure of the deviation of values of D from the mean value $\bar{D}$, and s_n^2 is the variance. A plot of the distribution function is the so-called *standard normal curve*, the area under which from $-\infty$ to $+\infty$ equals 1. The integral of the standard normal distribution function is the cumulative standard number distribution function $F(D)$:

Eq. (3) $$F(D)=\frac{1}{\sqrt{2\pi}}\int_{-\infty}^{D}\exp\left[-\frac{1}{2}\left(\frac{D-\bar{D}}{s_n}\right)^2\right]d\left(\frac{D-\bar{D}}{s_n}\right)$$

Plotting droplet size data on an arithmetic-probability graph paper will generate a straight line if the data follow normal distribution. Thus, the mean droplet diameter and standard deviation can be determined from such a plot.

Many droplet size distributions in natural droplet formation and liquid metal atomization processes conform to *log-normal distribution*:

Eq. (4) $$f(D) = \frac{dN}{dD} = \frac{1}{\sqrt{2\pi}Ds_g} \exp\left[-\frac{1}{2s_g^2} \left(\ln D - \ln \overline{D}_{ng} \right)^2 \right]$$

where $\overline{D}_{ng}$ is the number geometric mean droplet diameter and s_g is the geometric standard deviation. Plotting droplet size data on a log-probability graph paper will generate a straight line if the data follow log-normal distribution.

Log-normal distribution functions based on surface and volume, respectively, are:

Eq. (5) $$f(D^2) = \frac{1}{\sqrt{2\pi}Ds_g} \exp\left[-\frac{1}{2s_g^2} \left(\ln D - \ln \overline{D}_{sg} \right)^2 \right]$$

and

Eq. (6) $$f(D^3) = \frac{1}{\sqrt{2\pi}Ds_g} \exp\left[-\frac{1}{2s_g^2} \left(\ln D - \ln \overline{D}_{vg} \right)^2 \right]$$

where $\overline{D}_{sg}$ and $\overline{D}_{vg}$ are the geometric surface and volume mean droplet diameters, respectively. These diameters can be determined once the number geometric mean droplet diameter and the geometric standard deviation are known:

Eq. (7) $$\ln \overline{D}_{sg} = \ln \overline{D}_{ng} + 2s_g^2$$

Eq. (8) $$\ln \overline{D}_{vg} = \ln \overline{D}_{ng} + 3s_g^2$$

Eq. (9) $$\ln \text{SMD} = \ln \overline{D}_{ng} + 2.5s_g^2$$

In many atomization processes of normal liquids, droplet size distributions fairly follow *root-normal distribution* pattern:[264]

Eq. (10) $$f(D)=\frac{dN}{dD}=\frac{1}{\sqrt{2\pi}s_s}\exp\left[-\frac{1}{2s_s^2}\left((D/\mathrm{MMD})^{1/2}-1\right)^2\right]$$

where s_s is the standard deviation. This distribution function generates satisfactory fitting to the data from atomization experiments with a large number of gas-turbine fuel nozzles of different types, over a range of fuel viscosities, and at a variety of operation conditions.[264] In this distribution, a straight line can be generated by plotting $(D/\mathrm{MMD})^{0.5}$ *vs.* cumulative volume of droplets on a normal-probability scale. This distribution is simpler than the log-normal and Rosin-Rammler distributions.

Nukiyama, and Tanasawa[79] proposed a relatively simple function for adequate description of some actual droplet size distributions:

Eq. (11) $$f(D)=\frac{dN}{dD}=aD^p\exp\left[-bD^q\right]$$

where a, b and p are constants and q is the dispersion coefficient which is a constant for a given nozzle design. The value of q usually varies from 1/6 to 2, and is determined by trial-and-error. A graphical method to determine a and b has been proposed by Mugele and Evans.[423] In the *Nukiyama-Tanasawa distribution* equation, $p = 2$, such that:

Eq. (12) $$f(D)=\frac{dN}{dD}=aD^2\exp\left[-bD^q\right]$$

Rearranging this equation, it reduces to:

Eq. (13) $$\ln\left(\frac{1}{D^2}\frac{dN}{dD}\right) = \ln a - bD^q$$

Plotting $\ln[1/D^2(dN/dD)]$ *vs.* D^q based on experimental data and an assumed value of q, a straight line may be generated if the assumed q value is correct. Thus, the values of a and b can be determined from the plot so that the entire droplet size distribution can be characterized with the function.

Rosin-Rammler distribution function[428] is perhaps the most widely used one at present:

Eq. (14) $$V = 1 - \exp[-D/X]^q$$

where V is the fraction of total volume of droplets smaller than D, and X and q are constants. The exponent q is a measure of the spread of droplet sizes. A larger value of q corresponds to a more uniform droplet size. For many droplet generation processes, q ranges from 1.5 to 4, and for rotary atomization processes, q may be as large as 7. For those atomization processes that generate uniform droplet sizes, q is infinite. A straight line can be generated by plotting $\ln(1 - V)^{-1}$ *vs.* droplet diameter on a normal-probability scale. The value of q can be then obtained as the slope of the line, and the value of X equals the value of D corresponding to $1 - V = \exp(-1)$, i.e., $V = 0.632$. Rosin-Rammler distribution function assumes an infinite range of droplet sizes and therefore allows data to be extrapolated down to the range of very small droplet sizes where measurements are most difficult and least accurate.

Mugele and Evans[423] proposed the *upper-limit distribution* function based on their analyses of various distribution functions and comparisons with experimental data. This distribution function is a modified form of the log-normal distribution function, and for droplet volume distribution it is expressed as:

Eq. (15) $$\frac{dV}{dy} = \delta \exp\left[-\frac{1}{\sqrt{\pi}}\delta^2 y^2\right]$$

where $y = \ln\,[aD/(D_{max} - D)]$, δ is a factor related to the standard deviation of droplet size, a is a dimensionless constant, and D_{max} is the maximum droplet diameter. The Sauter mean diameter is formulated as:

Eq. (16) $$\text{SMD} = \frac{D_{max}}{1 + a\exp(1/4\delta^2)}$$

The upper-limit distribution function assumes a finite minimum and maximum droplet size, corresponding to a y value of $-\infty$ and $+\infty$, respectively. The function is therefore more realistic. However, similarly to other distribution functions, it is difficult to integrate and requires the use of log-probability paper. In addition, it usually requires many trials to determine a most suitable value for a maximum droplet size.

Some other distribution functions have also been derived from analyses of experimental data,[429][430] or on the basis of probability theory.[431] Hiroyasu and Kadota[317] reported a more generalized form of droplet size distribution, i.e., *χ-square distribution*. It was shown that the χ-square distribution fits the available spray data very well. Moreover, the χ-square distribution has many advantages for the representation of droplet size distribution due to the fact that it is commonly used in statistical evaluations.

It has been indicated[323] that for some distributions it is possible to find, at least, an empirical correlation between the mean droplet size and the standard deviation. Gretzinger and Marshall[102] have proposed such empirical equations relating the mean droplet size and the standard deviation for water-air system. Thus, once the mean droplet size is determined from a mathematical model, an empirical correlation, and/or experimental data, the entire droplet size distribution can be then predicted quantitatively.

In many applications, a mean droplet size is a factor of foremost concern. Mean droplet size can be taken as a measure of the quality of an atomization process. It is also convenient to use only mean droplet size in calculations involving discrete droplets such as multiphase flow and mass transfer processes. Various definitions of mean droplet size have been employed in different applications, as summarized in Table 4.1. The concept and notation of mean droplet diameter have been generalized and standardized by Mugele and Evans.[423] The arithmetic, surface, and volume mean droplet diameter (D_{10}, D_{20}, and D_{30}) are some most common mean droplet diameters:

Eq. (17)
$$D_{10} = \frac{\int_{D_{min}}^{D_{max}} D(dN/dD)dD}{\int_{D_{min}}^{D_{max}} (dN/dD)dD}$$

Eq. (18)
$$D_{20} = \left[\frac{\int_{D_{min}}^{D_{max}} D^2(dN/dD)dD}{\int_{D_{min}}^{D_{max}} (dN/dD)dD}\right]^{1/2}$$

Eq. (19)
$$D_{30} = \left[\frac{\int_{D_{min}}^{D_{max}} D^3(dN/dD)dD}{\int_{D_{min}}^{D_{max}} (dN/dD)dD}\right]^{1/3}$$

where D_{min} is the minimum droplet diameter. The arithmetic mean droplet diameter is the linear average diameter of all droplets in a sample. The volume mean droplet diameter represents the diameter of a droplet whose volume, multiplied by the total number of droplets, equals the total volume of the sample. A general form of mean droplet diameter can be written as:[423]

Eq. (20)
$$D_{ab} = \left[\frac{\int_{D_{min}}^{D_{max}} D^a (dN/dD) dD}{\int_{D_{min}}^{D_{max}} D^b (dN/dD) dD} \right]^{1/(a-b)}$$

or

Eq. (21)
$$D_{ab} = \left[\frac{\sum N_i D_i^a}{\sum N_i D_i^b} \right]^{1/(a-b)}$$

where a and b may take any values according to the effect considered, as listed in Table 4.1. Among these, the Sauter mean diameter SMD is perhaps the most widely used one. It represents the diameter whose ratio of volume to surface area is the same as that of the entire droplet sample. Analyses[1] showed that for combustion applications only SMD can properly indicate the fineness of a spray, thus SMD is to be used to describe atomization quality.

To characterize a droplet size distribution, at least two parameters are typically necessary, i.e., a representative droplet diameter, (for example, mean droplet size) and a measure of droplet size range (for example, standard deviation or q). Many representative droplet diameters have been used in specifying distribution functions. The definitions of these diameters and the relevant relationships are summarized in Table 4.2. These relationships are derived on the basis of the Rosin-Rammler distribution function (Eq. 14), and the diameters are uniquely related to each other *via* the distribution parameter q in the Rosin-Rammler distribution function. Lefebvre[1] calculated the values of these diameters for q ranging from 1.2 to 4.0. The calculated results showed that D_{peak} is always larger than SMD, and SMD is between 80% and 84% of D_{peak} for many droplet generation processes for which $2 \leq q \leq 2.8$. Thus, SMD always lies on the left-hand side of D_{peak}. The ratio MMD/SMD is always greater than one, and it changes only very little for $q \geq 3$; $D_{0.9}$ is more than double MMD for $q \leq 1.7$, but it is only 50% higher than MMD for $q = 3$.

Table 4.1. Definitions of Mean Droplet Diameters and Their Applications[423]

Quantity	Common Name	a	b	Definition	Application
D_{10}	Arithmetic Mean (Length)	1	0	$\frac{\sum N_i D_i}{\sum N_i}$	Comparison
D_{20}	Surface Mean (Surface Area)	2	0	$\left(\frac{\sum N_i D_i^2}{\sum N_i}\right)^{1/2}$	Surface Area Controlling
D_{30}	Volume Mean (Volume)	3	0	$\left(\frac{\sum N_i D_i^3}{\sum N_i}\right)^{1/3}$	Volume Controlling (Hydrology)
D_{21}	Length Mean (Surface Area-Length)	2	1	$\frac{\sum N_i D_i^2}{\sum N_i D_i}$	Absorption
D_{31}	Length Mean (Volume-Length)	3	1	$\left(\frac{\sum N_i D_i^3}{\sum N_i D_i}\right)^{1/2}$	Evaporation, Molecular Diffusion
D_{32}	Sauter Mean (SMD) (Volume-Surface)	3	2	$\frac{\sum N_i D_i^3}{\sum N_i D_i^2}$	Mass Transfer, Reaction
D_{43}	Herdan Mean (De Brouckere or Herdan) (Weight)	4	3	$\frac{\sum N_i D_i^4}{\sum N_i D_i^3}$	Combustion, Equilibrium

Table 4.2. Definitions of Representative Droplet Diameters

Symbol	Definition	Position in *V*-*D* plot	Relationship
$D_{0.1}$	10% of total volume of droplets are of smaller diameters than this value	V=10%	$\frac{D_{0.1}}{X} = (0.1054)^{1/q}$ $\frac{D_{0.1}}{\text{MMD}} = (0.152)^{1/q}$
$D_{0.5}$ (MMD)	*Mass median diameter* 50% of total volume of droplets are of smaller diameters than this value	V=50% Left-hand or right-hand side of D_{peak} for $q >$ or $<$ 3.2584	$\frac{\text{MMD}}{X} = (0.693)^{1/q}$ $\frac{\text{MMD}}{\text{SMD}} = (0.693)^{1/q}\Gamma\left(1-\frac{1}{q}\right)$
$D_{0.632}$	Characteristic diameter 63.2% of total volume of droplets are of smaller diameters than this value	V=63.2%	X (X in Rosin-Rammler distribution function)
$D_{0.9}$	90% of total volume of droplets are of smaller diameters than this value	V=90%	$\frac{D_{0.9}}{X} = (2.3025)^{1/q}$ $\frac{D_{0.9}}{\text{MMD}} = (3.32)^{1/q}$
$D_{0.999}$	*Maximum diameter* 99.9% of total volume of droplets are of smaller diameters than this value	V=99.9%	$\frac{D_{0.999}}{\text{MMD}} = (9.968)^{1/q}$
D_{peak}	*Peak diameter* Value of D corresponding to peak of droplet size frequency distribution curve	Peak point corresponding to $d^2v/dD^2=0$	$\frac{D_{peak}}{X} = \left(1-\frac{1}{q}\right)^{1/q}$ $\frac{D_{peak}}{\text{MMD}} = \left(1.4428-\frac{1.4428}{q}\right)^{1/q}$ $\frac{D_{peak}}{\text{SMD}} = \left(1-\frac{1}{q}\right)^{1/q}\Gamma\left(1-\frac{1}{q}\right)$

Since the ratio of any two representative diameters is a unique function of q, Rosin-Rammler distribution function can be rewritten as:

Eq. (22) $$V = 1 - \exp\left[-0.693\left(\frac{D}{\mathrm{MMD}}\right)^q\right]$$

or

Eq. (23) $$V = 1 - \exp\left[-\Gamma\left(1 - \frac{1}{q}\right)^{-q}\left(\frac{D}{\mathrm{SMD}}\right)^q\right]$$

where Γ denotes the gamma function. Since these formulations clearly show both the fineness and the spread of droplet sizes in a spray, their applications are strongly recommended.[1] Based on these equations and those relationships in Table 4.2, representative diameters such as SMD, MMD, $D_{0.1}$ and $D_{0.9}$ can be determined once q is determined from experimental data.

It should be indicated that a probability density function has been derived on the basis of maximum entropy formalism for the prediction of droplet size distribution in a spray resulting from the breakup of a liquid sheet.[432] The physics of the breakup process is described by simple conservation constraints for mass, momentum, surface energy, and kinetic energy. The predicted, most probable distribution, i.e., *maximum entropy distribution*, agrees very well with corresponding empirical distributions, particularly the Rosin-Rammler distribution. Although the maximum entropy distribution is considered as an ideal case, the approach used to derive it provides a framework for studying more complex distributions.

The ratio MMD/SMD is generally recognized as a good measure of droplet size range. In addition, various indices and factors have been defined to describe the spread of droplet sizes in a spray, for example, *droplet uniformity index* $\sum_i V_i(\mathrm{MMD} - D_i)/\mathrm{MMD}$[433] and *relative span factor* $(D_{0.9} - D_{0.1})/\mathrm{MMD}$, etc.

4.2.0 CORRELATIONS FOR DROPLET SIZES OF NORMAL LIQUIDS

In many atomization processes, physical phenomena involved have not yet been understood to such an extent that mean droplet size could be expressed with equations derived directly from first principles, although some attempts have been made to predict droplet size and velocity distributions in sprays through maximum entropy principle.[252][432] Therefore, the correlations proposed by numerous studies on droplet size distributions are mainly empirical in nature. However, the empirical correlations prove to be a practical way to determine droplet sizes from process parameters and relevant physical properties of liquid and gas involved. In addition, these previous studies have provided insightful information about the effects of process parameters and material properties on droplet sizes.

The process parameters influencing droplet sizes may include liquid pressure, flow rate, velocity ratio of air to liquid (mass flow rate ratio of air to liquid), and atomizer geometry and configuration. It has been clearly established that increasing the velocity ratio of air to liquid is the most important practical method of improving atomization.[211] In industrial applications, however, the use of mass flow rate ratio of air to liquid has been preferred. As indicated by Chigier,[211] it is difficult to accept that vast quantities of air, that do not come into any direct contact with the liquid surface, have any influence on atomization although mass flow rates of fluids include the effects of velocities.

The liquid properties of primary importance are density, viscosity and surface tension. Unfortunately, there is no incontrovertible evidence for the effects of liquid viscosity and surface tension on droplet sizes, and in some cases the effects are conflicting. Gas density is generally considered to be the only thermophysical property of importance for the atomization of liquids in a gaseous medium. Gas density shows different influences in different atomization processes. For example, in a fan spray, or a swirl jet atomization process, an increase in the gas density can generally improve atomization for a constant liquid flow rate.[116] However, beyond a certain limit, a further increase in the gas density may increase droplet size due to the reduced relative velocity resulted from the increased drag. In gas atomization of melts, using a lighter gas, for example, helium, to replace air can improve atomization for a constant gas pressure.

In the following sections, the correlations for droplet sizes generated by different types of atomizers will be summarized, and the effects of process parameters and material properties on droplet sizes will be discussed on the basis of the analytical and experimental studies available in published literature.

4.2.1 Pressure Jet Atomization

Various correlations for mean, minimum and maximum droplet sizes generated in pressure jet atomization using plain-orifice atomizers have been derived,[434]-[439] as listed in Table 4.3. In these correlations, d_0 is the diameter of discharge orifice, v_L is the kinematic viscosity of liquid, μ_G is the dynamic viscosity of gas, $\dot{V}_L$ is the volumetric flow rate of liquid, and C_f is the skin friction coefficient. The Jet Reynolds number and Weber number are defined as $\mathrm{Re}_L = U_L \rho_L d_0 / \mu_L$ and $\mathrm{We}_L = U_L^2 \rho_L d_0 / \sigma$, respectively. The primary parameters governing the mean droplet size are liquid injection pressure or velocity, physical properties of liquid (viscosity, density, surface tension), and ambient gas (viscosity, density), and atomizer geometry parameter such as discharge orifice diameter. Some other parameters that may influence the droplet size include relative velocity, ambient gas pressure and temperature, geometrical boundaries that confine the surrounding atmosphere (chamber walls), presence of nearby jets, chemical reactions (combustion), as well as nozzle and supply-line configurations.[220] In spray combustion applications, the density and surface tension of most commercial fuels differ only slightly from each other (Table 2.2) so that the significance of these properties is weak. However, the viscosity of different fuels spans a large range and may vary by nearly two orders of magnitude. Thus, the viscosity exhibits a significant effect on the mean droplet size.

Table 4.3. Correlations for Mean, Minimum and Maximum Droplet Sizes Generated in Pressure Jet Atomization by Plain-Orifice Atomizers

Correlations	Process Characteristics & Remarks	References
$SMD = \frac{500 d_0^{1.2} \nu_L^{0.2}}{U_L}$	Liquid jet from a plain circular orifice into quiescent air	Merrington & Richardson[434]
$MMD = 6 d_0 \, Re_L^{-0.15}$	Liquid jet from a diesel-type injector into quiescent air; $1000 < Re_L < 12000$	Panasenkov[435]
$SMD = 3330 d_0^{0.3} \mu_L^{0.07} \rho_L^{-0.648} \sigma^{-0.15} U_L^{-0.55} \mu_G^{0.78} \rho_G^{-0.052}$	Liquid jet from a diesel-type injector into quiescent air; effects of gas properties	Harmon[436]
$D_{0.999} = d_0 We_L^{-0.333} (23.5 + 0.000395 \, Re_L)$	Best-fit of previous experimental data for liquid jet disintegration	Miesse[220]
$SMD = 47 d_0 U_L^{-1} \left(\frac{\sigma g}{\rho_G} \right)^{0.25} \left[1 + 3.31 \frac{\mu_L}{(\rho_L \sigma d_0)^{0.5}} \right]$	Liquid jet from a diesel-type injector into quiescent air $D_{max} \approx (2 \sim 2.5) SMD$	Tanasawa& Toyoda[41]
$SMD = c \rho_A^{0.121} \dot{V}_L^{0.131} \Delta P_L^{-0.135}$ (μm), c=25.1, 23.9, 22.4 for pintle,hole,throttling pintle nozzles, resp.	Liquid jet from a diesel-type injector into quiescent air; ΔP_L (MPa), ρ_A (kg/m^3), $\dot{V}_L$ (mm^3/stroke)	Hiroyasu& Kadota[317]
$SMD = 6156 \nu_L^{0.385} (\sigma \rho_L)^{0.737} \rho_A^{0.06} \Delta P_L^{-0.54}$ (μm) $D_{max} = 1.75 D_{32}$ or $D_{0.999} = 1.75 SMD$	Liquid jet from a diesel-type injector into quiescent air; σ (N/m), ΔP_L (bar), ρ (kg/m^3), ν_L (m^2/s)	Elkotb[438]
$SMD = 47 d_0 \, Re_L^{-0.5} \left(\frac{\rho_G}{\rho_L} \right)^{0.26}$	Non-evaporating unsteady dense sprays: $l_0/d_0 = 4$, Re_L $(2.6 \sim 4) \times 10^4$, $\rho_G / \rho_L = (9.84 \sim 50.9) \times 10^{-3}$, $d_0 = 0.15 \sim 0.2$ mm, $\sigma = 2.7 \times 10^{-2}$ N/m, ν_L=2.2×10^{-6} m^2/s	Yanianeet al.[439]

Generally, an increase in liquid injection pressure (velocity) can promote jet breakup due to the increase in both the level of turbulence in the jet, and the aerodynamic drag forces between the jet and ambient gaseous medium. An increase in liquid viscosity resists the breakup of liquid jet and ligaments, and thus, delays the onset of atomization. Therefore, the mean droplet diameter is proportional to the liquid viscosity, gas viscosity, discharge orifice diameter[41] and volumetric flow rate of liquid, and inversely proportional to the liquid density, velocity and injection pressure,[41] with different proportional power indices representing the significance of each factor. However, different researchers reported distinctly different effects of liquid surface tension and gas density on droplet size. An increase in liquid surface tension typically increase the mean droplet size.[220][437][438] On the other hand, Harmon's[436] correlation suggested that an increase in liquid surface tension may reduce the mean droplet size. An increase in gas density can probably promote secondary breakup of large droplets, and reduce the maximum droplet size, improving the fineness,[41][436] and uniformity of entire droplet sizes, while the effect of gas density on the minimum droplet size seems to be little.[440] In contrast, some correlations[317][438][439] suggested that an increase in the gas density may increase the mean droplet size, plausibly due to the reduced relative velocity resulted from the increased drag. Moreover, Miesse's experimental results[220] revealed that the effects of flow conditions and physical properties on the maximum droplet diameter depend on the liquid Reynolds number. For example, an increase in jet velocity decreases the maximum droplet diameter for $Re_L < 11.9 \times 10^4$, but increases it for larger values of the Reynolds number. Similarly, an increase in liquid density decreases the droplet diameter for $Re_L < 2.975 \times 10^4$, but increases it for larger values of the Reynolds number.

In addition to these physical properties and process variables, the flow, flow direction, and shock wave of ambient air/gas relative to liquid jet may significantly influence the resultant droplet size distribution. In high-speed aerodynamic atomization, different flow arrangements have been used,[244] including: *(a)* injection of liquid jet into a subsonic or supersonic co-flow, or subsonic contra-flow of air/gas, *(b)* transverse injection of liquid jet into a subsonic or supersonic crossflow of air/gas, and *(c)* shattering of liquid jet by a traversing shock. Fine droplets can be obtained as a result of the intense shear at the liquid-gas interface by high-speed gas flow.

Recently, Razumovskii[441] studied the shape of drops, and satellite droplets formed by forced capillary breakup of a liquid jet. On the basis of an instability analysis, Teng et al.[442] derived a simple equation for the prediction of droplet size from the breakup of cylindrical liquid jets at low-velocities. The equation correlates droplet size to a modified Ohnesorge number, and is applicable to both liquid-in-liquid, and liquid-in-gas jets of Newtonian or non-Newtonian fluids. Yamane et al.[439] measured Sauter mean diameter, and air-entrainment characteristics of non-evaporating unsteady dense sprays by means of an image analysis technique which uses an instantaneous shadow picture of the spray and amount of injected fuel. Influences of injection pressure and ambient gas density on the Sauter mean diameter and air entrainment were investigated parametrically. An empirical equation for the Sauter mean diameter was proposed based on a dimensionless analysis of the experimental results. It was indicated that the Sauter mean diameter decreases with an increase in injection pressure and a decrease in ambient gas density. It was also shown that the air-entrainment characteristics can be predicted from the quasi-steady jet theory.

4.2.2 Pressure-Swirl and Fan Spray Atomization

Various correlations for mean droplet size generated using pressure-swirl and fan spray atomizers are summarized in Tables 4.4 and 4.5, respectively. In the correlations for pressure-swirl data, *FN* is the *Flow number* defined as $FN = \dot{m}_L/\Delta P_L \rho_L)^{0.5}$, l_0 and d_0 are the length and diameter of final orifice, respectively, l_s and d_s are the length and diameter of swirl chamber, respectively, A_p is the total inlet ports area, t_f is the film thickness in final orifice, θ is the half of spray cone angle, and We_f is the Weber number estimated with film

Table 4.4. Correlations for Mean Droplet Size Generated by Pressure-Swirl Atomizers

Correlations	Process Characteristics & Remarks	Refs.
$SMD = 7.3\sigma^{0.6}\nu_L^{0.2}\dot{m}_L^{0.25}\Delta P_L^{-0.4}$	Very small variations in σ and wide variations in μ_L; effects of atomizer geometry and air properties not included	Radcliffe [443]
$SMD = 4.4\sigma^{0.6}\nu_L^{0.16}\dot{m}_L^{0.22}\Delta P_L^{-0.43}$	Effects of atomizer geometry and air properties not included	Jasuja [83]
$SMD = \begin{cases} 133\dfrac{FN^{0.64291}}{\Delta P_L^{0.22565}\rho_L^{0.3215}}, & \Delta P_L < 2.8\text{ MPa} \\ 607\dfrac{FN^{0.75344}}{\Delta P_L^{0.19936}\rho_L^{0.3767}}, & \Delta P_L > 2.8\text{ MPa} \end{cases}$	For kerosene-type fuels; effects of air properties not included	Babu et al. [444]
$SMD = 10^{-3}\sigma(6.11 + 0.32\times10^5 FN\rho_L^{0.5} - 6.973\times10^{-3}\Delta P_L^{0.5} + 1.89\times10^{-6}\Delta P_L)$	Derived from experimental data for 25 different fuels using 6 different simplex nozzles of large Flow numbers; $We_f>10$; Strong effect of σ, no effect of μ_L; Discrepant with other data	Kennedy [445]
$MMD = 2.47\dot{m}_L^{0.315}\Delta P_L^{-0.47}\mu_L^{0.16}\mu_A^{-0.04}\sigma^{0.25}\rho_L^{-0.22}\left(\dfrac{l_0}{d_0}\right)^{0.03}\left(\dfrac{l_s}{d_s}\right)^{0.07}\left(\dfrac{A_p}{d_s d_0}\right)^{-0.13}\left(\dfrac{d_s}{d_0}\right)^{0.21}$	Derived from experimental data using large-capacity industrial pressure swirl atomizers of large Flow numbers with 50 different geometric configurations* $d_0\rho_L U_L^2/\sigma$=11.5×10³–3.55×10⁵, $d_0\rho_L U_L/\mu_L$=1.913×10³–21.14×10³, μ_L/μ_A=279–2235, ρ_L/ρ_A=694–964	Jones [446]
$SMD = 2.25\sigma^{0.25}\mu_L^{0.25}\dot{m}_L^{0.25}\Delta P_L^{-0.5}\rho_A^{-0.25}$	Consistent with theoretical value and other experimental data [83],[446],[447]	Lefebvre [199]
$SMD = 4.52\left(\dfrac{\sigma\mu_L^2}{\rho_A\Delta P_L^2}\right)^{0.25}(t_f\cos\theta)^{0.25} + 0.39\left(\dfrac{\sigma\rho_L}{\rho_A\Delta P_L}\right)^{0.25}(t_f\cos\theta)^{0.75}$	Effect of spray cone angle considered; film thickness is taken as a primitive variable; it may be estimated from $t_f = 2.7\left[\dfrac{d_0\dot{m}_L\mu_L}{\Delta P_L\rho_L}\right]^{0.25}$ [448]	Wang & Lefebvre [449]

*l_0/d_0-0.1–0.9, l_s/d_s=0.31–1.26, $A_P/(d_s d0)$=0.19–1.21, d_s/d_0-1.41–8.13

Table 4.5. Correlations for Mean Droplet Sizes Generated by Fan Spray Atomizers

Correlations	Process Characteristics & Remarks	References
$\mathrm{SMD} = 0.071\left(\dfrac{t_s L \sigma \mu_L^{0.5}}{\rho_L^{0.5} U_L^2}\right)^{1/3}$ in cm (cgs units)	Derived by empirically correcting a theoretical equation of inviscid flows for viscosity with fan spray data of wax; Valid for $3 \le \mu_L \le 25$ cP	Hasson & Mizrahi [450]
$D_m = \left[\dfrac{3\pi}{\sqrt{2}}\right]^{1/3} d_l \left[1 + \dfrac{3\mu_L}{(\rho_L \sigma d_l)^{1/2}}\right]^{1/6}, \quad K = t_s L$ $d_l = 0.9614\left[\dfrac{K^2\sigma^2}{\rho_G \rho_L U_R^4}\right]^{1/6}\left[1 + 2.6\mu_L\left(\dfrac{K\rho_G^4 U_R^7}{72\rho_L^2\sigma^5}\right)^{1/3}\right]^{1/5}$	Derived on the basis of a force balance including forces caused by gas pressure, liquid inertia, surface tension and viscosity; For fan sprays $K \approx 0.00315$ cm^2	Dombrowski & Johns [238]
$\mathrm{SMD} = 4.78 \times 10^{-6} A_L \left(\sigma / (\sin\theta \rho_L \dot{V}_L^2)\right)^{1/3}$	Derived from fan spray data of water and oil based on a simplified sheet breakup theory; Effect of liquid viscosity not included	Dombrowski & Munday [94]

thickness in final orifice. In the correlations for fan spray data, A_L is the cross-section area of liquid flow, and θ is the spray angle. The Flow number is a measure of the flow through an atomizer nozzle for a given supply pressure. Its value depends on the size of the orifice and the internal geometry of the nozzle. Both volume-flow-rate-based and mass-flow-rate-based Flow numbers are in use. The primary parameters influencing the mean droplet size include liquid injection pressure, physical properties of liquid (viscosity, density, surface tension) and ambient gas (viscosity, density), and atomizer geometry as described by the flow number and length to diameter ratios of swirl chamber and nozzle orifice.

In *pressure-swirl atomization*, the mean droplet size increases with an increase in liquid viscosity,[83][443][446][451][452] surface tension,[199][445][446][453] and/or liquid flow rate or Flow number.[83][443][446][454] The proportional power indices for viscosity, surface tension, and liquid flow rate are ~0.06–0.5, ~0.25, and ~0.22–0.75, respectively. The effect of liquid viscosity on mean droplet size diminishes with increasing Flow number or decreasing spray cone angle, while the effect of Flow number on mean droplet size also diminishes with increasing liquid injection pressure.[449] A high liquid injection pressure can promote atomization due to the resultant high liquid velocity, and thus the mean droplet size is inversely proportional to liquid injection pressure. The proportional power index may range from -0.275 to -0.44 for various atomization experiments.[83][443][447][449][451] Apparently, increasing Flow number generates coarse droplets.

As ambient air pressure is increased, the mean droplet size increases[455]-[458] up to a maximum and then turns to decline with further increase in ambient air pressure.[1] The initial rise in the mean droplet size with ambient pressure is attributed to the reduction of sheet breakup length and spray cone angle. The former leads to droplet formation from a thicker liquid sheet, and the latter results in an increase in the opportunity for droplet coalescence and a decrease in the relative velocity between droplets and ambient air due to rapid acceleration. At low pressures, these effects prevail. Since the mean droplet size is proportional to the square root of liquid sheet thickness and inversely proportional to the relative velocity, the initial rise in the mean droplet size can be readily explained. With increasing ambient pressure, its effect on spray cone angle diminishes, allowing disintegration forces become dominant. Consequently, the mean droplet size turns to decline. Since ambient air pressure is directly related to air density, most correlations include air density as a variable to facilitate applications. Some experiments[452] revealed that ambient air temperature has essentially no effect on the mean droplet size.

The mean droplet size reduces with decreasing length to diameter ratio of the final nozzle orifice.[459] However, the effect of the length to diameter ratio of swirl chamber on the mean droplet size is not straightforward. The mean droplet size reduces initially with an increase in the length to diameter ratio of swirl chamber due to the elimination of flow striations caused by finite number of swirl ports. Further increasing the length to diameter ratio of swirl chamber, the mean droplet size turns to increase due to the raised energy losses caused by the extra length while the contribution to the elimination of flow striations is damped out.[459] The mean droplet size may be reduced by increasing spray cone angle.

The variations of the mean droplet size and the droplet size distribution with axial distance in a spray generated by pressure swirl atomizers have been shown to be a function of ambient air pressure and velocity, liquid injection pressure, and initial mean droplet size and distribution.[460]

In *fan spray atomization*, the effects of process parameters on the mean droplet size are similar to those in pressure-swirl atomization. In general, the mean droplet size increases with an increase in liquid viscosity, surface tension, and/or liquid sheet thickness and length. It decreases with increasing liquid velocity, liquid density, gas density, spray angle, and/or relative velocity between liquid and surrounding air.

4.2.3 Air-Assist Atomization

Various correlations for mean droplet sizes generated by air-assist atomizers are given in Table 4.6. In these correlations, $\dot{m}_A$ is the mass flow rate of air, h is the height of air annulus, t_{f0} is the initial film thickness defined as $t_{f0} = d_o w/d_{an}$, d_o is the outer diameter of pressure nozzle, d_{an} is the diameter of annular gas nozzle, w is the slot width of pressure nozzle, C is a constant related to nozzle design, U_A is the velocity of air, and MMD_c is the modified mean droplet diameter for the conditions of droplet coalescence. Distinguishing air-assist and air-blast atomizers is often difficult. Moreover, many

Table 4.6. Correlations for Mean Droplet Size Generated by Air-Assist Atomizers

Correlations	Process Characteristics & Remarks	Refs.
$\mathrm{MMD} = 20\nu_L^{0.5}\dot{m}_L^{0.1}\left(1+\frac{\dot{m}_L}{\dot{m}_A}\right)^{0.5} \times h^{0.1}\sigma^{0.2}\rho_A^{-0.3}U_R^{-1.0}$ For conditions of droplet coalescence: $\mathrm{MMD_c} = \mathrm{MMD}\left[1+5.0\dot{m}_L^{0.1}\left(\frac{\dot{m}_L}{\dot{m}_A}\right)^{0.6}\right]$	*Internal mixing air-assist atomizer.* Derived from wax spray data in Ref. 461 using NGTE atomizer; Good agreement with fuel-air or fuel-steam spray data;[103] Discrepant with water-air spray data;[79][462] MMD is to be multiplied by an empirical correction factor for conditions of droplet coalescence.	Wigg[75]
$\mathrm{SMD} = 14\times10^{-6}d_0^{0.75}\left(\frac{\dot{m}_L}{\dot{m}_A}\right)^{0.75}$	*Internal-mixing air-assist atomizer.* Derived from water-air spray data at $\dot{m}_L$=30–100 kg/h, $\dot{m}_L/\dot{m}_A$=5–100 using immersion technique	Sakai et al.[76]
$\frac{\mathrm{SMD}}{t_{f0}} = \left[1+\frac{16850\mathrm{Oh}^{0.5}}{\mathrm{We}(\rho_L/\rho_A)}\right]\left[1+\frac{0.065}{(\dot{m}_A/\dot{m}_L)^2}\right]$ $\mathrm{Oh} = \left(\frac{\mu_L^2}{\rho_L t_{f0}\sigma}\right)^{0.5}$, $\mathrm{We} = \frac{\rho_A t_{f0} U_A^2}{\sigma}$	*External-mixing air-assist atomizers.* Derived from ethanol (glycerin)-air spray data with initial thickness of flat circular sheet up to 0.7 mm and varied air impingement angles; Sampled with oil-coated slides	Inamura & Nagai [77]
$\mathrm{SMD} = 51d_0\,\mathrm{Re}^{-0.39}\,\mathrm{We}^{-0.18}\left(\frac{\dot{m}_L}{\dot{m}_A}\right)^{0.29}$ $\mathrm{Re} = \frac{\rho_L U_R d_0}{\mu_L}$, $\mathrm{We} = \frac{\rho_L d_0 U_R^2}{\sigma}$	*External-mixing air-assist atomizers.* Derived from kerosene-air spray data with numerous nozzle configurations, including effects of air pressure; Sampled with coated slides	Elkotb et al.[78]
$\mathrm{SMD} = C\left(\frac{\rho_L^{0.25}\mu_L^{0.06}\sigma^{0.375}}{\rho_A^{0.375}}\right) \times\left(\frac{\dot{m}_L}{\dot{m}_L U_L+\dot{m}_A U_A}\right)^{0.55}$	*Pressure and air-assist atomizers.* Derived from calibrating fluid (MIL-F-70411)-air spray data using Parker Hannifin spray analyzer	Simmons [451]

features of external-mixing air-assist atomizers are in common with air-blast atomizers. Therefore, the correlations of droplet sizes derived for one type of atomizers may be relevant to the other type. The important parameters influencing the mean droplet size include velocities and mass flow rates of liquid and atomization gas/air, physical properties of liquid (viscosity, density, surface tension) and gas/air (density), and atomizer geometry as described by nozzle diameter, air annulus height, etc. The kinetic energy (dynamic pressure) of atomization gas is deemed to be the predominant factor governing the mean droplet size.[101][463] For low-viscosity liquids atomized using external-mixing air-assist atomizers, the relative velocity between liquid and air is an additional independent key factor governing atomization quality.[463]

In general, the mean droplet size is proportional to liquid viscosity and surface tension, and inversely proportional to air velocity, air pressure, relative velocity between air and liquid, and mass flow rate ratio of air to liquid, with different proportional power indices representing the significance of each factor. The effect of liquid flow rate on the mean droplet size is notable: an increase in liquid flow rate corresponds to a decrease in the mean droplet size at low air pressures, but an increase in the mean droplet size at high air pressures. Suyari and Lefebvre[463] attributed this behavior to the fact that the spray system operates essentially as a simplex pressure-swirl atomizer at low air pressures or velocities, while it operates in an air-blast mode at high air pressures or velocities. Thus, increasing liquid flow rate at low air pressures is equivalent to increasing liquid injection pressure in pressure-swirl mode, leading to a reduction in the mean droplet size, whereas increasing liquid flow rate at high air pressures lowers both the mass flow rate ratio of air to liquid, and the relative velocity between air and liquid, reducing liquid-air interaction and impairing atomization quality. Therefore, atomization quality can be improved by increasing liquid injection pressure in pressure-swirl mode, and by increasing air velocity or mass flow rate ratio of air to liquid in air-assist or air-blast mode. Suyari and Lefebvre[463] further indicated that for a constant air velocity, increasing liquid injection pressure from zero leads to an increase in the mean droplet size up to a maximum value, beyond which further increase in injection pressure causes the mean droplet size to decline. This maximum value corresponds to a minimum of the relative velocity or a minimum of the interaction level. The liquid injection pressure corresponding to the maximum value of mean droplet size increases with increasing air velocity. Whether an increase in liquid velocity may promote or hinder atomization depends on if it increases or decreases the relative velocity between liquid and air, whereas an increase in air velocity generally improves atomization quality.

4.2.4 Air-Blast Atomization

Various correlations for mean droplet size generated by plain-jet, prefilming, and miscellaneous air-blast atomizers using air as atomization gas are listed in Tables 4.7, 4.8, 4.9, and 4.10, respectively. In these correlations, ALR is the mass flow rate ratio of air to liquid, ALR $= m_A/m_L$, D_p is the prefilmer diameter, D_h is the hydraulic mean diameter of air exit duct, v_r is the kinematic viscosity ratio relative to water, a is the radial distance from cup lip, D_L is the diameter of cup at lip, U_p is the cup peripheral velocity, U_r is the air to liquid velocity ratio defined as U_r=U_A/U_p, L_w is the diameter of wetted periphery between air and liquid streams, A_A is the flow area of atomizing air stream, m is a power index, P_A is the pressure of air, and B is a composite numerical factor. The important parameters influencing the mean droplet size include relative velocity between atomization air/gas and liquid, mass flow rate ratio of air to liquid, physical properties of liquid (viscosity, density, surface tension) and air (density), and atomizer geometry as described by nozzle diameter, prefilmer diameter, etc.

Similarly to pressure-swirl atomization and air-assist atomization, the mean droplet size is proportional to liquid viscosity and surface tension, and inversely proportional to air velocity, air pressure, air density, relative velocity between air and liquid, and mass flow rate ratio of air to liquid, with different proportional power indices denoting the significance of each factor. The proportional power indices are (-0.15)–(-1.44) (with a frequently used value of -1) for air velocity, (-1)–0.72 (with typical values around -0.5) for air density, (-0.5)–0.5 (typically 0) for liquid density, 0–0.6 (with a typical value of 0.5) for surface tension, and 0–1.7 (frequently 0.5 or 1) for $(1 + \dot{m}_L/\dot{m}_A)$, respectively. For large mass flow rate ratios of air to liquid, the influence of liquid viscosity on the mean droplet size becomes negligibly small.[79] For liquids of low viscosities, the nozzle dimensions appear to have little influence on the mean droplet size.[80] For liquids of high viscosities, the mean droplet size is proportional to the nozzle diameter, with a proportional power index ranging from 0.16 to 0.6, and a typical value of 0.5.

Table 4.7. Correlations for Mean Droplet Sizes Generated by Plain-Jet Air-Blast Atomizers

Correlations	Process Characteristics & Remarks	Refs.
$SMD = 0.585\left(\frac{\sigma}{\rho_L U_R^2}\right)^{0.5} + 53\left(\frac{\mu_L^2}{\sigma\rho_L}\right)^{0.225}\left(\frac{\dot{V}_L}{\dot{V}_A}\right)^{1.5}$	Derived from spray data for water, gasoline, alcohol, and heavy fuel oil; Sampled with oil-coated glass slides; Dimensionally incorrect; No effects of nozzle sizes and shapes or air density	Nukiyama & Tanasawa [79]
$SMD = 0.95\frac{(\sigma\dot{m}_L)^{0.33}}{\rho_L^{0.37}\rho_A^{0.30}U_R}\left(1+\frac{\dot{m}_L}{\dot{m}_A}\right)^{1.70} + 0.13\left(\frac{\mu_L^2 d_0}{\sigma\rho_L}\right)^{0.5}\left(1+\frac{\dot{m}_L}{\dot{m}_A}\right)^{1.70}$	Derived from spray data for water, kerosene, and special solutions over a broad range of air and liquid properties using light-scattering technique	Lorenzetto & Lefebvre [80]
$SMD = 0.022\left(\frac{\sigma}{\rho_A U_A^2}\right)^{0.45}\left(1+\frac{1}{ALR}\right)^{0.5} + 0.00143\left(\frac{\mu_L^2}{\sigma\rho_L}\right)^{0.4}\left(1+\frac{1}{ALR}\right)^{0.8}$	Derived from spray data for residual fuel oils using light-scattering technique; Dimensionally incorrect; Discrete liquid jets into a swirling air stream (cross-flow breakup)	Jasuja [81]
$SMD = 0.48d_0\left(\frac{\sigma}{\rho_A U_R^2 d_0}\right)^{0.4}\left(1+\frac{1}{ALR}\right)^{0.4} + 0.15d_0\left(\frac{\mu_L^2}{\sigma\rho_L d_0}\right)^{0.5}\left(1+\frac{1}{ALR}\right)$	Derived from spray data for kerosene, gas oil, and blended oils using light-scattering technique; Dimensionally correct, excellent data correlation; Co-flowing air and liquids	Rizk & Lefebvre [82]
$D_{min} = \frac{4\sigma}{C_f\rho_A U_A^2}$, $C_f \approx 0.0196$	Liquid jet from a plain circular orifice into a coaxial co-flowing Mach 1.5 air flow	Issac et al. [244]

Table 4.8. Correlations for Mean Droplet Size Generated by Prefilming Air-Blast Atomizers

Correlations	Process Characteristics & Remarks	Refs.
$SMD = 3.33\times10^{-3}\frac{(\sigma\rho_L D_p)^{0.5}}{\rho_A U_A}\left(1+\frac{1}{ALR}\right) +13.0\times10^{-3}\left(\frac{\mu_L^2}{\sigma\rho_L}\right)^{0.425} D_p^{0.575}\left(1+\frac{1}{ALR}\right)^2$	Derived from dimensional analysis with various constants and indices being deduced from spray data for water and kerosene measured using light-scattering technique	Rizkalla & Lefebvre [1,84]
$SMD = 10^{-3}\left(1+\frac{1}{ALR}\right)^{0.5} \times\left[\frac{(\sigma\rho_L)^{0.5}}{\rho_A U_A}+0.06\left(\frac{\mu_L^2}{\sigma\rho_A}\right)^{0.425}\right]$	Derived from spray data for kerosene, gas oil, various blends of gas oil and residual fuel oil with nozzle of Bryan et al. [464] using light-scattering technique; No effect of nozzle dimensions; Less effect of ALR	Jasuja [83]
$\frac{SMD}{D_h} = \left[0.33\left(\frac{\sigma}{\rho_A U_A^2 D_p}\right)^{0.6}\left(\frac{\rho_L}{\rho_A}\right)^{0.1} +0.068\left(\frac{\mu_L^2}{\sigma\rho_L D_p}\right)^{0.5}\right]\left(1+\frac{1}{ALR}\right)$	Derived from spray data for water, kerosene, and special solutions of high viscosities with 3 geometrically similar atomizers, using light-scattering technique	El-Shanawany & Lefebvre [85]

Table 4.9. Correlations for Mean Droplet Size Generated by Miscellaneous Air-Blast Atomizers with Prefilming Effect

Correlations	Process Characteristics & Remarks	Refs.
$SMD = 6\times10^{-6} + 0.019\left[\frac{\sigma^{0.5}\nu_r^{0.21}}{\rho_A^{0.5}\left(aD_L + a^2\right)^{0.25}}\right] \times\left[1+0.065\left(\frac{\dot{m}_L}{\dot{m}_A}\right)^{1.5}\right]\left[\frac{\dot{V}_L}{U_p^2\left(0.5U_r^2 - U_r + 1\right)}\right]^{0.5}$	Derived from spray data for various oils of intermediate surface tension and a large range of viscosities, prefilming by spinning cups, using light-absorption technique; Prefilming type is superior to plain-jet type for generating fine droplets; Controlling over thin liquid sheets is a prerequisite to fine sprays.	Fraser et al. [73]
$SMD = \left(2.67\times10^4 U_L P_A^{-0.33} + 4.11\times10^6 \rho_A U_A P_A^{-0.75}\right)^{-1}$	Derived from spray data for water with splash plate injector using radiometer; $0.1<P_A<2.1$ MPa	Ingebo [465]
Single convergent nozzle: $MMD = 5.36\times10^{-3}\frac{\sigma^{0.41}\mu_L^{0.32}}{\left(\rho_A U_R^2\right)^{0.57} A_A^{0.36}\rho_L^{0.16}} + 3.44\times10^{-3}\left(\frac{\mu_L^2}{\sigma\rho_L}\right)^{0.17}\left(\frac{\dot{m}_A}{\dot{m}_L}\right)^m \frac{1}{U_R^{0.54}}$ Double concentric nozzle: $MMD = 2.62\frac{\sigma^{0.41}\mu_L^{0.32}}{\left(\rho_A U_R^2\right)^{0.72}\rho_L^{0.16}} + 1.06\times10^{-3}\left(\frac{\mu_L}{\rho_L\sigma}\right)^{0.17}\left(\frac{\dot{m}_A}{\dot{m}_L}\right)^m \frac{1}{U_R^{0.54}}$	Derived from spray data for wax melts with a single convergent nozzle (air converged and expanded through an annulus around a liquid nozzle) and a double concentric nozzle (a secondary air nozzle inserted axially in a liquid nozzle); Counted using a microscope; $m = \begin{cases} -1 \text{ for } \dot{m}_A/\dot{m}_L < 3 \\ -0.5 \text{ for } \dot{m}_A/\dot{m}_L > 3 \end{cases}$	Kim & Marshall [429]
$SMD = \frac{12\sigma}{\rho_L U_R^2/\left[1+1/\left(\alpha\frac{\dot{m}_G}{\dot{m}_L}\right)\right] + 4\frac{\sigma}{t_s}}$ $\alpha = \frac{1.62}{U_G^{1.3}\left(\frac{\dot{m}_G}{\dot{m}_L}\right)^{0.63}\mu_L^{0.3}}$	Derived from spray data for high-viscosity liquids (mixtures of glycerine and water) of $50\leq\mu_L\leq1140$ cP with a 2-D nozzle (air flows through discharge slots and impacts both sides of a flat liquid sheet from a discharge slot inbetween the air slots); Droplet size measured by Malvern 2600HSD Spray Analyzer; Effects of air slot thickness included	Knoll & Sojka [263]

Table 4.10. Correlations for Mean Droplet Size Generated by Miscellaneous Air-Blast Atomizers

Correlations	Process Characteristics & Remarks	References
$SMD = 5.0\left(\frac{\sigma\mu_L d_0^2}{\rho_A U_R^3 \rho_L}\right)^{0.25}$ for $WeRe<10^6$ $SMD = 37\left(\frac{\sigma\mu_L d_0^{0.5}}{\rho_A U_R^3 \rho_L}\right)^{0.4}$ for $WeRe>10^6$	Derived from spray data for water with cross-current air injection using high-speed photography technique; Applicable to fuel injection from plain-orifice atomizers into high-velocity cross-flowing air streams in various air-breathing propulsion engines (ramjets, turbojet afterburners)	Ingebo & Foster [466] Ingebo [467]
Convergent nozzle: $MMD = 2.6\times10^{-3}\left(\frac{\mu_A}{\rho_A U_A L_w}\right)^{0.4}\left(\frac{\dot{m}_L}{\dot{m}_A}\right)^{0.4}$ Impingement nozzle: $MMD = 1.22\times10^{-4}\left(\frac{\mu_A}{\rho_A U_A L_w}\right)^{0.15}\left(\frac{\dot{m}_L}{\dot{m}_A}\right)^{0.6}$	Derived from spray data for aqueous solutions of black dye with a convergent nozzle (liquid first contacts air at air nozzle throat) [79] and an impingement nozzle (a central circular air tube surrounded by an annular liquid duct); Collected in mineral oil and counted using a light microscope; Droplet size independent of liquid viscosity but dependent on air viscosity	Gretzinger & Marshall [102]
$\frac{MMD\rho_G U_R^2}{\sigma} = 0.61\left(\frac{U_R\mu_L}{\sigma}\right)^{2/3}\left(1+\frac{10^3\rho_G}{\rho_L}\right)\left(\frac{\dot{m}_L\rho_L\mu_G\sigma}{\mu_L^4}\right)^{1/12}$	Derived from spray data for wax melts using micromerograph and shifting technique; Cross-stream and co-stream injections of liquid into air; Relative velocity is important;Properties of fluids are less critical; Geometry and operation of the injection is of least importance.	Weiss & Worsham [259]
$MMD = 9\pi(16B)^{1/2}\left(\frac{\mu_L(\sigma/\rho_L)^{1/2}}{\rho_G U_G^2}\right)^{2/3}$	Semi-analytical correlation considering development of waves along liquid surface produced by high-velocity gas	Mayer [462]

The influence of liquid density on the mean droplet size is relatively small but complex. An increase in liquid density may reduce the mean droplet size due to a decrease in sheet thickness at the atomizing lip of a prefilming atomizer, or due to an increase in the relative velocity between liquid and air for a plain-jet atomizer. However, increasing liquid density may also increase the mean droplet size because a liquid sheet may extend further downstream of the atomizing lip of a prefilming atomizer so that the sheet breakup may take place at lower relative velocity between liquid and air.

For prefilming type of atomizers, minimum droplet sizes are obtained with nozzle designs that spread liquid into thinnest sheet before subjecting its both sides to air-blast action[86] and provide maximum contact between liquid and air.[468] From experimental data obtained over a wide range of process conditions and material properties, it was found[469] that the effect of liquid viscosity on the mean droplet size is independent of that of surface tension and air velocity. Therefore, the mean droplet size can be expressed as a sum of two terms: one dominated by surface tension, air velocity and air density, and the other by liquid viscosity, as suggested by Lefebvre:[469]

Eq. (24)

$$\frac{\text{SMD}}{L_c} = A\left(\frac{\sigma}{\rho_A U_A^2 D_p}\right)^{0.5}\left(1+\frac{\dot{m}_L}{\dot{m}_A}\right) + B\left(\frac{\mu_L^2}{\sigma \rho_L D_p}\right)^{0.5}\left(1+\frac{\dot{m}_L}{\dot{m}_A}\right)$$

where L_c is the characteristic dimension of air-blast atomizer, and may be set equal to D_h or D_p, and A and B are constants related to atomizer design and must be determined experimentally. This is considered as the basic equation for prefilming air-blast atomizers. For liquids of low viscosities, such as water and kerosene, the first term predominates, and the key factors influencing the mean droplet size are surface tension, air velocity and air density.[84] The mean droplet size increases with increasing surface tension and atomizer dimension with a proportional power index of ~0.5, and decreases with increasing air velocity, air density, and air to liquid ratio. For liquids of high viscosities, the second term becomes more significant, and thus the mean droplet size is less sensitive to variations in air velocity and density, but instead more dependent on liquid properties, particularly liquid viscosity. Some modified versions of this equation have been proposed accommodating the effect of liquid to air density ratio, liquid Reynolds number and air Mach number, as detailed in Table 4.8.

Some variant air-blast atomizer designs[102][259][429][465]-[467] have been developed and employed in certain important applications. These designs could not be fit into either of the two primary types of air-blast atomizers, i.e., plain-jet and prefilming configurations. Thus, they are considered as miscellaneous types of air-blast atomizers. The related correlations are given in Tables 4.9 and 4.10, respectively. These experimental observations showed general trends consistent with the findings in previous studies using the plain-jet and prefilming air-blast atomizers in terms of the beneficial effects of increasing the air to liquid ratio, air velocity, and air density. Typically, increasing the air to liquid ratio ALR and/or the dynamic force $\rho_A U_R^2$ can reduce the mean droplet size. However, with increasing air to liquid ratio, the mean droplet size approaches an asymptotic value.[272][273][429] Various studies recommended that the operation range for the air to liquid ratio ALR be from 0.1 to 10,[429] or 2 to ~5,[84]-[86] whereas Fraser et al.[73] suggested an upper limit of 1.5. Below the lower limit, the atomization quality deteriorates; above the upper limit, some portion of the air energy may be wasted so that the atomization efficiency may be lowered.

4.2.5 Rotary Atomization

Various correlations for mean and maximum droplet sizes generated by smooth flat vaneless disks, vaneless disks, and wheels are listed in Tables 4.11, 4.12 and 4.13, respectively. In these correlations, d is the diameter of disk/cup, ω and ω_{rps} are the rotational speed of disk/cup in radians/s and rps, respectively, θ is the semicone angle of rotating cup, n is the number of vanes, h is the vane height, w is the vane opening, and K is a constant. These correlations may be used for prediction purposes because each of them was derived from data involving only one well-defined mechanism of droplet formation. The important parameters influencing the mean droplet size include liquid flow rate, physical properties of liquid (viscosity, density, and surface tension), and rotational speed and diameter of disk. For disks or wheels fitted with vanes, the height, opening, and number of vanes are some additional geometry parameters of importance.

Table 4.11a. Correlations for Mean and Maximum Droplet Sizes Generated by Smooth Flat Vaneless Disks in Direct Droplet Formation Regime

Correlations	Process Characteristics & Remarks	References
$D_{0.999} = \frac{1.07}{\omega_{rps}}\left(\frac{\sigma}{d\rho_L}\right)^{0.5}$	Effect of liquid viscosity not included	Bär [16]
$\mathrm{SMD} = \frac{27}{\omega_{rps}}\left(\frac{\sigma}{d\rho_L}\right)^{0.5}\left(1+0.003\frac{\dot{V}_L}{d\nu_L}\right)$	All liquid properties included; SMD decreases with increasing ν_L	Tanasawa et al. [111]
$\mathrm{SMD} = 1.345d\mathrm{We}^{-0.5} = \frac{3.8}{\omega}\left(\frac{\sigma}{\rho_L d}\right)^{0.5}$ $\mathrm{We} = \rho_L d^3\omega^2/(8\sigma)$	Mean droplet size dependent only on Weber number	Matsumoto et al. [470] Walton & Prewett [471]

Table 4.11b. Correlations for Mean and Maximum Droplet Sizes Generated by Smooth Flat Vaneless Disks in Ligament Formation Regime

Correlations	Process Characteristics & Remarks	References
$D_{0.999} = \dfrac{0.87}{\omega_{rps}} \left(\dfrac{\sigma}{d\rho_L} \right)^{0.5}$	Effect of liquid viscosity not included	Walton & Prewett [471]
$\text{SMD} = \dfrac{0.177 \dot{V}_L^{0.2}}{\omega_{rps} d^{0.3}}$	Derived from water droplet size measurements; No liquid properties included; Applicable only to water	Oyama & Endou [472]
$\dfrac{\text{SMD}}{d_l} = (1.5\pi)^{1/3} \left(1 + 3 \dfrac{\text{We}}{\text{Re}^2} \right)^{1/6}$	$\text{Re} = \omega \rho_L d^2 / (4\mu_L)$; $\text{We} = \rho_L d^3 \omega^2 / (8\sigma)$	Matsumoto & Takashima [473]
$\text{SMD} = 0.119 \dfrac{\dot{V}_L^{0.1} \sigma^{0.5}}{\omega_{rps} d^{0.5} \rho_L^{0.4} \mu_L^{0.1}}$	All liquid properties and relevant parameters included; SMD inversely proportional to $\mu_L^{0.1}$	Tanasawa et al. [111]
$\text{SMD} = 0.26 \omega_{rps}^{-0.79} \dot{V}_L^{0.32} d^{-0.69} \rho_L^{-0.29} \sigma^{0.26} (1 + 1.027 \mu_L^{0.65})$	All liquid properties and relevant parameters included	Kayano & Kamiya [110]

Table 4.11c. Correlations for Mean Droplet Sizes Generated by Smooth Flat Vaneless Disks in Sheet Formation Regime

Correlations	Process Characteristics & Remarks	References
$D_{10} = \dfrac{0.76}{\omega_{rps}} \left(\dfrac{\sigma}{d\rho_L} \right)^{0.5}$	Very similar to correlations for Direct Droplet formation [16] and for droplet formation *via* Ligament Breakup [471]	Fraser & Eisenklam [424]
$\text{SMD} = 15.6 \dfrac{\dot{V}_L^{0.5}}{\omega_{rps}} \left(\dfrac{\sigma}{d^2 \rho_L} \right)^{0.4}$	Effect of liquid flow rate included	Tanasawa et al. [111]

Table 4.12. Correlations for Mean Droplet Size Generated by Vaneless Disks in Three Droplet Formation Regimes

Correlations	Regimes	References
$D_{10} = \left[\dfrac{3\dot{V}_L \mu_L}{2\pi\rho_L (\pi d \omega_{rps})^2 \sin\theta} \right]^{1/3}$	Direct Droplet Formation	Fraser et al. [73]
$D_{10} = \left[0.77 \dfrac{\dot{V}_L}{\omega_{rps} d} \left(\dfrac{\rho_L \omega_{rps}^2 d^3}{\sigma} \right)^{-5/12} \left(\dfrac{\rho_L \sigma d}{\mu_L^2} \right)^{-1/16} \right]^{0.5}$	Ligament Formation	Hinze & Milborn [112]
$D_{10} = 4.42 \left(\dfrac{\sigma}{\omega^2 \rho_L d} \right)^{0.5}$	Sheet Formation	Hege [474]

Table 4.13. Correlations for Mean Droplet Size Generated by Wheels Fitted with Vanes

Correlations	References
$\text{SMD} = 0.44 d \left(\dfrac{\dot{m}_L}{\rho_L \omega_{rps} d^2} \right)^{0.6} \left(\dfrac{\mu_L}{\dot{m}_L} \right)^{0.2} \left(\dfrac{\sigma \rho_L n h}{\dot{m}_L^2} \right)^{0.1}$	Friedman et al. [475]
$\text{SMD} = \dfrac{3.3\times10^{-9} K \dot{m}_L^{0.24}}{(\omega_{rps} d)^{0.83} (nh)^{0.12}}$, K=8.5×10^5–9.5×10^5	Herring & Marshall [476]
$\text{SMD} = 0.483 \omega_{rps}^{-0.6} \rho_L^{-0.5} \left(\dfrac{\mu_L \dot{m}_L}{d} \right)^{0.2} \left(\dfrac{\sigma}{nh} \right)^{0.1}$	Fraser et al. [477]
$\text{SMD} = 5240 \left(\dfrac{\dot{m}_L}{wn} \right)^{0.171} (\pi d \omega_{rps})^{-0.539} \mu_L^{-0.017}$ SMD [μm], $\dot{m}_L$ [g/s], w and d [cm], μ_L [poise]	Scott et al. [478]

A number of theoretical and experimental studies on centrifugal atomization[16][73][110]-[112][424][470]-[480] have been conducted in an effort to ascertain atomization mechanisms and establish correlations between droplet sizes and process parameters. The results of these studies suggested that, generally, the mean droplet size decreases with increasing rotational speed, liquid density, disk diameter, air density, and/or air velocity if exposed to an air flow. In contrast, an increase in liquid flow rate, viscosity, and/or surface tension generally increases the mean droplet size. The liquid flow rate appears to exert less influence on the droplet size than the rotational speed, suggesting the relative importance of controlling the rotational speed to obtain droplets with reproducible size characteristics. Uniform droplets can be produced in the range of $2.67 < \omega D(d\rho_L/\sigma)^{1/2} < 4.44$.[88]

Several research groups[73][112][470][480] presented experimental and semi-empirical equations for the transitional conditions from Direct Droplet Formation regime to Ligament Formation regime, or from Ligament Formation regime to Film/Sheet Formation regime. Matsumoto et al.[470] conducted a theoretical analysis and related three dimensionless parameters to the two transitions based on two simple models. The three dimensionless parameters are the Reynolds number, $\text{Re} = d^2\omega\rho_L/(4\mu_L)$, the Weber number, $\text{We} = d^3\omega^2\rho_L/(8\sigma)$, and the dimensionless flow rate, $Q^+ = 2\dot{V}_L/[\pi d^2(\mu_L\omega/\rho_L)^{0.5}]$. Matsumoto et al. derived correlations for the transition flow rates from their experimental results in terms of these dimensionless parameters.

Accordingly, for the transition from Direct Droplet to Ligament Formation regime and the reverse transition, the dimensionless transition flow rates are $Q_1^+ = 0.096\ \text{Re}^{0.95}/\text{We}^{1.15}$ and $\bar{Q}_1^+ = 0.073\ \text{Re}^{0.95}/\text{We}^{1.15}$, respectively.[470]

For the transition from Ligament to Film/Sheet Formation regime or the reverse transition, the equation of Hinze and Milborn,[112] $Q_2^+ = 0.340\ \text{Re}^{2/3}/\text{We}^{0.883}$, may be used to predict the dimensionless transition flow rate for liquids of low viscosities (less than a few poises). For more viscous liquids, the equation derived by Tanasawa et al.,[480] $Q_2^+ = 0.297\ \text{Re}^{6/5}/\text{We}$, is applicable for the calculation of the dimensionless transition flow rate.

4.2.6 Effervescent Atomization

The studies on the performance of effervescent atomizer have been very limited as compared to those described above. However, the results of droplet size measurements made by Lefebvre et al.[87] for the effervescent atomizer provided insightful information about the effects of process parameters on droplet size. Their analysis of the experimental data suggested that the atomization quality by the effervescent atomizer is generally quite high. Better atomization may be achieved by generating small bubbles. Droplet size distribution may follow the Rosin-Rammler distribution pattern with the parameter q ranging from 1 to 2 for a gas to liquid ratio up to 0.2, and a liquid injection pressure from 34.5 to 345 kPa. The mean droplet size decreases with an increase in the gas to liquid ratio and/or liquid injection pressure. Any factor that tends to impair atomization quality, and increase the mean droplet size (for example, decreasing gas to liquid ratio and/or injection pressure) also leads to a more monodisperse spray.

For a constant gas flow rate, a decrease in gas density leads to an increase in mean gas velocity and/or average gas flow area. The increases in both these quantities prove to be beneficial to atomization quality. In the former case, it accelerates the liquid flow through the injector orifice so that the liquid is discharged at a higher velocity. In the latter case, it reduces the area available for the liquid flow so that the liquid is squeezed into thinner films and ligaments as it flows through the injector orifice.

The decrease in the mean droplet size with increasing liquid injection pressure may be attributed to two effects. First, the high pressure-drop across the exit orifice makes the process more like a pressure atomization at high pressure. Second, the liquid is squeezed into fine ligaments as it flows through the injector orifice, and the ligaments are shattered into small droplets by the explosion downstream of the nozzle exit.

The experimental results of Lefebvre et al.[87] also revealed that the mean droplet size is virtually independent of injector orifice diameter. This is a desirable feature for spray combustion applications.

4.2.7 Electrostatic Atomization

Very few experimental data of droplet sizes in electrostatic atomization are available in published literature. Mori et al.[481] proposed the following correlation for the mean droplet size generated in electrostatic atomization:

Eq. (25) $$\mathrm{SMD} = 5.39 d_0 \overline{E}^{-0.255} \overline{V}^{0.277} \mathrm{Re}^{-0.124}$$

with

$$\overline{E} = \in E^2 d_0 / \sigma \text{ , } \overline{V} = \rho_L \dot{V}_L^2 / (\sigma d_0^3) \text{ , } \mathrm{Re} = \rho_L \dot{V}_L / (\mu_L d_0)$$

where d_0 is the outer diameter of liquid supply tube, $\in$ is the dielectric constant, and E is the intensity of electric field. Clearly, the mean droplet size decreases with increasing electric field intensity, dielectric constant, and/or liquid surface tension, but increases with increasing liquid flow rate, liquid density, liquid viscosity, and/or liquid supply tube diameter. The effects of liquid surface tension and liquid supply tube diameter appear less significant in electrostatic atomization.

Some researchers have suggested correlations based on theoretical analyses, as reviewed by Lee et al.[88] However, the equations have not been experimentally verified.

4.2.8 Ultrasonic Atomization

Analytical and empirical correlations for droplet sizes generated by ultrasonic atomization are listed in Table 4.14 for an overview. In these correlations, D_m is the median droplet diameter, λ is the wavelength of capillary waves, ω_0 is the operating frequency, a is the amplitude, U_{L0} is the liquid velocity at the nozzle exit in USWA, P_{max} is the maximum sound pressure, and U_s is the speed of sound in gas. Most of the analytical correlations are derived on the basis of the capillary wave theory. Experimental observations revealed that the mean droplet size generated from thin liquid films on an ultrasonically excited plate is proportional to the theoretical wavelength of capillary waves excited by the ultrasonic vibrations on the liquid surface. Lang[127] suggested $D = 0.34\lambda$, and Lobdell[484] derived a theoretical value of 0.36λ based on considerations of droplet formation from high-amplitude capillary waves, whereas Mizutani et al.[482] proposed 0.53λ for low flow rates, at which a thin liquid film forms before atomization. The wavelength of the capillary waves is controlled by the vibration frequency. Thus, the mean droplet size can be related to the operating frequency of the ultrasonic atomizer. In addition, the mean droplet size may be influenced by several other parameters, (such as flow rate and vibration amplitude) and physical properties of liquid (density, surface tension and viscosity), as suggested by many researchers on the basis of the Taylor instability[483] or experiments.[130][482]

Table 4.14. Analytical and Empirical Correlations for Droplet Sizes Generated by Ultrasonic Atomization

Correlations	Process Characteristics & Remarks	References
$D_m \approx 0.34\lambda = 0.34(8\pi\sigma / \rho_L \omega_0^2)^{1/3}$	Derived on the basis of capillary wave theory (semi-empirical)	Lang [127]
$\text{SMD} \approx 0.53\lambda = 0.53(8\pi\sigma / \rho_L \omega_0^2)^{1/3}$	Derived on the basis of capillary wave theory, for low flow rates	Mizutani et al.[482]
$D_m = \left(4\pi^3\sigma / \rho_L \omega_0^2\right)^{1/3}$ $D_m = 0.42\pi a$, for large $\rho_L \omega_0^2 a^3 / 2\sigma$	Derived on the basis of capillary wave theory in terms of Taylor instability, for thick films	Peskin & Raco [483]
$\text{SMD} = 0.158\left(\dfrac{\sigma}{\rho_L}\right)^{0.354} \mu_L^{0.303} \dot{V}_L^{0.139}$	For water-glycerin (methanol) solutions and distilled water with a horn type of ultrasonic atomizer at ω_0=26 kHz, $\dot{V}_L \leq 50$ l/h	Mochida [130]
$\text{MMD} = 0.53 d_0 Z^{-0.48} \text{Re}^{-0.69}$ $\text{Re} = \rho_L U_{L0} d_0 / \mu_L$, $Z = P_{max}^2 / \rho_G U_s^2 \rho_L U_{L0}^2$	Ultrasonic standing wave atomization of Drivanil 250/25: $0.01 \leq \text{Re} \leq 1$, $0.01 \leq \text{MMD}/d_0 \leq 0.3$, $2.5 Z^{0.5} \leq \text{Re} \leq 100 Z^{0.7}$, $10 \leq Z \leq 10000$	Bauckhage et al.[143]

Generally, the mean droplet size is proportional to liquid surface tension, and inversely proportional to liquid density and vibration frequency. The proportional power index is ~1/3 for the surface tension, about -1/3 for the liquid density, and -2/3 for the vibration frequency. The mean droplet size may be influenced by two additional parameters, i.e., liquid viscosity and flow rate. As expected, increasing liquid viscosity, and/or flow rate leads to an increase in the mean droplet size,[130][482] while the spray becomes more polydisperse at high flow rates.[482] The spray angle is also affected by the liquid flow rate, vibration frequency and amplitude. Moreover, the spray shape is greatly influenced by the direction of liquid flow (upwards, downwards, or horizontally).[482]

In ultrasonic standing wave atomization, increasing the amplitude of the sonotrodes and/or the gas pressure reduces the mass median diameter of the droplets generated.[143]

4.3.0 CORRELATIONS FOR DROPLET SIZES OF MELTS

As described above, a number of empirical and analytical correlations for droplet sizes have been established for normal liquids. These correlations are applicable mainly to atomizer designs, and operation conditions under which they were derived, and hold for fairly narrow variations of geometry and process parameters. In contrast, correlations for droplet sizes of liquid metals/alloys available in published literature[318][323][328]-[331][485]-[487] are relatively limited, and most of these correlations fail to provide quantitative information on mechanisms of droplet formation. Many of the empirical correlations for metal droplet sizes have been derived from off-line measurements of solidified particles (powders), mainly sieve analysis. In addition, the validity of the published correlations needs to be examined for a wide range of process conditions in different applications. Reviews of mathematical models and correlations for droplet sizes generated by twin-fluid atomization have been made by Mehrotra,[321][322] with emphasis on gas atomization of liquid metals/alloys using free-fall type of ring atomizers. In a recent review, Aller and Losada[485] have surveyed the mathematical models developed for commercial and some near-commercial atomization processes for metal powder production.

4.3.1 Gas Atomization

Analytical,[321][322][331][486] and experimental studies[162][164][318][319][323] on the atomization of molten metals have been performed in an effort to elucidate atomization mechanisms and control droplet size. These studies have led to the development of various correlations that may be used to predict the mean size and the size distribution of atomized droplets. Analytical, empirical, and semi-empirical correlations for droplet sizes of liquid metals/alloys are summarized in Tables 4.15 and 4.16, for gas atomization *via* jet breakup and film/sheet breakup, respectively. Among these, the empirical correlation developed by Lubanska[323] is probably the best known and most quoted one for *free-fall type of ring-like atomizers* due to its reasonable accuracy under various experimental conditions. Lubanska correlated a considerable amount of spray data for low melting-point alloys, Sn, Fe, Al, Cu, steel and wax, and proposed a modified form of Wigg's correlation.[75] In Lubanska's correlation, important process parameters and material properties are related to the mass median droplet size by the following equation:

Eq. (26) $$\frac{\text{MMD}}{d_0} = k_D \left[\left(1 + \frac{\dot{m}_L}{\dot{m}_G} \right) \frac{\nu_L}{\nu_G \text{We}} \right]^{1/2}$$

where d_0 is the melt delivery nozzle diameter, k_D is a constant whose value was found to be 40–50 for a wide variety of spray data (air-Sn, Fe, Al, Cu, steel, low melting-point alloys, and wax), and ν_L and ν_G

Table 4.15. Empirical Correlations for Droplet Sizes of Liquid Metals in Gas Atomization Jet Breakup

Correlations	Process Characteristics & Remarks	References
$\frac{\text{MMD}}{d_0} = k_D \left[\left(1 + \frac{\dot{m}_L}{\dot{m}_G}\right) \frac{\nu_L}{\nu_G \text{We}} \right]^{1/2}$ $\text{We} = \frac{\rho_L U_G^2 d_0}{\sigma}$, $k_D \approx 40\text{–}50$	Derived from gas-Sn (Fe, low melting point alloys) spray data with multiple gas jets from a free-fall, ring-like atomizer: d_0 =1/4–7/8"; Applicable to Cu, Al, steels, wax, oil	Lubanska [323]
$\frac{\text{MMD}}{d_0} = k_D \left[\left(1 + \frac{\dot{m}_L}{\dot{m}_G}\right) \frac{\nu_L}{\nu_G \text{We}} \right]^{m}$	Derived from air-Sn (Pb, Pb-Sn) spray data using free-fall atomizer with nozzle diameters of 1.3, 1.8, 2.4 mm, atomizing angles of 15,25, 35,45°, and different melt superheat	Rao & Mehrotra [331]
$\text{MMD} = \left[\frac{a\sigma}{\rho_L U_G^0 (1 - bd_0 \quad \mu_L)} \frac{\dot{m}_L}{\dot{m}_G} \right]^{1/2}$	Semi-empirical correlation for gas atomization with free-fall atomizer; Gas jet impacts melt stream vertically; Fitted well with data for Fe-,Ni-, Co-, Cu-based alloys; the values of *a* and *b* should be determined experimentally.	Tornberg [486]
$\text{MMD} \approx 43/(\text{SGC})^{1/2}$	For gas-atomized Al, Fe, Cu-Al and Sn-Sb	Dunkley [487]
(a) $D_{peak} = D_{10}\,(0.136 \times 10^{-5}\,\text{Re} + 0.294)$ (b) $D_{peak} = 0.382 d_0\, \xi^{-1} \text{We}^{-1.77}$ $\text{Re} = \frac{\rho_G U_{G0} d_{G0}}{\mu_G}$, $\text{We} = \frac{\rho_L U_{L0}^2 d_0}{\sigma}$	For air-atomized bronze at (a) constant metal nozzle diameter but different annular gas jet velocities, 182000<Re<238000, We=7.95; (b) varying metal nozzle diameters but constant annular gas jet velocity, 4.4<We<7.95, Re = 206000	Nichiporenko [488]
Schmitt [489]	Semi-empirical correlation for a ring-like atomizer	

Table 4.16. Analytical and Experimental Correlations for Metal Droplet Sizes in Gas Atomization via Film or Sheet Breakup

Correlations	Process Characteristics & Remarks	References
Subsonic gas flow: $D = \frac{11.85\varepsilon\sigma}{\chi\rho_G U_s^2}$, *Sonic/supersonic gas flow*: $D = \frac{11.85\varepsilon\sigma\chi_m}{\rho_G U_s^2}$, $\varepsilon = 0.25$; $\varepsilon = -0.12 + 0.57\vartheta + 0.38 d_0$ for tin; $\varepsilon = -0.14 + 0.66\vartheta + 0.48 d_0$ for lead	Derived from an analytical model for flat, infinitely thick liquid layer; Effects of gas compressibility included; Effects of gas/liquid ratio, liquid viscosity, and nozzle geometry not included; χ and χ_m can be determined from the universal curves for metals in [329] for subsonic gas flow and in [330] for sonic/supersonic gas flow	Bradley [329,330] Rao & Mehrotra [331]
$\frac{D_{0.5}\rho_G U_G^2}{\sigma} = \left(\frac{\rho_G}{\rho_L}\frac{U_G\mu_L}{\sigma}\right)^{2/3} \times \left(\frac{\rho_L\sigma d_0}{\mu_L^2}\right)^{1/6} f\left(\frac{\dot{m}_G}{\dot{m}_L}\right)$	Derived from spray data of a variety of metals with close-coupled atomizers	Miller & Giles [4]
Film breakup: $\text{MMD} \approx 2.9\left(\frac{\rho_L}{\rho_G}\right)^{0.5}\frac{U_L t_f}{U_G}$; For $15 \leq \rho_L\sigma^2/(\mu_L^2\rho_G U_G^2) \leq 5000$, $60 \leq \rho_L U_L t_f/\mu_L \leq 2200$; *Sheet breakup*: $\text{MMD} = 4.3\left(\frac{\rho_L U_L^2}{\rho_G U_G^2}\right)^{0.3}\frac{t_s}{\text{Re}^{0.22}}$, $\text{Re} = \frac{\rho_L U_L t_s}{\mu_L}$; For $720 \leq \rho_L\sigma^2/(\mu_L^2\rho_G U_G^2) \leq 5100$, $1.1\cdot10^5 \leq \rho_L t_s\sigma/\mu_L^2 \leq 3\cdot10^5$, $9.1\cdot10^{-4} \leq \mu_L U_L/\sigma \leq 1.2\cdot10^{-2}$	Derived from air-water (water-glycerol mixtures) spray data using an internal mixing air-blast atomizer with prefilming effect (a coaxial liquid injector placed at the throat on centerline of a converging-diverging air nozzle) similar to Ünal nozzle;[169] Liquid injector without or with a small deflecting plate at its exit plane creates conditions for Film or Sheet breakup; Observed by high-speed photography; Measured by video image analyzer; Calibrated by holography; Applicable to liquid metals due to dynamic similarity	Hespel et al. [328]

are the kinematic viscosity of liquid metal (melt) and atomization gas, respectively. The Weber number is defined as We $= P_L U_G^2 d_0/\sigma$, where U_G is the gas velocity at the impact of the gas jets with the liquid metal stream, that was derived (in Lubanska's work) from curves for velocity decay in sonic jets. Lubanska's correlation fitted well with the experimental data of Tamura and Takeda for copper,[490] Thompson's data for aluminum,[491] and Clare and Radcliffe's data for wax.[461] However, it failed to fit the experimental data of Rao and Mehrotra[331] for tin, lead, and lead-tin alloy using nozzles of different diameters and atomizing angles. Large discrepancies between the experiments and the predictions based on this correlation have also been reported for aluminum alloys,[492] suggesting the inadequacy of this correlation for the application to *close-coupled atomizers* with sonic-supersonic-sonic transitions. In addition, no experimental data and correlations are available for the size and size distribution of droplets generated with a linear atomizer which is composed of a long rectangular melt delivery nozzle and gas jet nozzle.[178]

Yule and Dunkley[5] made a detailed analysis of advantages and drawbacks of Lubanska's correlation. By comparing the predictions of Lubanska's and Wigg's equations to measured data, they showed that Wigg's equation[75] is in a reasonable fit to the data for MMD > 500 μm, and gives better fit than Lubanska's equation for MMD < 500 μm. Thus, they recommended to reconsider Wigg's equation:

Eq. (27) $$\text{MMD} = 20 \nu_L^{0.5} \dot{m}_L^{0.1} d_0^{0.1} \sigma^{0.2} \rho_G^{-0.3} U_R^{-1.0} \left(1+\frac{\dot{m}_L}{\dot{m}_G}\right)^{0.5}$$

because it has some theoretical reasoning behind it and appears to offer a better accuracy than Lubanska's modification to it. It should be noted that in the above equation, the length scale *h* defined in

Wigg's original equation is substituted by liquid nozzle diameter d_0. It is uncertain, however, if one could use the liquid nozzle diameter to substitute this length scale, although the prediction of this equation is not sensitive to the choice. In addition, there is evidence that the dependency of droplet size on surface tension is more pronounced in reality than that indicated by this equation. Thus, Yule and Dunkley pointed out that there is scope for exploring the use of other correlations which contain higher powers of surface tension.

To make Lubanska's correlation applicable to more general conditions, Rao and Mehrotra[331] proposed to substitute the square root by an exponent m which is not a constant, but a function of atomizing angle and independent of nozzle diameter (Table 4.17). In addition, in the modified Lubanska's correlation, the parameter k_D is also no longer a constant but varies with nozzle diameter and insensitive to atomizing angle (Table 4.17).

Table 4.17. Experimental Values of Parameters k_D and m* in Modified Lubanska's Correlation[331]

Nozzle Diameter (cm) → Atomizing Angle (°) ↓	0.24	0.18	0.13
15	6.31 (0.301)	8.19 (0.303)	11.72 (0.300)
25	6.26 (0.282)	8.23 (0.283)	11.67 (0.282)
35	6.29 (0.265)	8.32 (0.267)	11.60 (0.261)

*Values of m are in parenthesis

On the basis of a force balance model, Tornberg[486] derived a semi-empirical correlation for *free-fall type of atomizers*:

Eq. (28)
$$\mathrm{MMD} = \left[\frac{a\sigma}{\rho_L U_G^0 (1 - b d_0 \mu_L)} \frac{\dot{m}_L}{\dot{m}_G} \right]^{1/2}$$

where a and b are constants related to atomizer design, material system, and operation conditions, and U_G^0 is the atomization gas velocity at nozzle exit. This model encompassed a force applied to a melt to break up the liquid into droplets, and an internal friction force within the melt counteracting the atomization force. The atomization force included that created by the dynamic pressure on the melt stream, and the drag force created by gas surrounding the melt. The dynamic pressure was neglected in the model. The internal friction force is due to the liquid viscosity. The pressure from all these forces exerted on a droplet was then balanced by the surface tension force, leading to the above correlation. Although the physical meaning of this correlation is clear, it is apparently also system-dependent. Nevertheless, this correlation has been found to generate much better agreements with experimental data for free-fall type of atomizers than other available correlations.[493]

Dunkley[487] analyzed droplet size data for gas-atomized Al, Fe, Cu-Al and Sn-Sb, and developed an empirical correlation between the specific gas consumption, SGC, and mass median diameter of droplets for various atomization gases, pressures, and atomizers:

Eq. (29) $$\mathrm{MMD} \approx 43/(\mathrm{SGC})^{1/2}$$

where SGC is in m^3/kg, MMD is in μm, and the standard deviation of the correlation is within ±10 μm. This empirical correlation may also be employed as a useful rule of thumb for estimating gas requirements.

Some other correlations that have been specifically developed for liquid metals include those proposed by Nichiporenko,[488] Schmitt,[489] Thompson,[491] and Date et al.[494] Nichiporenko[488] correlated the diameter of particles (powder) of predominant fraction, D_{peak}, with the arithmetic mean particle diameter (linear average diameter), D_{10}, and the Reynolds number as well as the Weber number and metal nozzle diameter for air-atomized bronze under different experimental conditions. The definitions of the Reynolds and Weber numbers are given in Table 4.15, where U_{G0} and U_{L0} are the gas and liquid metal velocities at the gas and metal nozzle exit, respectively, d_{G0} is the equivalent diameter of the annular gas nozzle, and ξ is the predominant fraction corresponding to the peak of particle size frequency distribution curve. Schmitt[489] derived a correlation for the maximum droplet diameter on the basis of dimensional analysis and assuming that the disintegration process encompasses forces due to inertia, surface tension, and viscosity of melt as well as inertia of atomization gas. The correlation was used to explain his data for the atomization of a Ni-base alloy (IN 100) with argon using a ring-like atomizer. Thompson[491] derived an empirical equation to correlate atomization gas pressure, liquid metal temperature, nozzle diameter, and metal head to atomization rate. Date et al.[494] also related atomization gas pressure to the percent yield of each mesh fraction with orthogonal polynomials based on a statistical analysis of atomization data.

Many atomization experiments of liquid metals with free-fall type of atomizers *via* jet breakup[318][319][323][331][495][496] revealed that droplet/particle size distributions follow closely the logarithmic-normal distribution pattern. As discussed above, the logarithmic-normal distribution can be fully described by two parameters, i.e., the mean of the distribution and the standard deviation. The values of both these parameters can be readily determined from the plot of experimental data (cumulative weight percentage *vs*. droplet size) on a log-probability graph paper. Thus, the entire droplet size distribution can be characterized by two parameters, i.e., the mass median diameter $D_{0.5}$, and the geometric standard deviation of the distribution, σ_{LN}, which satisfies a relationship:

Eq. (30) $$\sigma_{LN} = D_{0.84} / D_{0.5} = D_{0.5} / D_{0.16} = (D_{0.84} / D_{0.16})^{1/2}$$

according to the mathematical characteristics of the log-normal distribution. In a sieve test, the mass median diameter $D_{0.5}$ corresponds to the opening of a screen mesh which lets through 50 wt.% of the atomized powder. Similarly, droplet sizes $D_{0.16}$ and $D_{0.84}$ correspond to the openings of the screen meshes which let through 16 wt.% and 84 wt.% of the atomized powder, respectively. In the log-probability graph, the 50 wt.% value of droplet diameter gives the mass median diameter. In addition, the standard deviation of the distribution can also be empirically correlated to the mass median diameter, as suggested by Lubanska:[323]

Eq. (31) $$D_{0.5} = 13\sigma_{LN}^3 \quad \text{or} \quad \sigma_{LN} = 0.425 D_{0.5}^{1/3}$$

Thus, both the mean droplet size and the size distribution may be predicted using these correlations [Eqs. (26), (27), (28), or (29) and Eqs. (30), (31)] for given process parameters and material properties. For a given atomizer design, the standard deviation of droplet size distribution has been found to increase with the melt flow rate, but appears to be less sensitive to the gas flow rate.[5] Moreover, the variation of the standard deviation is very atomizer- and melt-specific. An empirical correlation which fits with a wide range of atomization data has the following form:

Eq. (32) $$\sigma_{LN} = a D_{0.5}^b$$

where a and b are system-specific constants. The values of the standard deviation of droplet size distributions are given in Tables 2.3, 2.4 and 2.5 for different atomization processes. In spray forming processes, the mass median droplet diameter is typically in the range of 50 to 150 μm, and the log-normal standard deviation is generally in the range of 1.7 to 2.3, depending on process parameters and materials involved.[497]

Similarly to the trends in the atomization of normal liquids, droplet properties such as mean droplet size, and size distribution for liquid metals/alloys depend on a large number of parameters. These include thermophysical properties of melts and atomization gas, and operation variables such as liquid temperature and flow rate, gas pressure and flow rate, and atomizer geometry parameters and configurations. For a given atomizer design and under identical operation conditions, different atomization gases and/or metal/alloy systems may produce droplets of different mass median sizes and size distributions due to the difference in thermophysical properties of the gases (Tables 2.6 and 2.7) and/or metals/alloys (Tables 2.8, 2.9, 2.10 and 2.11). Gas velocity and gas to metal mass flow rate ratio are two important factors governing the resultant droplet sizes. As evident from the correlations (Tables 4.15 and 4.16), the mean droplet/particle size increases with increasing liquid viscosity, surface tension and/or gas density, and it decreases with increasing gas viscosity, liquid density, and/or gas velocity. The mean droplet/particle size reduces with increasing gas to liquid mass flow rate ratio and approaches an asymptotic value at large gas to liquid ratios. The dynamic force of atomization gas, in addition to the gas to liquid mass flow rate ratio, is also an important factor affecting the resultant metal droplet sizes.[498] As confirmed by many atomization experiments for melts,[80][82][322][323] the mean droplet/particle size generally increases with increasing liquid nozzle diameter for plain-jet air-blast type of atomizers, i.e., free-fall type of atomizers with an annular gas jet or discrete gas jets. However, the experiments of Nukiyama and Tanasawa[79] showed no effect under their experimental conditions and nozzle dimension. For other types of nozzle designs, such as those used by Gretzinger and Marshall,[102] the mean droplet size decreases with an increase in liquid nozzle dimension. Therefore, published data and correlations should be used with caution, and attention should be paid to the operation conditions, nozzle configurations and nozzle dimensions used for deriving the correlations. Moreover, the atomizing angle, or generally speaking, atomizer geometry parameters and configurations also have an important impact on the resultant droplet sizes and size distributions. Increasing the atomizing angle may reduce the mean droplet size (Table 4.17).[331]

In addition to the effects similar to those in normal liquid atomization, the thermophysical properties of melts and atomization gases also influence the metal droplet size by affecting the heat transfer between the two phases. Putimtsev[499] studied the atomization of Fe, Ni, Cu, Ag and Fe-Ni alloys in air, nitrogen, argon, and helium, and observed that the metal droplets atomized by gases of low density, low thermal conductivity, and/or high viscosity remain in liquid state for longer time, thus increasing the probability of secondary disintegration and generation of finer droplets. Metals with high thermal conductivity and/or low thermal capacity cool faster, inhibiting droplet deformation and disintegration and leading to large droplets/particles. For example, Ag and Cu have lower viscosity and surface tension than Fe and Ni. However, much smaller yield of finer droplets was obtained for Ag and Cu due to their high thermal conductivities. For the similar reason, increasing liquid temperature or melt superheat can reduce the mean droplet size.[331] Moreover, gas chemistry critically affects the size distribution, shape and thermal history of atomized metal droplets, as demonstrated by Liu et al.[154][174] and Zeng et al.[173][500] Retention of ligament shape due to the formation of oxide skin,[319] or change in droplet size distribution due to oxygen blended in inert atomization gas[154][174] are some examples of the effects of gas chemistry.

In gas atomization *via* film or sheet breakup (Table 4.16), the mean droplet size is proportional to liquid density, liquid viscosity, liquid velocity, and film or sheet thickness, and inversely proportional to gas density and gas velocity, with different proportional power indices denoting the significance of each factor. In recent experimental studies on liquid sheet and film atomization processes using a close-coupled atomizer, Hespel et al.[328] concluded that the control over the liquid layer thickness is of major importance to the control of the atomization quality.

4.3.2 Water Atomization

A limited number of empirical correlations have been developed for metal droplet sizes generated by water atomization, as listed in Table 4.18. In these correlations, c is a system-specific constant, ϑ is the atomizing angle, i.e., angle between water nozzle axis and metal delivery nozzle axis, k is a proportional constant specific to atomizer type, melt type and melt temperature, n is a parameter depending on atomizer type, ΔP_W is the water pressure, U_W is the water velocity, and $\dot{m}_W$ is the mass flow rate of water.

Some quantitative studies[498][501] on droplet size distribution in water atomization of melts showed that the mean droplet size increases with metal flow rate and reduces with water flow rate, water velocity, or water pressure. From detailed experimental studies on the water atomization of steel, Grandzol and Tallmadge[501] observed that water velocity is a fundamental variable influencing the mean droplet size, and further, it is the velocity component normal to the molten metal stream $U_W \sin \vartheta$, rather than parallel to the metal stream, that governs the mean droplet size. This may be attributed to the hypothesis that water atomization is an impact and shattering process, while gas atomization is predominantly an aerodynamic shear process.[5]

In the empirical correlation proposed by Kato et al.,[503] the mean droplet size is inversely proportional to the water pressure, with a power index of ~0.5 for conical shaped annular-jet atomizers, and 0.7–1.0 for V-shaped flat-jet atomizers. This suggests a lower efficiency of the annular-jet atomizers in terms of spray fineness at high water pressures. The data of Kato et al.[503] were obtained for water pressures lower than 10 MPa. Seki et al.[502] observed the similar trend in the water atomization of nickel and various steels at higher water pressures (>10 MPa). Since k is dependent on both atomizer type and melt properties, no satisfactory general correlation has been developed and it must be determined by experimental measurements. Qualitatively speaking, the value of k is larger for Cu, Fe, and other high surface tension metals which are difficult to atomize.[5] The values of k and n proposed by Seki et al.[502] are given in Table 4.18.

Table 4.18. Empirical Correlations for Mean Droplet Size of Liquid Metals in Water Atomization *via* Jet Breakup

Correlations	Remarks	References
$MMD = c/(U_W \sin \vartheta)$ $c = 2750,\ \dot{m}_W = 0.9\text{–}3.8$ kg/s, $\vartheta = 15°\text{–}30°$	Derived from spray data for water-4620 stee	Grandzol & Tallmadge [501]
$MMD = k\Delta P_W^{-n}$; MMD in μm, ΔP_W in MPa For annular jet: $n \approx 0.5$ [5]; $n = 0.58,\ k = 114$ [502] For V-jets: $n = 0.7\text{–}1.0$ [5]; $n = 0.56,\ k = 68$ [502]	Derived from data for ΔP_W less than 10 MP	Kato et al. [503] Seki et al. [502]
$MMD = kd_0\left(\frac{d_0\rho_L U_W}{\mu_L}\right)^{-0.57}\left(\frac{d_0\rho_L U_W^2}{\sigma}\right)^{-0.22} \times \left(\frac{\dot{m}_L}{\dot{m}_W}\right)^{0.043}$	For two-jet water atomizers	Kishidaka [504]

In the empirical correlation proposed by Kishidaka[504] for two-jet atomizers, melt nozzle diameter and physical properties, water velocity, and water to melt ratio are included. The constant k is again a function of atomizer geometry. The water velocity may be estimated with the following equation assuming loss-free water flow in the water nozzle(s):

Eq. (33) $$U_W \approx (2\Delta P_W / \rho_W)^{1/2}$$

Further studies are required to develop more comprehensive and general correlations for water-atomized metal droplets/particles.

Most water-atomized metal particles (powders) have been observed to follow the log-normal size distribution pattern. Relatively narrow size distributions of both fine and coarse particles may be generated by water atomization. A review of published data for droplet size distributions generated by gas and water atomization of a variety of liquid metals and alloys has been made by Lawley,[4] along with presentations of micrographs of surface morphology and internal microstructure of solidified particles.

4.3.3 Centrifugal Atomization

Correlations have been developed for metal droplet sizes generated by centrifugal atomization, as listed in Tables 4.19 and 4.20. Similarly to normal liquids, centrifugal atomization of melts may occur in three regimes: *(1)* Direct Droplet Formation, *(2)* Ligament Disintegration, and *(3)* Film/Sheet Disintegration, as depicted in Fig. 4.3.[320] Champagne and Angers[320] studied the atomization of Al, Cu, Fe, Zn, and steel, and proposed a number group X:

Eq. (34) $$X = X_n / X_d$$

with

Eq. (35) $$X_n = \dot{V}_L \omega^{0.6} d^{-0.68} \quad , \quad X_d = \sigma^{0.88} \mu_L^{-0.17} \rho_L^{-0.71}$$

This number group may be used to determine in which regime a centrifugal atomization takes place for given operation conditions and material properties. Since the group is not dimensionless and the plot was made using SI units, it should be used with caution. From Fig. 4.3,[320] it can be seen that for a given liquid metal/alloy at a given temperature (i.e., given material properties), a centrifugal atomization process may switch from Direct Droplet regime to Ligament regime, up to Film regime with increasing liquid flow rate, rotational speed, and/or decreasing electrode or disk diameter. For given operation conditions (liquid flow rate, rotational speed, and electrode or disk diameter), the reverse transition may occur, i.e., a centrifugal atomization process may switch from Film regime to Ligament regime, down to Direct Droplet regime with increasing surface tension and/or decreasing liquid viscosity and density. The transition from Direct Droplet regime to Ligament regime, or Ligament regime to Film/Sheet regime occurs at $X = 0.07$ or $X = 1.33$,[320] respectively.

Table 4.19. Correlations for Mean Droplet Sizes of Liquid Metals in Centrifugal Atomization

Correlations	Process Characteristics & Remarks	References
$\text{MMD} = \dfrac{K}{\omega}\left(\dfrac{1}{d}\right)^{1/2}$	For atomization in both Direct Droplet and Ligament regimes	Yule & Dunkley [5]
$\text{MMD} = \dfrac{1}{\pi\omega_{rpm}}\left(\dfrac{3\sigma}{d\rho_L}\right)^{1/2}$	Theoretical particle diameter	Schmitt [489]
$\text{MMD} = \left(\dfrac{a\sigma}{\rho_L\omega^2 d/2[1-(\omega d/2)^b\,\mu_L]}\right)^{1/2}$	Semi-empirical correlation for rotating disk atomization or REP; Fitted well with data of Merl76 alloy with $a = 4.8$, $b = 0.93$, $d = 90$ mm, $5000 \leq \omega \leq 24000$ rpm	Tornberg [486]

Table 4.20. Correlations for Droplet Sizes of Liquid Metals in Rotating Electrode Atomization (REP)

<table>
<tr><th>Correlations</th><th>Process Characteristics & Remarks</th><th>Refs.</th></tr>
<tr><td>$D_{10} = 1.77\times10^{6}\omega^{-1.02}d^{-0.48}\left(\frac{\sigma}{\rho_L}\right)^{0.53}\dot{V}_L^{-0.04}$
$D_{20} = 2.17\times10^{6}\omega^{-1.03}d^{-0.51}\left(\frac{\sigma}{\rho_L}\right)^{0.53}\dot{V}_L^{-0.03}$
$D_{30} = 2.50\times10^{6}\omega^{-1.03}d^{-0.53}\left(\frac{\sigma}{\rho_L}\right)^{0.52}\dot{V}_L^{-0.01}$
$D_{21} = 2.65\times10^{6}\omega^{-1.05}d^{-0.55}\left(\frac{\sigma}{\rho_L}\right)^{0.53}\dot{V}_L^{-0.01}$
$D_{31} = 2.96\times10^{6}\omega^{-1.03}d^{-0.56}\left(\frac{\sigma}{\rho_L}\right)^{0.52}\dot{V}_L^{0}$
$D_{32} = 3.3\times10^{6}\omega^{-1.03}d^{-0.58}\left(\frac{\sigma}{\rho_L}\right)^{0.50}\dot{V}_L^{0.02}$
$D_{43} = 3.65\times10^{6}\omega^{-1.0}d^{-0.58}\left(\frac{\sigma}{\rho_L}\right)^{0.46}\dot{V}_L^{0.06}$</td><td>Derived from data for REP of Armco iron, SAE 1090 steel, Cu, Al and Zn.

ω, d and $\dot{V}_L$ are the angular velocity (rad/s), diameter (m) and melting rate (m^3/s) of anode, respectively.

Droplet sizes are in µm.

No effect of viscosity</td><td>Champagne & Angers [320]</td></tr>
<tr><td>$D_{min} = 0.74\times10^{6}\omega^{-0.95}d^{-0.44}\left(\frac{\sigma}{\rho_L}\right)^{0.47}\dot{V}_L^{0.01}$
$D_{max} = 2.91\times10^{6}\omega^{-0.85}d^{-0.43}\left(\frac{\sigma}{\rho_L}\right)^{0.38}\dot{V}_L^{0.1}$</td><td>$D_{min}$ and D_{max} correspond to values above and below which 99% of the particles of a given distribution are found on a weight basis. No effect of viscosity</td><td>Champagne & Angers [320]</td></tr>
<tr><td>$D_{32} = 3.34\times10^{6}\frac{\sigma^{0.48}\mu_L^{0.02}\dot{V}_L^{0.03}}{\rho_L^{0.5}\omega^{1.02}d^{0.58}}$</td><td>Effect of viscosity considered within limited range of viscosity values</td><td>Champagne & Angers [320]</td></tr>
</table>

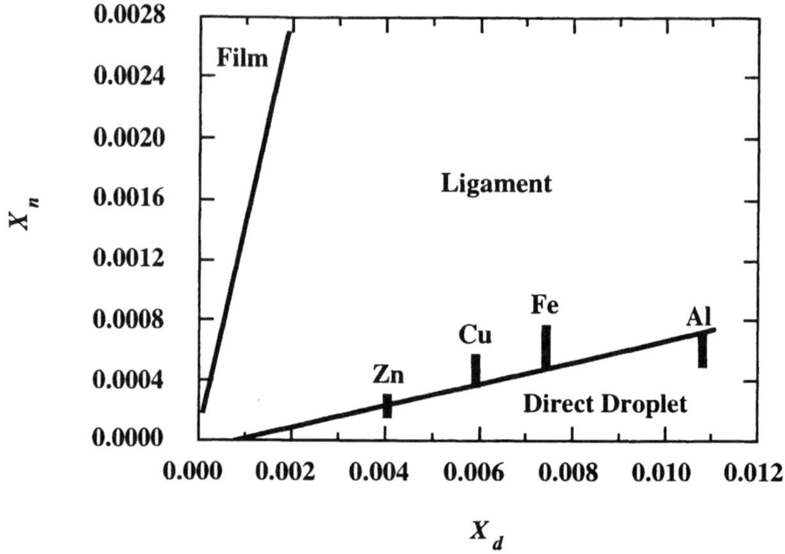

Figure 4.3. Regimes in centrifugal atomization of melts: Direct Droplet Formation, Ligament Disintegration, and Film/Sheet Disintegration.

In the Direct Droplet regime, droplets are in good spherical shape with smooth surfaces and almost free of satellites. Droplet size distribution is typically narrow. In the Ligament regime, droplet size distribution is typically bimodal due to fine satellite droplets, and droplet shape is more ellipsoidal than spherical. In the Film or Sheet regime, droplet size distribution may be as broad as that in twin-fluid atomization processes.[5] Generally, droplet size distribution becomes increasingly polydisperse as the droplet formation process moves from the Direct Droplet regime through the Ligament regime to the Film or Sheet regime. To avoid the Film/Sheet regime and prolong the Ligament regime to high flow rates, one can serrate the outer edge of the cup or disk. The optimum number of teeth, $N_{teeth,}$ has been proposed by Christensen and Steely[113] as:

Eq. (36) $$N_{teeth} = 0.215(\rho_L \omega^2 d^3 / \sigma)^{0.417} (\rho_L \sigma d / \mu_L^2)^{0.167}$$

In both the Direct Droplet and Ligament regimes, the mean droplet size is inversely proportional to the rotational speed ω and the square root of the electrode or disk diameter d:[5]

Eq. (37) $$\text{MMD} = K\omega^{-1} d^{-0.5}$$

This approximate relationship is similar to those for centrifugal atomization of normal liquids in both Direct Droplet and Ligament regimes. However, it is uncertain how accurately the model for K developed for normal liquid atomization could be applied to the estimation of droplet sizes of liquid metals.[5] Tornberg[486] derived a semi-empirical correlation for rotating disk atomization or REP of liquid metals with the proportionality between the mean droplet size, rotational speed, and electrode or disk diameter similar to the above equation. Tornberg also presented the values of the constants in the correlation for some given operation conditions and material properties.

From these correlations, it is clear that the mean droplet size may be controlled by changing rotational speed, and/or disk diameter for a given liquid metal. In addition, the droplet size increases with increasing liquid viscosity and/or surface tension, and/or decreasing liquid density. However, the effect of liquid viscosity is inconclusive. A large disk diameter and/or a high rotational speed are required to obtain a high metal velocity and fine droplets. The technical limitations to spin a large disk at a high rotational speed along with the additional effect of viscosity make it very difficult to produce fine powders *via* centrifugal atomization.[486]

4.3.4 Solidification and Spheroidization

In atomization of melts, the final droplet size also depends on the relative magnitude of the time t_b for a droplet to undergo deformation prior to secondary breakup, and the time t_{sol} required for cooling a superheated metal droplet to its melting temperature and for droplet solidification to complete. The latter time can be estimated using the following expression[319] for a small Biot number and negligible radiative heat transfer:

Eq. (38)
$$t_{sol} = \frac{D\rho_L}{6h_c}\left[c_{pL}\ln\left(\frac{T_D - T_G}{T_m - T_G}\right) + \frac{\Delta H_m}{T_m - T_G}\right]$$

where T_D is the initial droplet temperature and h_c is the heat transfer coefficient. Ranz-Marshall correlation[505] has been frequently used to determine the heat transfer coefficient:

Eq. (39)
$$h_c = \frac{k_G}{D}(2 + 0.6\,\mathrm{Re}^{0.5}\,\mathrm{Pr}^{0.33})$$

where Pr is the Prandtl number. If the breakup time is less than the total solidification time, secondary breakup may occur.

The final droplet/particle shape is determined by the time required for a deformed droplet to convert to spherical shape under surface tension force. If a droplet solidifies before the surface tension force contracts it into a sphere, the final droplet shape will be irregular. Nichiporenko and Naida[488] proposed the following dimensionally correct expression for the estimation of the spheroidization time, t_{sph}:

Eq. (40)
$$t_{sph} = \frac{3}{4}\frac{\pi^2\mu_L}{V_D\sigma}\left(\frac{1}{2}\right)^4\left(D_{end}^4 - D_{init}^4\right)$$

where V_D is the droplet volume, D_{init} is the smallest dimension of the initial, non-spherical droplet, and D_{end} is the final droplet diameter. A more detailed and general model for the estimation of the spheroidization of a droplet has been proposed by Rao and Tallmadge.[506] The model-predicted spheroidization times are not significantly different from those estimated with the equation of Nichiporenko and Naida.[488] Yule and Dunkley[5] warned that the published calculations of spheroidization time should be viewed with caution, and the use of the simplified form of the above equation under the assumption $D_{end}/D_{init} \approx 10$ is not generally recommended. They further pointed out the confusion in the concept of spheroidization time, and suggested that the mechanisms for primary and secondary breakup should be considered separately in terms of the deformation and spheroidization times.

4.4.0 CORRELATIONS FOR DROPLET DEFORMATION CHARACTERISTICS ON A SURFACE

One of the important deformation characteristics of droplets impinging on a surface is the *degree of flattening*, or *flattening ratio*, i.e., the ratio of final splat diameter (maximum spread diameter) D_s to the initial impact droplet diameter D_0. This is also termed *maximum spread factor*, mainly used in characterizing normal-liquid droplet deformation.[411][507] Some analytical models[157][401][508]-[512] have been developed to describe the impact deformation, heat transfer, and/or phase change (solidification or evaporation) of droplets, assuming normal impact on a smooth surface. Although these models can not describe the detailed physical phenomena associated with deformation process, they are relatively simple in general and can correlate important dimensionless deformation parameters to initial impact parameters.

For the deformation of droplets of normal liquids at low impact velocities on a horizontal plane surface without phase change, Tan et al.[513] developed a physical-mathematical model with a droplet falling from a certain height under the influence of gravity. They derived quantitative relations for the dimensions of the deformed droplet, including the effects of initial droplet diameter, height of fall, and thermophysical properties of liquid. In this model, the behavior of droplet deformation was assumed to be governed by the gravitational force, surface tension force, viscous friction force, and interaction with the solid surface. Using a high-speed cine camera, they observed that when a droplet fell onto a horizontal surface from different heights, it flattened out at the initial instant of contact with the surface, reached its maximum size, and then turned to contract until an equilibrium state was reached. The extent of the contraction depends primarily on the height of fall and the surface condition at impact. For a given liquid such as water with $D_0 = 6$ mm at room temperature, the maximum size of the deformed droplet, D_s, was expressed as a function of the height of fall, H, and the initial droplet size, D_0:

$$\text{Eq. (41)} \quad D_s = \begin{cases} D_0 & \text{for } H \leq 0 \text{ and } D_0 < 4\,\text{mm} \\ 3.18 D_0 - 0.009 & \text{for } H \leq 0 \text{ and } D_0 \geq 4\,\text{mm} \\ \exp(2.3 a_B) H^{b_B} & \text{for } H > 0 \end{cases}$$

with

$$a_B = \begin{cases} 158.75 D_0 - 2.33; \quad b_B = 0.17 & \text{for } 0 < H \leq 2\,\text{mm} \\ 158.75 D_0 - 1.82; \quad b_B = 0.36 & \text{for } 2 \leq H \leq 10\,\text{mm} \end{cases}$$

where a_B and b_B are coefficients related to initial droplet diameter. It should be noted that this model predicts that when the droplet diameter decreases to below a certain value, the droplet would not deform, but instead preserve its spherical shape on a surface. Under the conditions of creep flows at very low or zero Reynolds numbers, the interaction with substrate surface, against the surface tension of liquid, may become the driving force for droplet spreading.[342] Under such conditions, the above model would be no longer valid. To determine the relative importance of gravitational force to surface tension force, one may use the *Bond number*, $\mathrm{Bo} = \rho_L D_0^2 g/\sigma$, which is a ratio of the two forces.[514]

For the deformation of droplets of normal liquids on a heated, horizontal, flat surface with attendant phase change, i.e., evaporation, Chandra and Avedisian[411] derived the following equation based on a simplified model:

Eq. (42)
$$\frac{9/2(D_s/D_0)^4}{\mathrm{Re}}+\frac{3\left[(1-\cos\theta_{LS})(D_s/D_0)^2-4\right]}{\mathrm{We}}\approx 1$$

with

$$\theta_{LS}\approx \arctan\left[k_L(T_w-T_b)\upsilon_L/(\Delta H_e|A|)\right]$$

where θ_{LS} is the apparent advancing liquid-solid contact angle at the contact line, T_w is the substrate surface temperature, T_b is the boiling point of the liquid, υ_L is the kinematic viscosity of the liquid, and A is the Hamaker constant. It was suggested that the contact angle θ_{LS} be correlated to substrate surface temperature T_w to account for the effect of the surface temperature on droplet deformation. The influence of surface condition is carried entirely in the apparent contact angle.[385] It was found that this model predicts a maximum spread factor that is 20–40% higher than some experimentally measured data.[380]

Madejski's model and correlations[401] are perhaps the most frequently quoted ones in literature related to surface impact and solidification of melt droplets. In his full model, Madejski considered the effects of viscosity, surface tension, and solidification on droplet flattening, but ignored the compressibility of liquid metal. The flattening ratio was formulated as a function of Re, We and Pe to quantify the viscosity dissipation of inertial force of an impact droplet, conversion of impact kinetic energy to surface energy, and solidification processes, respectively. The relative importance of these three processes can be then determined according to the relative magnitude of the three dimensionless numbers. Madejski's full model consists of a complex integral-differential equation that has no analytical solution, and thus a general formula for the flattening ratio is difficult to find. However, the numerical fits made by Madejski resulted in some relatively simple correlations. In the general case of droplet spreading without solidification, the following equation was proposed for the calculation of the degree of flattening for We > 100 and Re > 100:

Eq. (43) $$\frac{3(D_s / D_0)^2}{\mathrm{We}} + \frac{1}{\mathrm{Re}}\left(\frac{D_s / D_0}{1.2941}\right)^5 = 1$$

Several other special cases have also been analyzed numerically, leading to simplified formulations for the flattening ratio, as detailed below in each model group.

Fukanuma and Ohmori[510] also presented an analytical model for the flattening ratio of a molten metal droplet on a surface as a function of time and compared the model prediction to experimental data for molten tin and zinc droplets. An expression for the flattening ratio was derived based on some simplified assumptions and approximations:

Eq. (44) $$\frac{6(D_s / D_0)^2}{\mathrm{We}} + \frac{1}{\mathrm{Re}}\left(\frac{D_s / D_0}{1.06}\right)^6 = 1$$

Bennett and Poulikakos[380] made a detailed review of the analytical models for droplet flattening ratio. They pointed out that almost all treatments of surface tension effects in the analytical models were fundamentally the same, with differences arising from varying degrees of simplifying assumptions. The splat energy dissipated through surface tension effects was shown to be proportional to the spread factor squared plus the initial droplet surface tension and contact angle effects as considered in some models. The treatments of viscous energy dissipation, however, were dissimilar. The various models yielded viscous energy dissipation as proportional to the spread factor raised to the fourth, fifth, sixth, or eighth power. Madejski's model provides the most theoretically sound treatment of viscous energy dissipation and has the most published experimental support. However, his model effectively ignored the initial surface tension energy of a droplet prior to impact and the contact angle effect during spreading. Bennett and Poulikakos refined Madejski's treatment of the surface tension effects using the treatment proposed by Collings et al.,[514] and Chandra and Avedisian,[411] and obtained the following general equation:

Eq. (45)
$$\frac{3[(1-\cos\theta)(D_s/D_0)^2-4]}{\mathrm{We}}+\frac{1}{\mathrm{Re}}\left(\frac{D_s/D_0}{1.2941}\right)^5=1$$

where θ is the equilibrium liquid-solid contact angle. It was indicated that the dynamic effect of the contact angle on the spread factor is substantial and this effect increases with decreasing Weber number. For large contact angles, i.e., a droplet does not wet the surface, the influence of the Weber number on the spread factor becomes significant.

Recently, Pasandideh-Fard et al.[366] developed a simple analytical model to predict the maximum spread factor on a flat solid surface:

Eq. (46)
$$D_s/D_0=\left(\frac{\mathrm{We}+12}{3(1-\cos\theta_a)+4(\mathrm{We}/\mathrm{Re}^{1/2})}\right)^{1/2}$$

where θ_a is the advancing contact angle. The advancing contact angle increases linearly with contact line velocity until an upper limit value is reached. Beyond that, the contact angle is independent of further increases in the velocity. A comparison of the model predictions with experimental measurements for a variety of droplet-surface combinations and over a wide range of Weber number ($26 < \mathrm{We} < 641$) and Reynolds number ($213 < \mathrm{Re} < 35339$) showed that the discrepancy is less than 15% in most cases and is largest at low Re. For $\mathrm{We} >> \mathrm{Re}^{1/2}$, the capillary effect becomes negligible so that $D_s/D_0 = 0.5\mathrm{Re}^{0.25}$.

In a subsequent study, Pasandideh-Fard et al.[367] derived a simple analytical model to estimate the maximum spread factor for a droplet with simultaneous deformation and solidification on a flat surface:

Eq. (47)

$$D_s / D_0 = \left(\frac{\mathrm{We} + 12}{3(1-\cos\theta) + 4(\mathrm{We}/\mathrm{Re}^{1/2}) + \mathrm{We}[3\mathrm{St}/(4\mathrm{Pe})]^{1/2}} \right)^{1/2}$$

where the Stefan number and the Peclet number are defined as $\mathrm{St} = c_{pL}(T_m - T_w)/\Delta H_m$ and $\mathrm{Pe} = u_0 D_0 / a_L$, respectively. In these equations, T_w is the initial substrate temperature and a_L is the thermal diffusivity of liquid droplet. In the above model, the following simplifying assumptions have been made: *(a)* The heat transfer is by one-dimensional heat conduction; *(b)* The substrate is isothermal; *(c)* The thermal contact resistance is negligible; *(d)* The Stefan number is small; and *(e)* The dimensionless deformation time is assumed to be 2.67. The magnitude of the term $\mathrm{We}[3\mathrm{St}/(4\mathrm{Pe})]^{1/2}$ in the above model determines whether solidification influences droplet spreading. The kinetic energy loss due to solidification will be too small to affect the extent of droplet deformation if $(\mathrm{St}/\mathrm{Pr})^{0.5} << 1$, where $\mathrm{Pr} = \mathrm{Pe}/\mathrm{Re} = \nu_L a_L$. Discrepancies between the model predictions and experiments may arise due to the assumptions made in deriving the model, i.e., the negligible thermal contact resistance and the isothermal substrate. Analysis of splat cooling revealed that the effect of thermal contact resistance is negligible only if $\mathrm{Bi} = hD_0/k_L > 30$. Thus, this model gives an upper bound on the thickness of solidified layer and the actual magnitude may be significantly lower. Hence, the model underestimates the maximum spread factor.

Overall, the available analytical models may be conveniently divided into four groups in terms of basic assumptions made and physical processes considered:

(1) *Viscous Dissipation Domain*: The decay of kinetic energy of an impacting droplet is due to viscous dissipation during flattening.

(2) *Surface Tension Domain*: The kinetic energy of an impacting droplet is converted to surface energy during spreading.

(3) *Solidification Domain*: Cooling and solidification of an impacting droplet occur simultaneously during flattening.

(4) *Partial Solidification prior to Impact*: Partial solidification of an impacting droplet prior to impact is considered.

For droplet spreading processes without attendant phase change, two primary domains have been identified, i.e., viscous dissipation domain and surface tension domain, which are discriminated by the principal mechanism responsible for arresting droplet spreading. In the viscous dissipation domain, the degree of flattening of a droplet is dominated by viscous energy dissipation, whereas in the surface tension domain, it is dominated by surface tension effects. According to Bennett and Poulikakos,[380] the condition under which surface tension effects dominate the termination of droplet spreading may be expressed as:

Eq. (48) $$\mathrm{We} < 2.80\,\mathrm{Re}^{0.457}$$

If the dominating domain is selected correctly, the error induced by the simplification will be no more than about 20%. However, even well into the viscous dissipation domain, the effects of the surface tension are still significant, while in the surface tension domain, the effects of viscous dissipation disappear far more rapidly as one moves away from the borderline. In other words, the viscous energy dissipation contribution to the spread factor rapidly declines within the surface tension-dominated domain, while significant residual surface tension effects extend well into the viscous energy dissipation domain.

The correlations derived from the analytical models, numerical modeling, and experimental results are listed in Table 4.21. The dimensionless numbers used to describe the droplet deformation processes are summarized in Table 4.22. In these tables, T_{Leid} is the Leidenfrost temperature, ρ_S is the density of solidified material, U is the freezing constant, u_s is the speed of sound in liquid, a_S is the thermal diffusivity of solidified material, ΔT is the difference between droplet temperature and melting/liquidus temperature for solidification, or substrate surface temperature and vapor temperature for evaporation, $\dot{T}$ is the splat cooling rate, and ΔH_e is the latent heat of evaporation.

Table 4.21a. Correlations for Final Splat Diameter Created by a Droplet Impinging on a Solid Surface

Correlations	Models	References
Group 1: *Decay of Kinetic Energy via Viscous Dissipation* $D_s / D_0 = 1.16\text{Re}^{0.125}$	Analytical	Jones[508]
$D_s / D_0 = 1.2941(\text{Re} + 0.9517)^{0.2}$, for $\text{We}^{-1}=0$ $D_s / D_0 = 1.2941\text{Re}^{0.2}$, for $\text{We}^{-1}=0$, Re>100	Analytical-Numerical	Madejski [401]
$D_s / D_0 = 1.0\text{Re}^{0.2}$	Numerical	Trapaga & Szekely [515]
$D_s / D_0 = 0.83\text{Re}^{0.2}$	Numerical	Hamatani et al. [516]
$D_s / D_0 = 0.82\text{Re}^{0.2}$	Numerical	Watanabe et al. [517]
$D_s / D_0 = 1.04\text{Re}^{0.2}$	Numerical	Liu et al. [18]
$D_s / D_0 = 1.06\text{Re}^{0.17}$, for $\text{We}^{-1}=0$	Analytical	Fukanuma & Ohmori [510]
$D_s / D_0 = 0.925\text{Re}$	Numerical	Bertagnolli et al. [518]
$D_s / D_0 = 0.5\text{Re}^{0.25}$	Analytical	Pasandideh-Fard et al. [366]

Table 4.21b. Correlations for Final Splat Diameter Created by a Droplet Impinging on a Solid Surface

Correlations	Models	References
Group 2: *Conversion of Kinetic Energy to Surface Energy* $D_s / D_0 = (\mathrm{We}/3)^{0.5}$, for Re^{-1}=0, We>100	Analytical-Numerical	Madejski [401]
$\mathrm{We} = 3(D_s / D_0)^2 + 8/(D_s / D_0) - 11$, no limit on We	Analytical	Dykhuizen [390]
$D_s / D_0 = 0.816\mathrm{We}^{0.25}$	Analytical	Cheng [509]
$D_s / D_0 = (\mathrm{We}/6 + 2)^{0.5}$	Analytical	Ueda et al. [519]
$D_s / D_0 = [(\mathrm{We}/3)/(1-\cos\theta_{LS})]^{0.5}$, for large Re $D_s / D_0 = (\mathrm{We}/6)^{0.5}$, for large Re and θ_{LS}=180°	Analytical	Collings et al. [514]
$D_s / D_0 \approx [(\mathrm{We}/3+4)/(1-\cos\theta_{LS})]^{0.5}$, for Re^{-1}=0 $D_s / D_0 \approx (\mathrm{We}/3+4)^{0.5}$, for Re^{-1}=0, $T_w > T_{\mathrm{Leid}}$ (θ_{LS}=180°)	Analytical	Chandra & Avedisian [411]
$D_s / D_0 = 1 + 0.463\mathrm{We}^{0.345}$, on hot surface	Experimental	Senda et al. [422]

Table 4.21c. Correlations for Final Splat Diameter Created by a Droplet Impinging on a Solid Surface

Correlations	Models	References
Group 3: *Simultaneous Cooling and Solidification* $D_s / D_0 = 1.4991(\rho_L / \rho_S)^{0.395}(\mathrm{Pe}/U^2)^{0.1975}$, for $\mathrm{Re}^{-1}=\mathrm{We}^{-1}$=0	Analytical-Numerical	Madejski [401]
$D_s / D_0 = 0.82(\mathrm{Pe}/U^2)^{0.25}$	Analytical	Dykhuizen [390]
$D_s / D_0 = \begin{cases} 0.8546\chi^{0.5}\mathrm{Re}^{0.25}[1+0.34\beta\mathrm{Re}^{0.5}\ln(0.3\mathrm{Re})] & \text{smooth} \\ 0.8546\chi^{0.5}\mathrm{Re}^{0.25}[1-0.06\omega\alpha^{0.5}\mathrm{Re}^{0.5} + 0.34\beta\mathrm{Re}^{0.5}\ln(0.3\mathrm{Re})] & \text{rough} \end{cases}$	Analytical	Sobolev et al. [511]
$D_s / D_0 = \left(\dfrac{\mathrm{We}+12}{3(1-\cos\theta) + 4(\mathrm{We}/\mathrm{Re}^{0.5}) + \mathrm{We}[3\mathrm{St}/(4\mathrm{Pe})]^{0.5}} \right)^{0.5}$	Analytical	Pasandideh-Fard et al. [367]
Group 4: *Partial Solidification prior to Impact*	Analytical-Numerical	San Marchi et al. [157]

Table 4.22a. Dimensionless Numbers Used to Describe Droplet Deformation Processes with Simultaneous Heat Transfer and Phase Change (Solidification or Evaporation)

Applications	Numbers	Definitions	Remarks	References
Deformation & Spreading	Reynolds Number	$\text{Re} = \rho_L u_0 D_0 / \mu_L$	Compare inertial force to viscous force	Madejski [401]
	Weber Number	$\text{We} = \rho_L u_0^2 D_0 / \sigma$	Compare inertial force to surface tension force	Madejski [401]
	Froude Number	$\text{Fr} = u_0 / (D_0 g)^{0.5}$	Compare inertial force to gravitational force	Tsurutani et al. [334]
	Mach Number	$\text{Ma} = u_0 / u_s$	Compare impact velocity to speed of sound in liquid	Hofmeister et al. [410]
	Bond Number	$\text{Bo} = \rho_L D_0^2 g / \sigma$	Compare gravitational force to surface tension force	Collings et al. [514]
Splashing	Capillary Number	$\text{Ca} = \text{We/Re} = u_0 \mu_L / \sigma$	Compare inertial-viscous force to surface tension force	Yarin & Weiss [357]
	Ohnesorge Number	$\text{Oh} = \mu_L / (\rho_L \sigma D_0)^{0.5}$	Compare internal viscosity force to surface tension force	Walzel [398]

Table 4.22b. Dimensionless Numbers Used to Describe Droplet Deformation Processes with Simultaneous Heat Transfer and Phase Change (Solidification or Evaporation)

Applications	Numbers	Definitions	Remarks	References
Heat Transfer	Peclet Number	$\text{Pe} = u_0 D_0 / a_S$	Compare heat removal by flow to heat extraction by conduction	Madejski [401]
	Impact Number	$\text{I} = \dfrac{\pi \rho_L D_0^2 u_0^3}{12 A_c h (T_L - T_w)}$	Compare kinetic energy transfer rate to heat extraction rate	Matson et al. [409]
Solidification	Stefan Number	$\text{St} = c_{pL} \Delta T / \Delta H_m$	Compare sensible to latent heat in liquid-solid phase change	Matson et al. [409]
	Freezing Number	$\text{F} = [\Delta H_m (1 - f_s) + c_{pL} \Delta T] \dfrac{\pi u_0 \rho_L D_0^2}{6 A_c h (T_L - T_w)}$	Compare solidification time to impact time	Matson et al.[409]
	Nucleation Number	$\text{N} = 0.5 \Delta H_m u_0 / (D_0 c_{pL} \dot{T})$	Compare nucleation time to impact time	Dykhuizen [390]
Evaporation	Jakob Number	$\text{Ja} = c_{pL} \Delta T / \Delta H_e$	Compare sensible to latent heat in liquid-vapor phase change	Tio & Sadhal [414]

4.4.1 Viscous Dissipation Domain

One of the earliest analytical models for the calculation of flattening ratio of a droplet impinging on a solid surface was developed by Jones.[508] In this model, the effects of surface tension and solidification were ignored. Thus, the flattening ratio is only a function of the Reynolds number. Discrepancies between experimental results and the predictions by this model have been reported and discussed by Bennett and Poulikakos.[380]

In case that the decay of impact kinetic energy due to viscous dissipation is the predominant mechanism in droplet flattening, Madejski's full model reduces to:

Eq. (49) $$D_s / D_0 = 1.2941(\mathrm{Re} + 0.9517)^{0.2}$$

This is sometimes referred to as Madejski's flow model in literature. In a subsequent study, Madejski[520] carried out calculations using the same model for a large range of Reynolds and Weber numbers. It was indicated that the model is applicable to plasma spraying of ceramic powder for surface coating. For Al_2O_3 powder, for example, a flattening ratio of 5–6 can be achieved for $1000 \leq \mathrm{We} \leq 10000$ and $4000 \leq \mathrm{Pe} \leq 8000$. Other application fields include: (1) Boiler tubes at mist flow and post-critical conditions ($50 \leq \mathrm{We} \leq 300$, $200 \leq \mathrm{Re} \leq 20000$); (2) Last stages of steam turbines at mist flow conditions after condensation shock ($20 \leq \mathrm{We} \leq 50$, $50 \leq \mathrm{Re} \leq 200$); (3) Rain erosion of aircraft ($50000 \leq \mathrm{We} \leq 300000$, $40000 \leq \mathrm{Re} \leq 200000$); (4) Aircraft icing ($500 \leq \mathrm{We} \leq 3500$, $400 \leq \mathrm{Re} \leq 1400$); (5) Splat quenching of alloys (mainly aluminum alloys) ($50 \leq \mathrm{We} \leq 500$, $500 \leq \mathrm{Re} \leq 10000$).

An analysis very similar to Madejski's study[401] has been conducted by Fiedler and Naber[521] for normal liquids in combustion applications from which solidification phenomenon is absent. The derived results are almost identical to Madejski's model. Markworth and Saunders[522] presented a substantial improvement upon the velocity field used by Madejski, and demonstrated a correspondingly improved model prediction.

Trapaga and Szekely[515] conducted a detailed numerical analysis of droplet flattening process on a flat surface. A fit to the numerical results yielded a correlation similar to Madejski's flow model, but with a slightly smaller proportional coefficient. Hamatani et al.[516] employed a numerical algorithm to model the impact of alumina droplets assuming a constant droplet temperature during deformation and investigated the effects of droplet size, velocity and viscosity on the deformation process. They also derived a correlation similar to Madejski's flow model, but with an even smaller coefficient. A very similar correlation has been reported by Watanabe et al.[517]

Overall, if the decay of kinetic energy of an impacting droplet is due to viscous dissipation during flattening, the corresponding models in this group predict that a large flattening ratio can be obtained for large droplets with high density and/or low viscosity at high impact velocity. For large Weber numbers, the droplet flattening behavior is dependent only on the Reynolds number. A notable example is thermal spray applications, in which surface tension forces are shown to be unimportant to droplet flattening processes.[401][508]

4.4.2 Surface Tension Domain

For droplets of high surface tension, the droplet flattening process may be governed by the transformation of impact kinetic energy to surface energy. In case that this mechanism dominates, the flattening ratio becomes only dependent on the Weber number, as derived by Madejski by fitting the numerical results of the full analytical model:

Eq. (50) $$D_s / D_0 = (\mathrm{We}/3)^{0.5}$$

The values of the Weber number pertinent to this correlation were suggested to be sufficiently small so that viscous and solidification effects can be neglected. Another analytical expression, derived from Madejski's full model after simplification under the conditions that surface tension dominates the droplet flattening, has no limitation on the value of the Weber number:[390]

Eq. (51) $$\mathrm{We} = 3(D_s / D_0)^2 + 8/(D_s / D_0) - 11$$

As pointed out by Dykhuizen,[390] the assumptions made in Madejski's model in terms of surface energy may be questionable. The model ignored the surface energy associated with the liquid-solid and gas-solid interfaces, and included only the liquid-gas interface. Thus, the flattening process in the model is independent of the liquid contact angle. In reality, however, the droplet flattening behavior is intimately related to the contact angle in flattening processes at low Weber numbers.[380][391] Therefore, Madejski's model is valid only for the condition of 90° contact angle.[390] Under such special condition, the ignored two surface energy terms will vanish. In general, the contact angle is a complicated function of contact-line speed.[391] The contact line is a curve formed when an interface between two immiscible fluids (for example, gas and liquid) intersects a solid. The effect of the contact line on droplet spreading has been investigated by Haley and Miksis[523] on the basis of lubrication theory. They considered several slip coefficients and relationships between contact-line speed and contact angle, and found that the spreading rates strongly depend on the relationships, although the qualitative features of the droplet spreading are similar for the different relationships.

Cheng[509] studied dynamic spreading of droplets impacting onto a solid surface without wetting and solidification. A significantly different expression was derived for the maximum degree of flattening under the conditions where surface tension is the only retarding force to droplet spreading:

Eq. (52) $$D_s / D_0 = 0.816\mathrm{We}^{0.25}$$

The correlations derived by Chandra and Avedisian,[411] and Collings et al.[514] are similar to Madejski's equation[401] except that the former include the effect of contact angle. However, there are some problems associated with these correlations, and discrepancies between experimental results and the predictions by these correlations have been reported, as discussed in detail by Bennett and Poulikakos.[380] For example, Chandra and Avedisian's model[411] predicted a maximum spread factor that is 20–40% higher than some experimentally measured results. This discrepancy was attributed to the deficient model for viscous energy dissipation. The model of Collings et al.[514] predicted a maximum spread factor 100% too small. The grossly large discrepancy may indicate that the treatment of the equilibrium contact angle in their model was incorrect.[380] These also demonstrated the importance of contact angle consideration and inclusion of the initial droplet surface tension in analytical models for predicting the flattening ratio.

4.4.3 Solidification Domain

In Madejski's full model,[401] solidification of melt droplets is formulated using the solution of analogous Stefan problem. Assuming a disk shape for both liquid and solid layers, the flattening ratio is derived from the numerical results of the solidification model for large Reynolds and Weber numbers:

Eq. (53) $$D_s / D_0 = 1.4991(\rho_L / \rho_S)^{0.395} (\mathrm{Pe} / U^2)^{0.1975}$$

Dykhuizen[390] introduced a simplified model:

Eq. (54) $$D_s / D_0 = 0.82(\mathrm{Pe} / U^2)^{0.25}$$

which compares well with Madejski's solidification model.

To use these models, the freezing constant, *U*, must be determined. One choice is the solution of the Stefan problem of solidification, as described by Madejski:[401]

Eq. (55)

$$\frac{U}{2}\sqrt{\pi} = \frac{T_0}{\left[\operatorname{erf}\frac{U}{2} + \sqrt{\frac{(kc_p\rho)_S}{(kc_p\rho)_W}}\right]\exp\left(\frac{U^2}{4}\right)} - \frac{T_p\sqrt{(kc_p\rho)_L/(kc_p\rho)_S}}{\operatorname{erfc}\left(\frac{U}{2}\sqrt{\frac{a_S}{a_L}}\right)\exp\left(\frac{U^2}{4}\frac{a_S}{a_L}\right)}$$

where $a = k/c_p\rho$ is the thermal diffusivity, $T_0 = k_S(T_m - T_w)/(a_S\rho_S\Delta H_m)$ and $T_p = k_S(T_L - T_m)/(a_S\rho_S\Delta H_m)$ are the dimensionless substrate temperature and dimensionless droplet temperature, respectively. Subscripts L, S, and W denote liquid droplet, solidified material, and substrate, respectively. This equation is derived for 1-D conduction-limited solidification of a molten layer at an initial temperature T_L that comes in contact with a substrate at an initial temperature T_w. The equation dictates that splat solidification is dependent on substrate thermal properties, conflicting with many experimental data. Thus, Madejski assumed that the substrate remained isothermal during the cooling process that requires the freezing constant be calculated from the following equation:

Eq. (56) $$T_0 = \sqrt{\pi}\frac{U}{2}\operatorname{erf}\frac{U}{2}\exp(U^2/4)$$

To rationalize the isothermal assumption, Dykhuizen[390] discussed two related physical phenomena. First, heat may be drawn out of the substrate from an area that is much larger than that covered by a splat. Thus, the 1-D assumption in the Stefan problem becomes invalid, and a solution of multidimensional heat conduction may make the interface between a splat and substrate closer to isothermal. Second, the contact resistance at the interface is deemed to be the largest thermal resistance retarding heat removal from the splat. If this resistance does not vary much with substrate material, splat solidification should be independent of substrate thermal properties. Either of the phenomena would result in a heat-transfer rate that is less dependent on the substrate properties, but not as high as that calculated by Madejski based on the isothermal assumption. Thus, for most thermal spray conditions, Madejski's solidification model predicts a splat that is smaller than those predicted by his flow and surface-tension models. This suggested that solidification process acted faster than viscous dissipation process, and hence, were the predominant mechanism in limiting splat flattening. On the contrary, Jones[508] predicted a solidification rate that is significantly lower due to thermal contact resistance, dismissing the solidification mechanism in favor of viscous dissipation mechanism. Jones' scaling argument is often quoted for thermal spray conditions where the existence of a contact resistance is much more important than in millimeter-sized droplet experiments. The contact resistance has been calculated by Moreau et al.[524] on the order of 10^{-6} Km^2/W for plasma-sprayed molybdenum on various substrates, and measured by Fantassi et al.[406] as 2×10^{-6} Km^2/W, which are in general agreement with the values (10^{-4} to 10^{-6} Km^2/W) assumed by Clyne,[155] and (2×10^{-5} Km^2/W) by Jones.[508]

In addition, Madejski's solidification model did not account for any time delay for nucleation of solid phase due to contact resistance and undercooling of liquid phase. The nucleation delay due to undercooling is common in rapid solidification processes, and may be an order of magnitude longer than impact time.[155] Nucleation delay becomes increasingly important as the droplet size decreases and the impact velocity increases. Thus, Dykhuizen[390] suggested to introduce an additional dimensionless number to determine when contact resistance and nucleation delay are important. Dykhuizen further proposed this number be formulated as a ratio of nucleation time $(\Delta T + 0.5\Delta H_m/c_{pL})/\dot{T} \approx (0.5\Delta H_m/c_{pL})/\dot{T}$ to impact time D_0/u_0 (Table 4.22). In the present book, this number will be referred to as *Nucleation number*. The Nucleation number is identical to the Freezing number proposed by Matson et al.[409] per se. If the value of this number is small, nucleation delay needs not to be considered. In millimeter-sized droplet experiments, the Nucleation number is small, but it is five orders of magnitude larger in thermal spray applications[390] where micrometer-sized droplets impact substrate at high speeds. Hence, extrapolation of experimental results for millimeter-sized droplets to thermal spray applications may not be easily justified.

Bennett and Poulikakos[380] also emphasized the important effect of droplet undercooling on solidification, although they indicated that solidification does not contribute significantly to terminating droplet spreading and the effect of solidification on arresting the spreading of a superheated droplet is likely to be secondary compared to the effects of viscous dissipation and surface tension. They further indicated that consideration of the undercooling encountered in rapid solidification dictates that the reduction in freezing time due to higher solidification speed is offset by the delay in nucleation, although rapid solidification that is typical of splat-quenching can produce much faster crystal growth kinetics than predicted by the Stefan solution.

Sobolev et al.[511] conducted a series of analytical studies on droplet flattening, and solidification on a surface in thermal spray processes, and recently extended the analytical formulas for the flattening of homogeneous (single-phase) droplets to composite powder particles. Under the condition Re >> 1, the flattening ratios on smooth and rough surfaces are formulated as:

Eq. (57)

$$D_s / D_0 = \begin{cases} 0.8546\chi^{0.5}\,\mathrm{Re}^{0.25}[1+0.34\beta\mathrm{Re}^{0.5}\ln(0.3\mathrm{Re})] & \text{smooth} \\ 0.8546\chi^{0.5}\,\mathrm{Re}^{0.25}[1-0.06\omega\alpha^{0.5}\mathrm{Re}^{0.5} & \\ \qquad\qquad +0.34\beta\,\mathrm{Re}^{0.5}\ln(0.3\mathrm{Re})] & \text{rough} \end{cases}$$

where χ is the ratio of the remaining droplet mass after mass loss due to splashing to the initial droplet mass, $\alpha = \varepsilon/(D_0/2)$, ε is the roughness height, $\beta = V_s/u_0$, V_s is the solidification speed of the splat lower part due to the heat removal through the substrate, $\omega = V_s(D_0/2)/(u_0\delta)$, and δ is the thickness of the splat lower part.

4.4.4 Partial Solidification Prior to Impact

Madejski's solidification model did not account for partial solidification of a droplet prior to impact. San Marchi et al.[157] modified some of the assumptions in Madejski's model and addressed the effects of different solid fractions of a droplet prior to impact on its flattening and solidification behavior. The modeling results showed that under the conditions typical of thermal spray processes, the impact kinetic energy of a droplet governs its spreading process, and increasing the solid fraction (or reducing the amount of liquid) reduces the extent of the droplet spreading. Partial solidification of a droplet prior to impact reduces the kinetic energy of the remaining liquid through reducing the volume or mass of the liquid. Thus, with increasing solid fraction, the kinetic energy of the liquid decreases, leading to a decrease in the spreading extent. The results also showed that the effect of the solid fraction on the decrease in the spreading extent is essentially independent of material systems considered. This further demonstrated that the partial solidification of a droplet prior to impact influences its spreading behavior primarily through reducing the liquid volume or mass. Overall, the partial solidification prior to impact does not affect the droplet deformation and solidification on substrate as much as might be expected. A 10% solid at impact results in a reduction in splat size of less than 4%. This is mainly because of the predominant effect of the impact kinetic energy on the flattening and solidification behavior. In addition, the solid fractions considered in this study were less than 0.4, limiting the generalization of the results. The effect of the solid fraction on the final splat morphology has been experimentally investigated and discussed in Ref. 409.

Further extensions of Madejski's model[401] may include *(a)* turbulence effect, *(b)* Rayleigh instability or Taylor instability and droplet breakup, *(c)* vibrational energy, and *(d)* influence of solidification on flow.[514] Some issues related to the deformation and solidification of droplets on a flat substrate in splat quenching have been addressed in Refs. 380 and 514. To date, analytical models addressing droplet impingement on a semi-solid surface have not been found in available literature.

5

Theoretical Calculations and Numerical Modeling of Droplet Processes

Droplet generation and deformation processes involve complex physical phenomena, such as liquid-gas, or liquid-surface interactions, primary and/or secondary breakup of liquid, droplet dynamics, and in many applications, heat transfer and phase change. To date, no general theoretical treatment of droplet processes is available, except for few simple processes under restricted conditions. Therefore, numerical modeling and simulation have been increasingly employed for analysis of droplet processes and optimization of process designs. Computer modeling is usually preferred over experiments for several reasons. For example, computer modeling typically can be carried out more quickly and with less lead-time and expense than experiments. In addition, it can be much better controlled and used to explore a much wider range of conditions and systems, some of which are physically inaccessible.

In this chapter, basic theoretical calculations and numerical modeling of droplet generation and deformation processes of both normal liquids and melts will be discussed in detail. The review of modeling efforts will outline the current status and recent developments in numerical models and computational methods for the droplet processes. The information will also be useful for understanding various mechanisms governing droplet processes as well as effects of process parameters on droplet properties.

5.1.0 ENERGY REQUIREMENTS AND EFFICIENCY

To generate droplets from a bulk liquid, a certain amount of energy is required to make some area of the liquid surface unstable. The surface then may rupture into fragments, which subsequently disintegrate into droplets. The energy requirement is specific to the technique used. For example, energy is needed to compress the gas used in gas atomization, and to melt and superheat a solid material in atomization of melts. In powder production process *via* atomization, the energy used for melting and superheating is wasted, because it is removed from droplets during flight in spray chamber through heat transfer to surrounding gas, and no heat energy recovery is usually made.

Theoretically, the energy, E, required to generate droplets from a liquid or melt can be roughly estimated by that needed to create the surface area of droplets, i.e., the product of the surface area of droplets and the surface tension of the liquid:

Eq. (1) $$E = \sigma \sum_{i=1}^{i=N} \pi D_i^2$$

where i represents the ith droplet, N is the total number of droplets generated from the liquid of mass m_L in unit time, and D_i is the diameter of the ith droplet. The mass of the liquid atomized in unit time can be formulated as:

Eq. (2) $$m_L = \rho_L \frac{\pi}{6} \sum_{i=1}^{i=N} D_i^3$$

In the above estimation of the energy requirement for atomization, the energy needed to overcome viscous force during liquid breakup is neglected. Under such assumption, the theoretical energy requirement for atomizing unit liquid mass can be calculated using the following equation:[5]

Eq. (3) $$E / m_L = \frac{6\sigma}{\rho_L D_{32}}$$

The theoretical energy efficiency, η, is defined as:

Eq. (4) $$\eta = \frac{\text{theoretical input energy}}{\text{actual input energy to atomizer}}$$

Neglecting the energy for overcoming viscous force during liquid breakup, a simple equation for the theoretical energy efficiency has been derived by Yule and Dunkley[5] for a pressure-swirl atomizer:

Eq. (5) $$\eta = 600\sigma / (\Delta P D_{32})$$

where the actual input energy is the product of the pressure drop ΔP across the atomizer nozzle and the volume flow rate of liquid, which can be expressed as a sum of the volumes of N droplets created in unit time. For swirl jet sprays, the theoretical energy efficiencies are 0.22%, 0.22%, 0.49%, and 0.95% for water, oil, solder, and aluminum, at ΔP of 10, 1, 6.7, and 2.4 MPa, and for D_{32} of 20, 59, 60, 150 μm, respectively.[5] The efficiencies are very small because only a very small portion of the kinetic energy supplied to the liquid is converted into surface energy on the droplets generated. A large percentage of the kinetic energy is consumed by accelerating the liquid and droplets.

Atomization is a process of fast transition of a continuous bulk liquid into a dispersion of the liquid in an ambient gas phase. It involves the change in the partition energies of flowing fluids. The initial energies of the fluids entering an atomizer are partitioned mainly between the pressure-volume and kinetic energies. During atomization, some of the energies are converted into interfacial energy and other dissipative forms of energies such as turbulence, heat, and acoustic energy. Generally, atomization efficiency may be defined as the ratio of interfacial energy output to the total energy input. Thermodynamic analyses may be used to evaluate the energy partition that sets limits to the efficiency of producing interfacial energy, i.e., atomization efficiency.

Thermodynamic analyses of the conversion efficiency of pressure-volume and kinetic energies to interfacial energy during atomization have been conducted by Li and Tankin,[252] and Shellens and Brzustowski.[432] The pressure of the ambient gas and the entropy density of droplets are considered in the analyses. It is showed that increasing the pressure and entropy density leads to an increase in the proportion of bulk energy in the total energy of droplets and a corresponding decrease in atomization efficiency. Cohesive forces become dominant as droplet size decreases, suggesting that there exists a limit to the smallest size of droplets that can be generated by atomization.

Detailed descriptions of the practical aspects of atomizer performance, plant design, and operation for powder production through atomization of liquid metals have been given by Yule and Dunkley.[5]

5.2.0 MODELING OF DROPLET PROCESSES OF NORMAL LIQUIDS

Computer modeling and simulation are aimed at identifying control parameters, establishing detailed correlations between design parameters and spray characteristics, and providing better understanding of real phenomena in droplet processes and guidelines for optimization of processes and designs. The objective of spray combustion modeling, for example, is to develop computer codes that can provide sufficiently accurate descriptions of the physical and chemical processes during spray combustion, and to predict thermal, combustion, and energy efficiencies and emission of pollutants. Due to the highly transient and complex nature, atomization and spray are difficult to model and analyze. Theoretical models of liquid breakup processes from first principles are currently limited to predicting mean droplet size and velocity distributions.[252][432] These models need to be extended to predict local droplet size and velocity distributions. In spray modeling, it is desired to predict and control the conditions at the initial plane of the spray for given atomizer geometry and input liquid and gas flow rates. The long-term goal is to design spray systems from first principles. This will require computations of nonlinear physical processes and coupling of different physical and chemical processes. The ultimate objective of theoretical efforts is to tailor droplet size distributions, and to generate specified distributions, such as mono-size, unimodal, multi-modal, skew, narrow-size, or other specific distributions.

Detailed modeling study of practical sprays requires solving the dynamics of interactions of droplets with surrounding gas. To calculate the mass, momentum, and energy exchanges between the droplets and gas, one must account for the distributions of droplet sizes, velocities, and temperatures. It is also necessary to consider the distortions, breakups, collisions, coalescences, and turbulent dispersions of the droplets, and in many cases the phase changes (vaporization or solidification) of the droplets. Modeling primary breakup of liquid requires information on the liquid and gas flows, such as velocity, shear stress, and turbulent kinetic energy distributions inside atomizer passages as initial conditions for liquid breakup region. In addition, disturbances on liquid surfaces need to be ascertained, along with their wavelengths, amplitude growth rates, and frequencies. To date, these phenomena have been the subjects of intensive theoretical analyses and numerical modeling studies. Comprehensive computer models and numerical methods have been developed and implemented with varied degrees of success. Currently, spray modeling is yet ineffective in dense sprays. An extensive review has been made by Reitz[525] on the current status and recent developments in computer modeling of sprays. A survey has been presented by Chigier[526] on the most recent developments in models and correlations of many droplet systems.

5.2.1 Theoretical Analyses and Modeling of Liquid Jet and Sheet Breakup

The subject of liquid jet and sheet atomization has attracted considerable attention in theoretical studies and numerical modeling due to its practical importance.[527] The models and methods developed range from linear stability models to detailed nonlinear numerical models based on boundary-element methods[528][529] and Volume-Of-Fluid (VOF) method.[530]

The classical stability theories of liquid breakup developed by Taylor in 1940[205] and others are linear theories that involve wave amplitude growth and changes in wavelength. On the basis of this theory framework, Reitz and Bracco[229] conducted an analysis of liquid jet stability to address the droplet formation process by jet breakup. The analysis considered the growth of initial perturbations on the liquid surface and included the effects of liquid inertia, surface tension, viscous, and aerodynamic forces on the jet. A dispersion equation was derived to relate the growth rate of an initial perturbation of infinitesimal amplitude to its wavelength. The relationship also included the physical and dynamical parameters of the liquid jet and the surrounding gas. In this analysis, a column of liquid was assumed to be infinite in the axial direction. Governing equations for mass and momentum conservation were solved with a kinematic jump boundary condition at the interface, i.e., a normal stress balance and a tangential stress balance. The normal stress balance accounted for surface tension, dynamic pressure (inertia), viscous (normal) force, and recoil forces due to the vapor leaving the interface of an evaporating jet.[243] In the tangential stress balance, the gas was typically assumed to be inviscid, i.e., slip condition was presumed at the liquid-gas interface. This assumption may not allow for the presence of shear at the interface in the boundary layer of the gas flow.

Typical solutions of the dispersion equation predicted wave growth rates versus wave number as a function of the Weber and Ohnesorge numbers.[232] The analysis results showed that there is a maximum wave growth rate at a particular wavelength, and the fastest growing (or most probable) waves on the liquid surface are responsible for the breakup. This maximum wave growth rate increases and the corresponding wavelength decreases rapidly with increasing Weber number. Liquid viscosity reduces the wave growth rate, but has a secondary stabilizing effect. Gas compressibility and liquid vaporization destabilize low-speed liquid jets, whereas vaporization stabilizes high-speed liquid jets.[243] Surface evaporation slows the breakup of surface waves and produces larger droplets.

This jet stability theory has been found to provide a reasonably complete description of the mechanisms of low-speed jet breakup and a framework for organizing jet breakup regimes. For high-speed liquid jets, the initial state of the jets appears to be progressively more important and less understood. Reitz[232] suggested that the surface wave mechanism of the second wind-induced regime could still apply to a jet in the atomization regime. However, direct verification of this mechanism by means of experiments has not yet been possible because the dense spray surrounding the jet obscures the breakup details.

The jet stability theory has also been used to model liquid jet breakup by considering the unstable growth of waves on a distorting liquid surface.[232] In this model, surface waves grow and break up when a breakup time is reached, producing droplets with sizes that are related to the wavelength of the most unstable wave. The breakup time is related to the wave growth rate. The model is implemented by injecting *parcels of liquid*, called *blobs* or *bulk liquid masses*, which have a characteristic size equal to the nozzle orifice diameter. The blobs distort and new blobs are formed from the liquid that is stripped from the parent blobs as a result of the wave breakup process. It is assumed that the sizes of small droplets are proportional to the wavelength of the fastest-growing or most unstable surface wave, while the sizes of large droplets are determined from the volume of the liquid contained in one surface wave, assuming one droplet is formed each period with a presumed disturbance frequency.

The most comprehensive theoretical treatment of jet atomization is perhaps that by Ponstein.[531] However, the number of published studies addressing the effect of swirl on liquid jet atomization has been very limited. Kang and Lin[532] studied the spatial instability of a swirling liquid jet including the effect of nonaxisymmetric disturbances. Lian and Lin[242] investigated the convective instability of a viscous liquid jet interacting with a swirling inviscid gas and found that gas swirl hinders jet atomization. Methods for solving the problems of secondary atomization, droplet size distribution and properties of multiple jets have been suggested by different researchers, but none of these problems has been solved satisfactorily.[220]

All these previous studies on liquid jet atomization were mainly based on the linear theory of hydrodynamic stability. The problem associated with the linear stability analyses lies in the assumption that the perturbations are infinitesimal. In reality, unstable perturbations will grow to finite quantities after a finite interval of time. Liquid breakup, droplet formation and secondary breakup are all nonlinear phenomena that can no longer be treated by the linearizing method of small disturbances since the amplitude of the disturbance at breakup is of the same order as the jet radius.[220] Such nonlinear phenomena may be analyzed by, for example, bifurcation theory that predicts bursting and chaotic formation of droplets, as suggested by Chigier.[211] In addition, liquid breakup may also be analyzed on the basis of thermodynamic theories that relate the input kinetic energies of gas and liquid to the final state of droplets generated.[252][432]

Alternative methods of analysis have been examined and evaluated. Shokoohi and Elrod[533] solved the Navier-Stokes equations numerically in the axisymmetric form. Bogy[527] used the Cosserat theory developed by Green.[534] Ibrahim and Lin[535] conducted a weakly nonlinear instability analysis. The method of strained coordinates was also examined. In spite of the mathematical or computational elegance, all of these methods suffer from inherent complexity. Lee[536] developed a 1-D, nonlinear direct-simulation technique that proved to be a simple and practical method for investigating the nonlinear instability of a liquid jet. Lee's direct-simulation approach formed the basis of a comprehensive treatment of the jet instability and atomization presented by Chuech et al.[537] Ibrahim[241] further extended the direct-simulation analysis of Chuech et al. to include the effect of swirl on jet atomization. It was found that liquid swirl enhances jet atomization.

Under most practical conditions, relatively small satellite droplets may form, interspersed among the main droplets, during the disintegration of a liquid jet. Because the formation of the satellite droplets is a nonlinear phenomenon, a linear theory can not adequately describe the formation mechanism of the satellite droplets. Ibrahim[538] studied the modification of the interfacial profile of a liquid jet caused by swirl on the basis of the nonlinear direct-simulation technique of Chuech et al.[537] The simulation results suggested that the effect of gas swirl contradicts that of liquid jet swirl. Thus, swirling gas would hinder liquid jet atomization. In addition, liquid swirl causes the satellite droplets to merge forward with the main droplets.

Recently, Spangler, Hilbing and Heister[528][529] performed nonlinear modeling of liquid jet breakup in the wind-induced regimes. They developed a boundary element method to solve for the nonlinear evolution of a liquid jet at unsteady inflow conditions and under the influence of both surface tension and aerodynamic interactions with surrounding gas. For longer waves, the aerodynamic effects were shown to cause a "swelling" of the liquid surface in the trough region. The model predicted the formation of main (parent) and satellite (product) droplets for fixed wavelength perturbations in the first wind-induced regime, and the evolution of a "spiked" surface at the periphery of the jet for conditions corresponding to the second wind-induced regime. These researchers further examined the effects of disturbance wave number, liquid Weber number, and liquid-gas density ratio on the jet breakup. They also identified transition conditions for various flow regimes. It was shown that the size of droplets is influenced by changes in the perturbation wavelength, perturbation magnitude and Weber number. Satellite droplet velocities are lower than main droplet velocities due to the sequential shedding of droplets from the orifice. Based on these modeling results, one can predict the likelihood of droplet recombination downstream of the initial pinching event and thus control the size of droplets generated by liquid jet breakup.

Linear stability theories have also been applied to analyses of liquid sheet breakup processes. The capillary instability of thin liquid sheets was first studied by Squire[258] who showed that instability and breakup of a liquid sheet are caused by the growth of sinuous waves, i.e., sideways deflections of the sheet centerline. For a low viscosity liquid sheet, Fraser et al.[73] derived an expression for the wavelength of the dominant unstable wave. A similar formulation was derived by Li[539] who considered both sinuous and varicose instabilities. Clark and Dombrowski[540] and Reitz and Diwakar[316] formulated equations for liquid sheet breakup length.

Current breakup models need to be extended to encompass the effects of liquid distortion, ligament and membrane formation, and stretching on the atomization process. The effects of nozzle internal flows and shear stresses due to gas viscosity on liquid breakup processes need to be ascertained. Experimental measurements and theoretical analyses are required to explore the mechanisms of breakup of liquid jets and sheets in dense (thick) spray regime.

5.2.2 Modeling of Droplet Formation, Breakup, Collision and Coalescence in Sprays

Substantial modeling efforts have been made by numerous investigators to study droplet formation, secondary breakup and coalescence in various processes. Meakin[35] made a comprehensive review of droplet formation processes *via* deposition growth and coalescence on a surface. Practical examples of condensation stages and droplet pattern formed during the condensation of water vapor on a cold surface, deposition of tin vapor on a hot surface, and deposition of GaAs on Ga-stabilized GaAs surfaces were illustrated. Although the condensation of water vapor may include breath figures or dropwise condensation, the main focus of this review was on the advances in the studies of breath figures and related phenomena. Computer models and simulation results were described in detail for all four stages in the breath figure process: *(a)* droplet growth without coalescence, *(b)* droplet growth and coalescence without renucleation, *(c)* droplet growth and coalescence with renucleation, and *(d)* droplet growth, coalescence and renucleation with large droplet removal. It was demonstrated that the use of scaling ideas was the most successful theoretical approach that had been applied to almost all the theoretical studies of droplet coalescence phenomena.

Detailed modeling study of practical sprays has a fairly short history due to the complexity of the physical processes involved. As reviewed by O'Rourke and Amsden,[310] two primary approaches have been developed and applied to modeling of physical phenomena in sprays: *(a) spray equation* approach and *(b) stochastic particle* approach. The first step toward modeling sprays was taken when a statistical formulation was proposed for spray analysis.[541] Even with this simplification, however, the mathematical problem was formidable and could be analyzed only when very restrictive assumptions were made. This is because the statistical formulation required the solution of the spray equation determining the evolution of the probability distribution function of droplet locations, sizes, velocities, and temperatures. The spray equation resembles the Boltzmann equation of gas dynamics[542] but has more independent variables and more complex terms on its right-hand side representing the effects of nucleations, collisions, and breakups of droplets.

Two numerical methods have been used for the solution of the spray equation. In the first method, i.e., the *full spray equation method*,[543][544] the full distribution function f is found approximately by subdividing the domain of coordinates accessible to the droplets, including their physical positions, velocities, sizes, and temperatures, into computational cells and keeping a value of f in each cell. The computational cells are fixed in time as in an Eulerian fluid dynamics calculation, and derivatives of f are approximated by taking finite differences of the cell values. This approach suffers from two principal drawbacks: *(a)* large numerical diffusion and dispersion errors[545] associated with convection through the fixed Eulerian mesh, and *(b)* enormous computer storage requirements.

In the second method, i.e., the *particle method*,[546][547] a spray is discretized into computational particles that follow droplet characteristic paths. Each particle represents a number of droplets of identical size, velocity, and temperature. Trajectories of individual droplets are calculated assuming that the droplets have no influence on surrounding gas. A later method,[548] that is restricted to steady-state sprays, includes complete coupling between droplets and gas. This method also discretizes the assumed droplet probability distribution function at the upstream boundary, which is determined by the atomization process, by subdividing the domain of coordinates into computational cells. Then, one parcel is injected for each cell.

An important advance in numerical methods for sprays was made by Dukowicz[549] who suggested that the ideas of the Monte Carlo method could be combined with the particle methods for spray calculations. In the method of Dukowicz, i.e., the *stochastic particle method*, the distribution of droplets at the upstream boundary is sampled stochastically by a relatively small number of computational particles. The droplet distribution function is obtained by averaging over a long time in steady-state calculations, or over many calculations in unsteady problems. The stochastic particle method can calculate unsteady sprays and account for the full coupling between droplets and gas due to mass, momentum, and energy exchanges. The method is robust and economical, and provides a framework for including some important new physical effects in spray calculations. In particular, this method has led to significant progress in discovering the mechanisms governing droplet sizes.

The first major extension of the stochastic particle method was made by O'Rourke[550] who developed a new method for calculating droplet collisions and coalescences. Consistent with the stochastic particle method, collisions are calculated by a statistical, rather than a deterministic, approach. The probability distributions governing the number and nature of the collisions between two droplets are sampled stochastically. This method was initially applied to diesel sprays[317] where it was found that coalescences caused a seven-fold increase in the mean droplet size. The importance of droplet collisions in diesel-type sprays has been corroborated by many subsequent studies.[228][551]

The second major extension of the stochastic particle method was supplied by Reitz and Diwakar[316] who developed a method for calculating droplet breakup. Droplet breakup is an important phenomenon in both hollow-cone and full-cone sprays typically formed in direct-injected stratified charge engines. Droplet sizes downstream of the injector are found to be determined primarily by the competition between coalescences and breakups, and influenced by vaporization and condensation. Reitz and Diwakar[316] also proposed a numerical method for calculating atomization using a droplet breakup model. In this method, droplets are injected with a diameter that equals the nozzle exit diameter. The breakup of these large droplets is then accomplished by the breakup model. This method for calculating atomization is based on an assumption that the dynamics and breakup of a liquid jet are indistinguishable from those of a train of droplets with equal diameter. Although this assumption requires further experimental validation, the method promises to remove one of the major weaknesses associated with spray modeling, i.e., the uncertainty in the specification of upstream boundary conditions.

To corroborate the findings of Reitz and Diwakar[316] concerning the importance of droplet breakup, O'Rourke and Amsden[310] developed an alternative model for calculating droplet aerodynamic breakup in spray calculations using the stochastic particle method. This model, called TAB (*Taylor Analogy Breakup*) model, is based on the analogy, suggested by Taylor,[245] between a spring-mass system and an oscillating-distorting droplet in a gas. The restoring force of the spring is analogous to the surface tension forces of the droplet. The external force on the mass is analogous to the aerodynamic forces of the gas on the droplet. The damping forces due to liquid viscosity were added to the analogy by O'Rourke and Amsden. This TAB model provides a means of predicting distortion, stretching and aerodynamic breakup of liquids such as droplets or ligaments in sprays.

The TAB model has several advantages. First, as pointed out by Taylor,[245] the model predicts that there is no unique critical Weber number for breakup, and whether or not a droplet breaks up depends on the history of its velocity relative to the gas. The Weber number is a dimensionless measure of the relative importance of the gas aerodynamic forces that distort a droplet and the surface tension forces that restore sphericity. Second, the effects of liquid viscosity are included. Although these effects are negligible for large droplets, liquid viscosity can significantly affect the oscillations of small droplets. Third, the model predicts the state of oscillations and distortions of droplets. Thus, once information is available on how distortions and oscillations affect the exchange rates of mass, momentum, and energy between the droplets and gas, this information can be incorporated in the model. Fourth, the model predicts droplet sizes that are more consistent with experimentally determined mechanisms of liquid jet breakup.[552][553] If this model is further used for calculating liquid jet breakup, an additional advantage is that the model predicts the velocities of the product droplets normal to the path of the original parent droplet. These normal velocities determine an initial spray angle, $\tan(\theta/2) = \sqrt{3}/3(\rho_G/\rho_L)^{1/2}$, that is in good agreement with some measured spray angles.[552] By giving an initial oscillation to the large, injected droplets, the initial spray angle can be varied and the effects of nozzle geometry changes can be included in numerical calculations. This eliminates the need for inputting the spray angle, i.e., the spray angle is automatically calculated in the TAB model.

The major limitation of the TAB model is that it can only keep track of one oscillation mode, while in reality there are many oscillation modes. Thus, more accurately, the Taylor analogy should be between an oscillating droplet and a sequence of spring-mass systems, one for each mode of oscillations. The TAB model keeps track only of the fundamental mode corresponding to the lowest order spherical zonal harmonic[554] whose axis is aligned with the relative velocity vector between the droplet and gas. This is the longest-lived and therefore the most important mode of oscillations. Nevertheless, for large Weber numbers, other modes are certainly excited and contribute to droplet breakup. Despite this limitation, good agreement between the model-predicted and experimentally observed breakup times was achieved, as reported in Ref. 310.

The equations in the TAB model contain four dimensionless constants that are determined by some theoretical and experimental results. The model predicts and continuously connects breakup times experimentally observed for the *bag* and *stripping* breakup regimes. The bag mode occurs when the Weber number is slightly large than a critical value, while the stripping mode takes place for the Weber numbers much larger than this critical value. The Taylor analogy equations do not predict product droplet sizes; thus an additional equation for the product droplet sizes is included in the TAB model and motivated by an energy conservation argument. A droplet breaks up into smaller droplets with a specified distribution of sizes when droplet distortion reaches a critical value. The product droplet sizes are determined by a Weber number criterion. For example, the Weber number based on the Sauter mean diameter of the product droplets is equal to 12 in the bag breakup regime.

A computer code, called KIVA, has been developed by Amsden et al.[555] at Los Alamos National Laboratory for modeling multidimensional fluid flows with chemical reactions and fuel sprays. The TAB model has been implemented numerically in the KIVA code,[555] along with the wave breakup model of Reitz[232] to describe the breakup process. A comparison of the computational results showed[310] that the two models[310][316] predicted different droplet sizes near the injector because they used different breakup times. Downstream, the two models generated similar results at lowest backpressure; at higher backpressures, however, the TAB model predicted larger droplet sizes than the calculations of Reitz and Diwakar[316] as well as the experiments of Hiroyasu and Kadota.[317] The differences in the implementation of the wave and TAB models have a significant impact on fuel vapor distribution in high-pressure combustion sprays since small product droplets vaporize very rapidly,[232] particularly when breakup time is much longer than numerical time step.

A modified version of the TAB model, called *dynamic drop breakup* (DDB) model, has been used by Ibrahim et al.[556] to study droplet distortion and breakup. The DDB model is based on the dynamics of the motion of the center of a half-drop mass. In the DDB model, a liquid droplet is assumed to be deformed by extensional flow from an initial spherical shape to an oblate spheroid of an ellipsoidal cross section. Mass conservation constraints are enforced as the droplet distorts. The model predictions agree well with the experimental results of Krzeczkowski.[311]

The substantial effect of secondary breakup of droplets on the final droplet size distributions in sprays has been reported by many researchers, particularly for overheated hydrocarbon fuel sprays.[557] A quantitative analysis of the secondary breakup process must deal with the aerodynamic effects caused by the flow around each individual, moving droplet, introducing additional difficulty in theoretical treatment. Aslanov and Shamshev[557] presented an elementary mathematical model of this highly transient phenomenon, formulated on the basis of the theory of hydrodynamic instability on the droplet-gas interface. The model and approach may be used to make estimations of the range of droplet sizes and to calculate droplet breakup in high-speed flows behind shock waves, characteristic of detonation spray processes.

Droplet collision and coalescence phenomena become increasingly important in dense sprays.[316] For solid-cone sprays, droplet coalescence is particularly important at high gas densities. For hollow-cone sprays, droplet breakup dominates because droplet coalescence is minimized by the expanding spray geometry. *Binary collisions* between droplets may lead to coalescence or separation, depending on operation conditions.[558] For a fixed orientation of colliding droplets, coalescence occurs at low collision Weber numbers. Separation of colliding droplets occurs in reflexive mode as the Weber number is increased beyond a critical value. O'Rourke[550] and O'Rourke and Bracco[559] developed a method for calculating droplet collisions in sprays on the basis of the kinetic theory of gases and Langmuir's work.[560] O'Rourke[550] assumed the probable number of collisions to follow the Poisson distribution, and proposed three possible outcomes of binary collisions for increasing collision Weber number: *(a)* permanent coalescence, *(b)* collision followed by breakup, and *(c)* shattering. It was indicated that the outcome of a binary droplet collision depends on forces acting on the coalesced pair of droplets. At low relative velocities or low Weber numbers, surface tension forces dominate over liquid inertia forces so that the droplets coalesce permanently. At high Weber numbers, the liquid inertia forces predominate so that the droplets separate, possibly forming satellite droplets, a phenomenon referred to as *grazing collision*. At very high relative velocities, the droplets may shatter. Droplets of largely unequal sizes are more likely to coalesce than droplets of nearly equal sizes. The possible formation of satellite droplets due to collision and the shattering collision have not yet been included in the currently available models.

An attempt has been made by Tsouris and Tavlarides[561] to improve previous models for breakup and coalescence of droplets in turbulent dispersions based on existing frameworks and recent advances. In both the breakup and coalescence models, two-step mechanisms were considered. A *droplet breakup function* was introduced as a product of droplet-eddy collision frequency and breakup efficiency that reflect the energetics of turbulent liquid-liquid dispersions. Similarly, a *coalescence function* was defined as a product of droplet-droplet collision frequency and coalescence efficiency. The existing coalescence efficiency model was modified to account for the effects of film drainage on droplets with partially mobile interfaces. A probability density function for secondary droplets was also proposed on the basis of the energy requirements for the formation of secondary droplets. These models eliminated several inconsistencies in previous studies, and are applicable to dense dispersions.

5.2.3 Theories and Analyses of Spray Structures and Flow Regimes

The degree of development of liquid and air (gas) flows inside an atomizer governs the velocity profiles and turbulence levels at the nozzle exit, that in turn have a significant influence on the spray structure formed. Current theories and analyses of spray structures are based on the assumptions and/or measurements of droplet sizes and velocity distributions at a designated initial cross-section plane of a spray where breakup is assumed to be completed. A *Lagrangian sectional approach* for calculating droplet size distributions of vaporizing fuel sprays in turbulent jets has been developed on the basis of detailed and numerous local droplet size and velocity distributions in the sprays.[562] In this approach, the effects of dispersion, acceleration, coalescence and evaporation can be calculated at any locations downstream. Accordingly, a spray is divided into three regions along the downstream direction: near-field, far-field, and fully developed far-field regions.

The region from the end of the liquid breakup region down to the far-field region is defined as *near-field region*. In this region, droplet velocities are significantly higher or lower than local gas velocities, so that the droplets are decelerated or accelerated by the gas flow during momentum exchanges. Inter-droplet distances are small initially but increase subsequently as the droplets disperse with increasing distance downstream. Droplet size and velocity are not well correlated. Initially, droplet velocity varies dramatically and then the variance decreases as gas turbulence decays with increasing distance downstream. The probability of droplet collision and subsequent coalescence is high due to high droplet number density, significant transverse velocity components, impinging droplet trajectories, and turbulent fluctuation components of both the droplets and gas. Coalescence may reduce the length of the near-field region due to the momentum exchanges between droplets during collisions.

The region where droplets and gas are approaching equilibrium is defined as *far-field region*. This region may be characterized by the small velocity difference (less than 10%) between droplets and gas. The far-field region develops with distance downstream to fully-developed far-field region. In the *fully developed far-field region*, the relative velocity between droplets and gas vanishes. Thus, momentum exchanges between the two phases also cease. The droplets are well dispersed in the gas with no collision or coalescence occurring, and spray becomes dilute with large inter-droplet distances. Droplet size is well correlated to its velocity. Droplet temperature is in equilibrium with local gas temperature so that no heat exchanges occur between the two phases.

Since a spray is a particle-laden turbulent jet, the fully-developed far-field region of a spray is analogous to the fully-developed region of a turbulent jet. In a turbulent jet, the profiles of mean velocity components, turbulent fluctuations, shear stress, and kinetic energy of turbulence tend towards self-similarity and thus can be described with a single self-similar profile independent of down-stream distance by normalizing the respective quantities and the distance with characteristic quantities and length. Different quantities may achieve the self-similarity at different distances downstream from the nozzle exit. The fully-developed region of a turbulent jet is the region where all quantities have attained equilibrium and self-similar-ity of profiles. Similarly, the distance for different droplets in a spray to achieve the equilibrium with local gas velocity will vary, depending on the magnitude of droplet sizes. The fully-developed region represents asymptotic conditions that may not be achieved in a particular spray.

The denseness of sprays may be measured by the ratio of the inter-droplet distance to droplet diameter. For the ratios larger than 30, a spray is dilute, and evaporation rates are almost same as those of single isolated droplets evaporating in the same atmosphere. As the ratio decreases to 15, small departures from dilute conditions are encountered. For the ratio of 10, a spray is dense with substantial departures from dilute conditions. For the ratios less than 10, a spray becomes so dense that the gas phase may be saturated before evapo-ration is complete.

Sprays and other dispersed flows are normally divided into *dense* and *dilute* spray regimes. Faeth[563] described the characteris-tics of each of the regimes. A jet spray may be divided into four regimes: churning, thick, thin, and very thin, as defined by O'Rourke.[550] The *churning regime* is defined as the flow region where the volume fraction of the liquid equals or exceeds that of the gas, and the liquid is no longer dispersed in a continuous gas phase. This regime occurs very close to the nozzle exit where the intact injected liquid starts to break up into blobs and droplets. For this regime, the equations of mass, momentum and energy conservation for compressible two-phase flows have been derived by considering the phases to be co-existing continua.[564] However, it is still uncertain about the form and ranges of applicability of the coupling terms between the liquid and gas phases.

In the *thick (dense) regime* next to the churning regime, the volume fraction of droplets is significant in the two-phase flow. In spite of the significant volume fraction, droplets are still recognizable as discrete droplets or liquid blobs in a continuous phase. Droplet-droplet effects such as collisions and coalescences are extremely important. The rates of mass, momentum, and energy exchanges between the droplets and gas are influenced by neighboring droplets.

Downstream in the spray, droplet size is reduced due to evaporation and breakup. Far from the nozzle, droplets become isolated and have negligible mass and volume relative to the gas phase. This regime is termed *very thin (dilute) regime*. Although the mass, momentum, and energy exchanges between the droplets and gas continue in this regime, the state of the gas is not altered appreciably by the exchanges. Therefore, the gas flow can be computed ignoring the volume occupied by the droplets, and the droplets can be tracked using the spray equation.[541] Isolated droplet correlations can be used to calculate the exchanges, and collisions and other interactions between droplets can be ignored.

The intermediate regime between the thick and very thin regimes is referred to as *thin regime*. In this regime, the droplets occupy negligible volume but have significant mass compared to the gas due to the large liquid to gas density ratio. The droplets influence the state of the gas, that in turn influences the exchange rates between other droplets and the gas. For example, the wake produced by droplets at the tip of a transient spray leads to a reduced relative velocity between the subsequent droplets and the gas because these droplets experience drafting forces. The reduced drag on the subsequent droplets allows them to overtake the leading droplets at the tip of the spray. The process repeats until entering the very thin regime.

5.2.4 Modeling of Droplet-Gas and Droplet-Droplet Interactions

Droplet-gas and droplet-droplet interactions in a spray affect droplet size and velocity. For modeling of mass, momentum, and energy exchanges between droplets and gas in a spray, the knowledge of droplet-gas and droplet-droplet interactions is required. It is common practice to calculate the rates of mass, momentum, and energy exchanges by means of engineering calculations that, however, may not adequately describe the complex, unsteady characteristics of the exchange processes. Laminar convective models have been used, but are not applicable to turbulent flows. Very sophisticated single droplet models and highly developed codes by Raju and Sirignano[565] and Dandy and Dwyer[566] include the time-dependent equations for the exchanges at droplet-gas interfaces and are being extended to arrays of single droplets and clusters. Advances in measurement techniques and spray calculations will lead to correlations of probability density functions that will allow more comprehensive and realistic descriptions of the transfer processes.

Accurate representations of the drag coefficients of droplets over a wide range of flow conditions are a necessary prerequisite to the calculation of gas-droplet flows. Yung et al.[567] performed numerical modeling of unsteady motion of a spherical fluid droplet under the influence of gravity to examine the steady state droplet velocity and drag coefficient. Numerical results for these two quantities were presented in graphical form for a water droplet moving through air and an air bubble rising in water under conditions from creep flows to moderate Reynolds numbers.

Due to computer storage and run time limitations, it is not yet possible to accurately model the details of flows around each individual droplet in a spray. Thus, empirical or semi-empirical correlations are typically used to model the exchange processes between droplets and gas. Correlations for drag coefficients have been suggested by many researchers.[45][559][568]-[571] For thin sprays, the drag coefficient is often calculated as a function of the droplet Reynolds number using the single solid-sphere correlation.[568] For thick sprays, O'Rourke and Bracco[559] suggested a correlation that includes the effect of local void (air/gas) fraction on the drag coefficient.[550] The effects of droplet oscillation and distortion on the drag coefficient were investigated by Hwang and Reitz[315] and Liu et al.[572] who proposed a quantitative correlation to relate the magnitude of droplet deformation to its drag coefficient. Using this dynamic drag model, the calculated drag coefficient of a distorting droplet lies between that of a rigid sphere and that of a flat disk whose drag coefficient at high Reynolds numbers is about 3.6 times higher than that of a sphere. Renksizbulut and Yuen[573] suggested a correlation for drag coefficient to account for the effect of droplet vaporization. They showed that droplet vaporization reduces its heat transfer rate and drag coefficient, whereas Gonzalez et al.[574] found that the effect of droplet vaporization on its drag is small in diesel-type sprays.

Henderson[575] presented a set of new correlations for drag coefficient of a single sphere in continuum and rarefied flows (Table 5.1). These correlations simplify in the limit to certain equations derived from theory and offer significantly improved agreement with experimental data. The flow regimes covered include continuum, slip, transition, and molecular flows at Mach numbers up to 6 and at Reynolds numbers up to the laminar-turbulent transition. The effect on drag of temperature difference between a sphere and gas is also incorporated.

Turbulence in a spray may have a significant effect on the drag coefficient.[576] The effect of turbulence is quantified by relative turbulence intensity. Torobin and Gauvin[577] found that for relative Reynolds numbers of the order of 10^3, relative turbulence intensities of the order of 20% can cause an order of magnitude decrease in the drag coefficient. Warnica et al.[578] presented data of drag coefficients for Reynolds numbers less than 100 and relative turbulence intensities as high as 60%. The data suggested that within the experimental accuracy achievable, the drag coefficient for a sphere can be used for a droplet.

Table 5.1. Correlations for Drag Coefficient of a Single Sphere in Continuum and Rarefied Flows[575]

Valid Regime	Correlations
Subsonic Flow $Ma \le 1$	$C_D = 24\left[\mathrm{Re} + S\left\{4.33 + \left(\dfrac{3.65 - 1.53\dfrac{T_w}{T}}{1+0.353\dfrac{T_w}{T}}\right)\times \exp\left(-0.247\dfrac{\mathrm{Re}}{S}\right)\right\}\right]^{-1}$ $+\exp\left(-\dfrac{0.5\mathrm{Ma}}{\sqrt{\mathrm{Re}}}\right)\left[\dfrac{4.5+0.38\left(0.03\,\mathrm{Re}+0.48\sqrt{\mathrm{Re}}\right)}{1+0.03\,\mathrm{Re}+0.48\sqrt{\mathrm{Re}}}+0.1\mathrm{Ma}^2+0.2\mathrm{Ma}^8\right]$ $+\left[1-\exp\left(-\dfrac{\mathrm{Ma}}{\mathrm{Re}}\right)\right]0.6S$
Supersonic Flow $Ma \le 1.75$	$C_D = \dfrac{0.9+\dfrac{0.34}{\mathrm{Ma}_\infty^2}+1.86\left(\dfrac{\mathrm{Ma}_\infty}{\mathrm{Re}_\infty}\right)^{1/2}\left[2+\dfrac{2}{S_\infty^2}+\dfrac{1.058}{S_\infty}\left(\dfrac{T_w}{T}\right)^{1/2}-\dfrac{1}{S_\infty^4}\right]}{1+1.86\left(\dfrac{\mathrm{Ma}_\infty}{\mathrm{Re}_\infty}\right)^{1/2}}$
Supersonic Flow $1 < Ma < 1.75$	$C_D(\mathrm{Ma}_\infty, \mathrm{Re}_\infty) = C_D(1., \mathrm{Re}) + \dfrac{4}{3}(\mathrm{Ma}_\infty - 1.)\left[C_D(1.75, \mathrm{Re}_\infty) - C_D(1., \mathrm{Re})\right]$

C_D — Drag coefficient; C_D(1.,Re) is the drag coefficient calculated using the correlation for subsonic flow with Ma=1.; $C_D(1.75, \mathrm{Re}_\infty)$ is the drag coefficient calculated using the correlation for supersonic flow with Ma_∞=1.75;

Ma — Mach number based on relative velocity between gas and sphere;

Re — Reynolds number based on sphere diameter, relative ve locity, and freestream density and viscosity;

S — Molecular speed ratio, $S = \mathrm{Ma}\,(\gamma/2)^{0.5}$, where γ is the ratio of specific heats;

T — Temperature of gas in freestream;

T_w — Temerature of sphere;

∞ — Refers to freestream conditions.

Interactions of a shock wave with a single droplet and multiple droplets of various liquids have been investigated in many numerical modeling studies.[579]-[582] It was found[579] that for a flow across a shock, the *Basset history force* acting on a droplet can be many times larger than the viscous drag in the immediate shock region. It was demonstrated how this force influences the motion of particles of different densities or sizes across the shock. For the theoretical description of the motion of particles commonly used in flow measurement tracer techniques, however, it is justified to neglect the Basset integral in the *Basset-Boussinesq-Oseen equation*.[568]

Resistance functions have been evaluated in numerical computations[583] for low Reynolds number flows past spherical particles, droplets and bubbles in cylindrical tubes. The undisturbed fluid may be at rest or subject to a pressure-driven flow. A *spectral boundary element method* was employed to calculate the resistance force for torque-free bodies in three cases: *(a)* rigid solids, *(b)* fluid droplets with viscosity ratio of unity, and *(c)* bubbles with viscosity ratio of zero. A lubrication theory was developed to predict the limiting resistance of bodies near contact with the cylinder walls. Compact algebraic expressions were derived to accurately represent the numerical data over the entire range of particle positions in a tube for all particle diameters ranging from nearly zero up to almost the tube diameter. The resistance functions formulated are consistent with known analytical results and are presented in a form suitable for further studies of particle migration in cylindrical vessels.

Correlations for heat transfer coefficient between a single sphere and surrounding gas have been proposed by many researchers (Table 5.2), for example, Whitaker,[584] and Ranz and Marshall,[505] among others. The correlation recommended by Whitaker is accurate to within ±30% for the range of parameter values listed. All properties except μ_s should be evaluated at T_∞. For freely falling liquid droplets, the Ranz-Marshall correlation[505] is often used. The correlations may be applied to mass transfer processes simply by replacing Nu and Pr with Sh and Sc, respectively, where Sh and Sc are the *Sherwood number* and *Schmidt number*, respectively. Modifications to the Ranz-Marshall correlation have been made by researchers to account for the effects of droplet oscillation and distortion,[585] droplet vaporization and mass transfer,[555] as well as compressibility of gas.[154][156] The influence of superheating of diffusing vapor around a droplet on its heat transfer rate was examined by Priem et al.[586] As with the drag coefficient, turbulence also affects the heat transfer coefficient. The effect of turbulence is quantified by the relative turbulence intensity. A modified form of the Ranz-Marshall correlation has been proposed on the basis of experimental data for water, ethanol and methanol droplets at relative turbulence intensities up to 12%.[587]

Table 5.2. Correlations for Heat Transfer Coefficient of a Single Sphere in Gas

Correlations	Remarks	References
$Nu = 2 + (0.4\,Re^{1/2} + 0.06\,Re^{2/3})\,Pr^{0.4}\left(\frac{\mu}{\mu_s}\right)^{1/4}$ $0.71 < Pr < 380,\ 3.5 < Re < 7.6\times10^4,\ 1 < \frac{\mu}{\mu_s} < 3.2$	All properties (except μ_s) evaluated at T_∞	Whitaker [584]
$Nu = 2 + 0.6\,Re^{1/2}\,Pr^{1/3}$	Freely falling liquid droplet	Ranz & Marshall [505]
$Nu = 2 + 0.6\,Re^{1/2}\,Pr^{1/3}\left[25\left(\frac{x}{D}\right)^{-0.7}\right]$ $3mm \le D \le 6mm,\ 10 \le x/D \le 600$	Effects of droplet oscillation & distortion included	Yao & Schrock [585]

Nu, Pr, Re — Nusselt, Prandtl, and Reynolds numbers, respectively,
D — Droplet diameter,
T_∞, T_s — Gas temperature in freestream and sphere surface temperature, respectively,
x — Falling distance measured from rest,
μ, μ_s — Gas viscosity evaluated at T_∞ and T_s, respectively.

A model for multi-droplet interaction effects has been developed by Silverman and Sirignano[588] using a new approach for dense-spray modeling. The model was employed to study the effects of droplet interactions on the evaporation and motion of a dense spray in a hot gaseous environment. A statistical description of droplets in a cloud and correlation functions for the effects of interactions between neighboring droplets were used to extract correction factors for various parameters that are affected by droplet interactions (for example, drag coefficient, Nusselt number and Sherwood number). The correction factors enable the calculations of the drag coefficient, evaporation rate and heat transfer of a droplet in a cloud based on models for a single droplet. It was shown that the interaction effects are important during a large fraction of the droplet lifetime as the droplet size decreases. The multi-droplet interactions cause the drag coefficient of a droplet in a dense spray to be lower and hence its velocity higher than that for an isolated droplet. Thus, spray penetration is very much greater than that predicted for isolated droplets at the same initial conditions. For droplets of 100 μm in diameter, the evaporation rate decreases and the droplet lifetime increases due to the multi-droplet interactions. For droplets of 40 μm in diameter, however, the evaporation rate increases and the droplet lifetime decreases, despite slower heat transfer to an interacting droplet than to an isolated droplet under the same gas-phase conditions and relative velocity of the droplets. It is the higher relative velocity of a 40 μm interacting droplet that compensates for this effect, resulting in a shorter droplet lifetime.

5.2.5 Modeling of Multiphase Flows and Heat and Mass Transfer in Sprays

A spray is a turbulent, two-phase, particle-laden jet with droplet collision, coalescence, evaporation (solidification), and dispersion, as well as heat, mass and momentum exchanges between droplets and gas. In spray modeling, the flow of gas phase is simulated typically by solving a series of conservation equations coupled with the equations for spray process. The governing equations for the gas phase include the equations of mass, momentum and energy conservation, as well as turbulence model equations. The simulation can be in the *Lagrangian*,[549][555] *Eulerian*,[543] or *Lagrangian-Eulerian*[555] *reference frame*. Source terms are required for the coupling of the gas and droplet/particle phases. Detailed descriptions of the conservation equations, spray equation, exchange terms, and computational techniques have been given by O'Rourke,[550] Williams,[541] and Amsden et al.[555]

Two (or three) phase flow in a spray is typically turbulent with resultant dispersion of droplets, turbulence damping by small droplets,[589] and turbulence generation by large droplets.[590] Large-scale turbulent structures, with sizes of the order of spray diameter, influence liquid breakup, dispersion, and mixing. Dispersion of droplets in a spray is intimately related to turbulence.[591] Large droplets with *Stokes numbers much larger than one* are projected along their initial trajectories without (or with little) deflection by the turbulent gas flow. On the contrary, small droplets with *Stokes numbers much smaller than one* follow the gas flow closely without (or with small) slip velocity. These small droplets spiral and are centrifuged corresponding to their sizes when entering large-scale eddies. Droplets with *Stokes numbers of the order of one* are partially rotated when entering eddies and before exiting from the eddies. Dispersion of droplets in a spray may become preferential, and in some cases may be higher than that of the gas phase alone. Thus, physical processes such as *droplet clustering* and *Stokes-number dispersion*[591] need to be considered in spray modeling. However, droplet dispersion cannot be predicted accurately without significant improvements in understanding and modeling of turbulence phenomenon. Currently, the most popular approach to predicting droplet dispersion in turbulence is the Monte-Carlo method in which the turbulence field is represented by a random number generator. The gas velocity is considered as the sum of a time-averaged velocity and a fluctuating velocity. The fluctuating velocity is selected from a Gaussian distribution with a variance proportional to the turbulence kinetic energy. The droplet motion is then integrated with this velocity field until it passes through an eddy. Reitz and Diwakar[316] found that the effect of turbulence modulation is most important in the jet development region close to the nozzle in a high-pressure spray. The calculated velocity decay is faster if turbulence modulation is not included, and the primary effect of the turbulence modulation is a shift in the virtual origin of the spray. The dispersion of droplets by turbulence and the gas phase turbulence modulation have been reviewed by Crowe et al.[591] and Faeth,[589] respectively.

Most current multidimensional spray simulations have adopted the thin or very thin spray assumptions,[550] i.e., the volume occupied by the dispersed phase is assumed to be small. This can be justified if a simulation starts some distance downstream of the nozzle exit, where the gas volume fraction is large enough, or if the computational cells are relatively large. Accordingly, two major classes of models have been used in spray modeling: *locally homogeneous flow* (LHF) models and two-phase-flow or *separated-flow* (SF) models.

In the *LHF models*, it is assumed that droplets are in dynamic and thermodynamic equilibrium with gas in a spray. This means that the droplets have the same velocity and temperature as those of the gas everywhere in the spray, so that slip between the phases can be neglected. The assumptions in this class of models correspond to the conditions in very thin (dilute) sprays. Under such conditions, the spray equation is not needed and the source terms in the gas equations for the coupling of the two phases can be neglected. The gas equations, however, need to be modified by introducing a mixture density that includes the partial density of species in the liquid and gas phases based on their mass fractions. Details of the LHF models have been discussed by Faeth.[589]

In the *SF models*, all of the terms in the droplet and gas conservation equations are retained. Therefore, the SF models are the more general models for spray calculations. The models account for mass, momentum and energy exchanges between droplets and gas. To formulate the exchange terms, the nature of the conditions at droplet-gas interface is of importance. The exchange processes are typically modeled by means of semi-empirical correlations.

To date, spray modeling has been largely dependent on solving the governing equations of multiphase flows and specifying initial and boundary conditions. This typically involves solving the governing equations for 2-D, turbulent, particle-laden flows with simultaneous chemical reactions and geometrically complex boundary conditions. Chemical reactions for individual species are treated by Arrhenius-type kinetics[555] and vaporization is calculated using the Frossling correlation[1] or the comprehensive model of Priem et al.[586] Detailed reviews of advanced droplet vaporization models have been given by Sirignano.[14][15] For specifying initial and boundary conditions, an initial plane needs to be designated where liquid breakup is complete and droplets are spherical. The typical outputs of spray modeling calculations include detailed spatial and temporal distributions of droplet sizes, velocities, angles of flight, number densities, temperatures, and mass flux, along with turbulent characteristics of gas flow, such as distributions of mean and fluctuating velocities, shear stresses, and kinetic energy of turbulence. Numerical computations have focused on seeking steady state solutions of 2-D, time-averaged, turbulent flow equations. New models/methods that allow computations of 3-D, time-dependent flows with resolution of large scales of turbulence include those for large eddy[592] and vortex dynamics[593] simulations. These models/methods have been applied to single phase jet flows, and need adaptation to two-phase jet spray flows.

The state-of-the-art in computations of sprays is the relation of the Reynolds-averaged Navier-Stokes equations with k-ε turbulence closure (where k is turbulent kinetic energy and ε is its dissipation rate) and particle-source-in-cell coupling to a Lagrangian condensed phase for droplets.[548] The particle-source-in-cell method proposed by Crowe et al.[548] relates the liquid phase to gas phase and is very sensitive to accurate determination of the joint distribution functions of droplet sizes, 3-D components of velocities, temperatures, and evaporation rates in sprays. The method is most suited for steady-state spray computations and is currently used in many commercial spray codes. The developments of new computer programs in massively parallel processing will allow more coupling of the full Reynolds-averaged Navier-Stokes equations with droplet dynamics.

Young[594] derived a set of fundamental equations for gas-droplet multiphase flows in which small liquid droplets polydisperse in a continuous gas-phase consisting of either a pure vapor of the same chemical species as the droplets or a mixture of the pure vapor and an inert gas. Some problems associated with previous formulations were resolved by more judicious definitions of interphase transfer terms. A consistent model was included to represent the surface energy and entropy of the droplets in order to maintain consistency with droplet growth models in which the droplet temperature depends on its radius due to capillary effect. An equation for the rate of entropy creation due to departures from equilibrium was also derived on the basis of linear irreversible thermodynamics. It was indicated that some commonly used mathematical models of droplet growth may not be physically realistic in certain circumstances.

Creismeas[595] calculated the evolution of a water droplet spray inside a complex geometry by means of an Eulerian-Lagrangian model. In this modeling, the transient, incompressible, turbulent Navier-Stokes equations were solved using a projection method. To accommodate the complex geometry, a rectangular mesh was defined and each mesh node was assigned to a node function whose value depends on the location of the node. Droplets were regarded as individual entities and treated using the Lagrangian approach. Mass and heat transfer coefficients were introduced for individual droplets to model mass and heat transfer phenomena. The modeling results may be used to determine the spray characteristics in the vicinity of obstacles. Landau-Levich model of a diffusion boundary layer was suggested for modeling of mass transfer in continuous and dispersed phases of two immiscible liquids during droplet motion in extractors.[64]

Aerosol production and transport over the oceans are of interest in studies concerning cloud physics, air pollution, atmospheric optics, and air-sea interactions. However, the contribution of sea spray droplets to the transfer of moisture and latent heat from the sea to the atmosphere is not well known. In an effort to investigate these phenomena, Edson et al.[12] used an interactive Eulerian-Lagrangian approach to simulate the generation, turbulent transport and evaporation of droplets. The k-ε turbulence closure model was incorporated in the Eulerian-Lagrangian model to accurately simulate stable, near-neutral, and unstable boundary layers of the air-sea interactions. The Langevin equation was modified to account for the effects of the gravity and inertia of the droplets. It was showed that although the influence of the droplets was small under the conditions in this modeling, the potential for substantial modification of the surface energy budget exists for high-wind conditions over the oceans. The model system may be used to investigate interactions between evaporating droplets and turbulent fields of temperature and humidity in general.

In spray combustion processes, the multi-phase flows in internal combustion piston engines are unsteady, compressible (low Mach number), turbulent, cyclic and non-stationary, both spatially and temporally. The distribution of liquid and vapor fuel in an engine greatly influences the combustion characteristics. Fuel injection introduces additional complexity to the process physics and consequently to the modeling of the dense multi-phase flows. In recent years, the use of computer models to simulate fuel injection has become common practice. The models range from highly empirical equations[90] to detailed multidimensional models requiring the solution of basic conservation equations.[596] Intermediate approaches have also been adopted using integral continuity and momentum equations and relying heavily on experimental data. These models have been used to characterize and predict fuel spray penetration, trajectory, mixing and combustion within an engine combustion chamber.[438] The predictions are then used for engine design and analysis of data acquired from engines in operation. Comparisons of model predictions and experiments[597] have shown that the model predictions can respond properly to changes in nozzle geometry, injection pressure and direction, and air properties with reasonable accuracy.

Some early spray models were based on the combination of a discrete droplet model with a multidimensional gas flow model for the prediction of turbulent combustion of liquid fuels in steady flow combustors and in direct injection engines. In an improved spray model,[438] the full Reynolds-averaged Navier-Stokes equations were solved along with energy equation and k-ε turbulence model including compressibility effect for air flow velocity, pressure, density and enthalpy. Fuel droplet trajectories, size and temperature histories were calculated by solving the equations of motion of discrete droplets as well as heat and mass transfer equations. In the equations of motion, the pressure gradient and Basset terms were neglected due to their very small magnitudes. The Saffman lift and Magnus forces were also neglected because the droplets are not in a high shear region of the air flow under the flow conditions considered. The model included the effects of droplet size distribution and turbulence mixing. The interaction between a droplet and air was modeled by averaging the overflow processes on a scale smaller than the droplet size and using typical correlations for droplet drag, heat and mass transfer. The drag coefficient correlation for a non-evaporating droplet and the heat transfer coefficient correlation were multiplied by a correction factor to account for the decrease of droplet mass during the evaporation period. The injection nozzle was divided into five introduction locations and the injection period was discretized into a number of equal intervals of 1° crank angle. The quantity of fuel injected during each period was calculated from a detailed atomization model. The number of droplets for each size class was determined by dividing the total number of droplets equally on the five introduction locations and using the droplet size distribution. The initial droplet velocity varied corresponding to variations in fuel and combustion chamber pressures. Each droplet was assigned with a different initial location, size and velocity. A size increment of 10 μm was used to cover the entire range of droplet size. The flow field in the combustion chamber was subdivided into grid elements and each element was treated as a control volume. The net efflux rate of droplet mass from a computational element, which is a source term in the continuity equation, was obtained by assuming overall trajectories traversing that element. At any instant of time, the flow field was first assumed and predictions for a time increment were then obtained by solving the gas flow equations using a marching integration algorithm. From the solutions of the gas flow equations at each time step, the fuel concentration was calculated to determine spray contours. The solutions were marched forward in time until the desired period had been covered. The calculated results suggested that the characteristics of the mixing and the ensuing combustion processes in diesel engines are mainly influenced by droplet size distribution, physical and chemical properties of fuel, and medium conditions. The droplet size distribution is one of the important factors affecting the characteristics of spray combustion and subsequent emission formation.

A detailed review has been made by Reitz[525] on recent advances in the development of a comprehensive analytical model of spray combustion as a design tool for industrial applications. The model is based on the KIVA code.[555] In the model, the conservation equations are solved for the transient dynamics of vaporizing fuel sprays interacting with flowing multi-component gases and undergoing mixing, ignition, chemical reactions and heat transfer. The model encompasses sub-models for those processes that occur on the time and length scales much smaller than can be resolved by a single macro-model. The modifications made to the KIVA code include sub-models for atomization[232] and droplet drag,[572][598] droplet-droplet and droplet-wall impingement, heat transfer, ignition, combustion, soot formation and emissions.

As described previously, in the *atomization sub-model*,[232] droplet parcels are injected with a size equal to the nozzle exit diameter. The subsequent breakups of the parcels and the resultant droplets are calculated with a breakup model that assumes that droplet breakup times and sizes are proportional to wave growth rates and wavelengths obtained from the liquid jet stability analysis. Other breakup mechanisms considered in the sub-model include the Kelvin-Helmholtz instability, Rayleigh-Taylor instability,[206] and boundary layer stripping mechanisms. The TAB model[310] is also included for modeling liquid breakup.

In the *droplet drag sub-model*,[572][598] the effects of droplet distortion and oscillation due to droplet-gas relative motion on droplet drag coefficient are taken into account. Dynamical changes of the drag coefficient with flow conditions can be calculated with this sub-model. Applications of the sub-models to diesel sprays showed that spray tip penetration is relatively insensitive to droplet drag coefficients whereas droplet size distribution is greatly influenced by droplet drag.

The spray model in the KIVA code[555] could not assess the physical phenomena in the dispersion process of spray impingement on a wall since all impinging droplets, called parcels, are treated as sticking on the wall.[422] To address this problem, Naber and Reitz[599] developed *droplet-wall impingement sub-models* for droplet-wall interactions during spray impingement on a wall, an important phenomenon in direct injection engines. The models include a reflect sub-model for droplet rebound and a liquid jet analogy sub-model for wall jet. Impinging droplets may rebound or form liquid films on a wall. During cold starting of engines, droplet impingement velocities are relatively low and the models predict droplet rebound as the dominant mode. The droplet rebound may be a mechanism for the back penetration of fuel vapor into the central region of a combustion chamber where gas temperature is high enough for ignition.

During the ignition delay period, the injected fuel is atomized and mixes with air after evaporation. Kong and Reitz[598] applied the *multistep Shell ignition model*, developed by Halstead et al.[600] for the autoignition of hydrocarbon fuels at high pressures and temperatures, to diesel combustion processes. The Shell model gives good predictions of diesel ignition.

Up to the time point of ignition, combustion may be controlled by the chemical kinetic rates. The combustion rate turns to be controlled by turbulent mixing after ignition. Several turbulent combustion models, such as the *one-step reaction model*,[601] the *coherent flame model*,[602] and the *laminar-and-turbulent characteristic-time model*,[229] have been developed that account for the effects of turbulence on mean reaction rates. In these models, the combustion time scale was related to the turbulent kinetic energy and its dissipation rate on the basis of the k-ε turbulence model. The model of Reitz and Bracco[229] adopted the *eddy-breakup concept* and was demonstrated to perform well along with the Shell ignition model in various engine applications. Details of the combustion model have been described by Kong and Reitz.[598]

For the prediction of NOx formation, the extended Zeldovich mechanism described by Heywood[603] was implemented. The *soot emission model* is a modified version of previously published models for soot formation and oxidation. Details of the soot emission model have been discussed by Han et al.[604]

5.3.0 MODELING OF DROPLET PROCESSES OF MELTS

As for normal liquids, modeling of droplet processes of melts provides tremendous opportunities to improve the understanding of the fundamental phenomena and underlying physics in the processes. It also provides basic guidelines for optimization and *on-line* control of the processes. This section is devoted to a comprehensive review of process models, computational methods, and numerical modeling results of the droplet processes of melts. The emphasis of this section will be placed on the droplet processes in spray atomization for metal powder production, and spray forming for near-net shape materials synthesis and manufacturing. Details of these processes have been described in Ref. 3.

Several basic process stages are common in these droplet processes: *(a)* metal and gas delivery, *(b)* atomization, and *(c)* spray. Droplet cooling, solidification and microstructure evolution may start as early as in the atomization stage and continue in the spray stage. In spray forming and thermal spray processes, droplet deposition, cooling and consolidation on target substrate is an important final stage. A more detailed review of modeling studies on this stage has been given in Ref. 3, and hence this section will concentrate on more common phenomena, i.e., droplet-substrate interactions.

The fundamental issues to be addressed in the process modeling include spray enthalpy, gas consumption, spray mass distribution, microstructure of solidified droplets, and droplet-substrate interactions. The effects of atomization gas chemistry, alloy composition and operation conditions on the resultant droplet properties are also to be investigated in the process modeling.

5.3.1 Modeling of Melt Flows and Heat Transfer in Near-Nozzle Region

In spray atomization and spray forming processes, a liquid metal is normally transported from a melting crucible (tundish) to an atomizer through a delivery nozzle. During the flow in the delivery nozzle, the melt temperature decreases gradually due to the heat exchange with the nozzle wall. Premature solidification of the melt in the delivery nozzle (a phenomenon referred to as *freeze-up*) may occur when the melt temperature decreases down to or below its melting temperature. The freeze-up may be caused by: *(a)* inadequate melt superheat; *(b)* excessive residence time of the melt in the delivery nozzle; and/or *(c)* recirculation of undercooled droplets at the nozzle exit. The selection of appropriate melt superheat requires that it can offset the heat loss of the melt during the flow in the delivery nozzle. The heat loss and the residence time of the melt are closely related to its flow and heat transfer behavior in the nozzle.

The mass flow rate of a liquid metal through a delivery nozzle orifice may be calculated with the following equation:[605]

Eq. (6) $$\dot{m}_L = cA_L\rho_L(2gh + 2\Delta P_L / \rho_L)^{1/2}$$

where c is the nozzle discharge coefficient, A_L is the cross-sectional area of the liquid nozzle orifice, g is the gravitational acceleration, h is the metal head, and ΔP_L is the overpressure. The nozzle discharge coefficient is unity for an ideal flow and decreases with increasing melt viscosity. For $l_0/d_0 \leq 4$, $c \approx 0.62$; for $l_0/d_0 > 4$, $c \approx 0.82$.[605] The overpressure may be caused by applying a pressure differential between melting chamber and spray chamber for free-fall atomizers, or by the suction at the nozzle base for close-coupled atomizers.[325] In absence of an overpressure, the metal head should be maintained constant by pouring molten metal from a melting tundish into a pouring tundish during atomization in order to maintain a constant metal flow rate.

Modeling of melt flow and heat transfer in the near-nozzle region is required to predict the process parameters necessary for preventing the premature solidification of liquid metals/alloys and to determine the optimum atomizer parameters. The modeling results of the metal delivery stage also provide initial conditions for the modeling of the ensuing atomization stage.

Liu et al.[606][607] developed a numerical model and computer code on the basis of the boundary layer theory and the modified van Driest and Cebeci mixing length turbulence model to simulate the flow and heat transfer of molten metals in a delivery nozzle. Numerical calculations were conducted for liquid Al, Cu, Mg, Ni, Ti, W, In, Sn, Bi, Pb, Zn and Sb to study the effects of process parameters and material properties on the *minimum melt superheat* that is necessary to prevent the freeze-up during delivery of the molten metal prior to atomization. To facilitate the application of the numerical model and to provide direct insight into the complex relationship among the minimum melt superheat, process parameters and material properties, *processing maps* were developed on the basis of the modeling results, as shown in Fig. 5.1.

The processing maps can be used to select the process conditions that are necessary to avoid the freeze-up. From the processing maps it can be seen that within the range of low overpressure values (for example, $\Delta P_L \leq 60$ kPa), increasing the overpressure can effectively decrease the minimum melt superheat, especially for a large nozzle length to diameter ratio (for example, $l_0/d_0 = 17$) and for materials of low densities (for example, Al and Mg). This effect diminishes with further increase in overpressure. The minimum melt superheat can be decreased by reducing the nozzle length to diameter ratio, by selecting a smooth delivery nozzle with low thermal conductivity and thick nozzle wall, by imposing a high overpressure, and/or by enhancing atomization gas temperature. The overpressure, which is necessary to avoid the freeze-up (referred to as *minimum overpressure*), decreases with decreasing nozzle length to diameter ratio or decreasing melting-temperature to gas-temperature ratio.

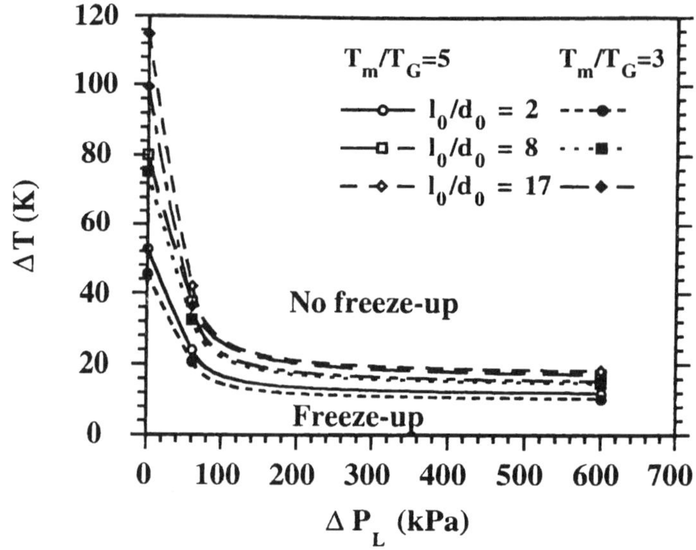

(a)

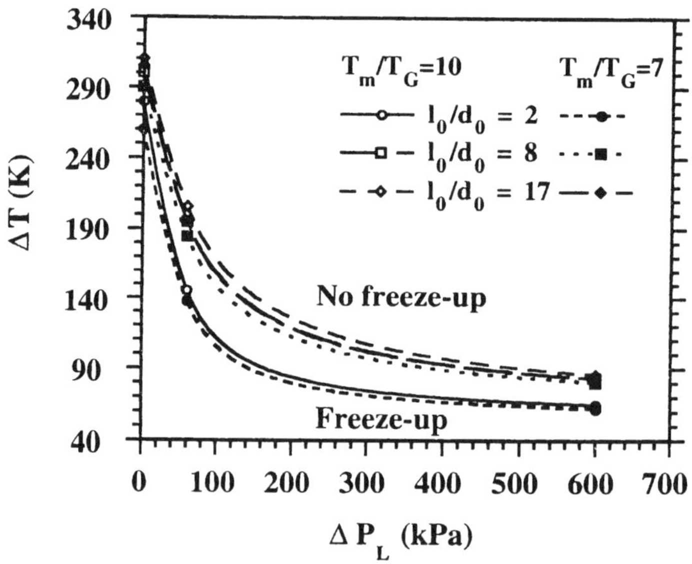

(b)

Figure 5.1. Processing maps for (*a*) Al and (*b*) Ti. *(Reprinted from Ref. 606. © 1995, with kind permission from Elsevier Science Ltd., The Boulevard, Langford Lane, Kidlington 0X5 1GB, UK.)*

The modeling results showed that for the materials studied, the minimum melt superheat ranges from $0.005T_m$ to $0.19T_m$, depending on process parameters and material properties. The dependence is quantitatively expressed in a correlation derived from a regression analysis of the numerical results:

Eq. (7)
$$\frac{\Delta T}{T_m} = 0.15\left[\frac{\mu_L U_{heat}^2}{k_L c_{pL} \rho_L \Delta P_L}\frac{T_m}{T_G}\right]^{0.1}\left[\frac{l_0}{d_0}\right]^{0.3}$$

where ΔT is the melt superheat, T_m is the equilibrium melting temperature of metal, μ_L, k_L, c_{pL}, and ρ_L are the kinematic viscosity, thermal conductivity, thermal capacity, and density of liquid metal, respectively, U_{heat} is the overall heat transfer coefficient between liquid metal at tube wall and atomization gas, ΔP_L is the overpressure, T_G is the gas temperature, and l_0/d_0 is the nozzle length to diameter ratio.

In this correlation, the material properties are evaluated at the melting temperature. The left hand side of the correlation is the dimensionless minimum melt superheat. The right hand side of the correlation is also dimensionless, and represents a combination of the Prandtl number, *Euler number*, Reynolds number and Nusselt number, as well as temperature and length ratios T_m/T_G and l_0/d_0. The correlation is accurate within 10%. In addition, considering the effects of the surface roughness of nozzle wall, the pre-basal coefficient in the regression expression has been increased by 25% in order to predict a safe estimate of the minimum melt superheat.

The above correlation can be used directly to calculate the process conditions that are necessary to avoid the freeze-up. From this correlation, it is evident that materials with high thermal conductivity, high thermal capacity and/or large density allow small minimum melt superheat, whereas materials with high melting temperature and/or high viscosity require large minimum melt superheat. For materials with high surface tension, the minimum melt superheat calculated with this correlation represents an upper bound.

In addition to the above process parameters, atomizer geometry and configuration also influence the minimum melt superheat by altering gas flow pattern and distribution. The freeze-up may occur even for the melt superheats that are higher than the predicted minimum value, if recirculation gas flow forms in the exit region of the delivery nozzle. The recirculation gas flow in this region may drag relatively cool or undercooled droplets upwards, and deposit them on the tip of the delivery nozzle, where they eventually solidify. The solidified metal may continue to increase in thickness during atomization, eventually choking the melt flow, a phenomenon that has been described previously as freeze-up. Therefore, the minimum melt superheat calculated with the above correlation is only a necessary condition, rather than a sufficient condition. The sufficient condition to avoid the freeze-up must be determined by examining the flow field below the delivery nozzle. If the geometry and configuration of the delivery nozzle and atomization gas nozzles are arranged in such a manner that the recirculation gas velocity is not large enough to drag cool droplets up to the delivery nozzle tip, then employing the melt superheat calculated with the above correlation may avoid the freeze-up. Accordingly, for free-fall atomizers, in order to minimize the probability of the freeze-up, it is necessary to design the atomizer geometry and select the position of the delivery nozzle in such a way as to minimize the formation of the recirculation gas flow in the region below the delivery nozzle. However, for close-coupled atomizers, the recirculation gas flow is utilized to generate a thin liquid film and sheet prior to atomization, as discussed previously in the liquid film-sheet breakup mechanism.

The temperature of a liquid metal stream discharged from the delivery tube prior to primary breakup can be calculated by integrating the energy equation in time. The cooling rate can be estimated from a cylinder cooling relation for the liquid jet-ligament breakup mechanism (with free-fall atomizers), or from a laminar flat plate boundary layer relation for the liquid film-sheet breakup mechanism (with close-coupled atomizers).

5.3.2 Modeling of Gas Flows in Near-Nozzle Region

In spray atomization and spray forming processes, the distribution of velocity and temperature of atomization gas in the near-nozzle region determines the impact kinetic energy and cooling effect on the molten metal stream, and hence significantly influences the generation, initial size, shape, radial distribution and flight direction of droplets. On the other hand, the melt flow and heat transfer during delivery and atomization determine its temperature, viscosity and surface tension, which in turn affect the continuous operation of the delivery nozzle and the droplet size distribution. Therefore, the flows and heat transfer during metal and gas delivery and in the vicinity of the atomizer are critical to the atomization efficiency and gas consumption, and have a significant impact on the distributions of droplet size, spray enthalpy and spray mass.

For the delivery of atomization gas, different types of nozzles have been employed, such as straight, converging, and converging-diverging nozzles. Two major types of atomizers, i.e., free-fall and close-coupled atomizers, have been used, in which gas flows may be subsonic, sonic, or supersonic, depending on process parameters and gas nozzle designs. In sonic or supersonic flows, the mass flow rate of atomization gas can be calculated with the following equation based on the compressible fluid dynamics:

Eq. (8)
$$\dot{m}_G = A_G P_{G0}\left(\frac{2}{\gamma+1}\right)^{(\gamma+1)/2(\gamma-1)}\left(\frac{\gamma}{RT_{G0}}\right)^{1/2}$$

where A_G is the cross-sectional area of the gas nozzle, P_{G0} and T_{G0} are the gas pressure and temperature in the supply manifold, respectively. The values of the gas constant R and the isotropic factor of gas γ can be found in Table 2.7 for various gases.

The power requirement for pressurizing atomization gas may be estimated using the following expression:

Eq. (9)
$$\text{Power} = \dot{m}_G R M_G T_{G0} \ln(P_{G0} / P_{chamber})$$

where $P_{chamber}$ is the gas pressure in spray chamber, and the values of the molecular weight of gas, M_G, are given in Table 2.7 for various gases.

While the gas flow in the nozzle can be calculated on the basis of isentropic expansion of stagnation properties, the analysis of gas flows in the near-nozzle region can be performed by means of numerical simulation/modeling using computational fluid dynamics (CFD) codes. Such codes can generally solve incompressible and/or compressible flows, and the k-ε two-equation model is frequently used as turbulence model. The modeling of the flow and temperature fields of atomization gas in the near-nozzle region provides basic guidelines for optimization of atomizer parameters. The modeling results of this stage also provide initial conditions for the modeling of the ensuing atomization stage.

Recently, there has been an increasing number of numerical studies on the gas delivery stage[161][163][324][325][496][608] and some experimental measurements in the near-nozzle region.[160][162][169][170][177][327][608]-[610] Extensive theoretical,[611] numerical,[161] and experimental[170][175] studies on high-speed gas jet flows in the near-nozzle region have been conducted to investigate velocity profiles, pressure distributions, shock waves and flow structures.

Espina et al.[611] applied the method of characteristics based on the Prandtl-Meyer theory to analyze gas flows near a close-coupled atomizer. The analysis was aimed at understanding the complex flows, determining the values of aspiration pressure, and locating the shock waves. The theoretical calculations explained the measured aspiration behavior and provided a potential tool for evaluation of different atomizer geometries, gases and pressure ratios. The theoretical analysis also provided a means for examining and optimizing the geometric design of gas atomizers and correlating the geometric design parameters to atomization efficiency. In addition, several interesting conclusions were drawn from the theoretical calculations. First, the energy dissipated in the flow interactions between discrete jets requires a higher operational pressure than that required by an annular jet to achieve the maximum of aspiration. This reduction in the operational pressure when using an annular-jet atomizer may decrease the gas mass flow rate required for the atomization of molten metals. Second, the maximum of aspiration in an annular-jet atomizer is of lesser magnitude than in a discrete-jet atomizer. This reduction in the aspiration may reduce the liquid metal flow rate during atomization. Third, the good agreement between the calculated and experimental results suggested that the errors caused by neglecting turbulent effects are small in this type of flows.

See and Johnston[319] studied the gas velocity profiles generated by ring atomizers with multiple discrete jets. In the atomization experiments of molten lead by See et al.,[318] a ring atomizer with four discrete jets was used to impinge the liquid stream on the centerline at different angles. The following empirical equations developed originally for a single submerged gas jet were used to calculate the gas velocity at the breakup point:

Eq. (10) $$\frac{U_{Gc}}{U_{G0}} = 6.2\left(\frac{d_{G0}}{l}\right) \quad \text{for} \quad 7 < \frac{l}{d_{G0}} < 100$$

Eq. (11) $$\log\left(\frac{U_{Gc}}{U_{Gr}}\right) = U_{G0}\left(\frac{r}{l}\right)^2 \quad \text{for} \quad 7 < \frac{l}{d_{G0}} < 100$$

where U_{Gc} is the gas velocity on the jet center line at a distance l from the gas nozzle exit, U_{G0} is the initial gas velocity at the gas nozzle exit, d_{G0} is the gas nozzle diameter, and U_{Gr} is the gas velocity at a radius r from the jet axis.

Liu[325][612] performed numerical modeling of gas flows in the near-nozzle region for both the close-coupled and free-fall atomizer configurations. The flow fields involved subsonic and supersonic flows, turbulence, shock waves, free shear flows, and recirculation flows. In view of the complex gas flows during atomization, the full Reynolds-averaged Navier-Stokes equations were solved along with turbulence transport equations.[613] The effect of turbulence was modeled using the standard Boussinesq approximation. Two turbulence models, i.e., combined Thomas/Baldwin-Lomax model and k-ε two-equation model, were used, switching from the former to the latter after some iterations. These models can accurately predict both wall bounded flows and free shear flows.[613]

The calculated gas velocity distributions for a *free-fall atomizer* configuration in different nozzle geometries are shown in Fig. 5.2. The results suggested that for a constant gas pressure, the recirculation gas flow can be minimized by arranging the geometry of a free-fall atomizer. For a *close-coupled atomizer* configuration, the calculated Mach number contour is shown in Fig. 5.3 and the calculated gas velocity vectors and stream lines are illustrated in Fig. 5.4. The gas flow is choked at the nozzle exit (Ma = 1), and then expands into the atomization chamber while the Mach number reaches a maximum of about 6. The supersonic gas flow separates at the corner of the metal delivery nozzle, leading to the formation of an annular free jet and a recirculation region between the delivery nozzle tip base and the realignment point where the centerline velocity vanishes. The gas velocity in the recirculation region is relatively low, and the gas flows up towards the delivery nozzle tip and then turns outwards along the delivery nozzle base plane, as clearly demonstrated by the stream lines in Fig. 5.4. The gas pressure attains a maximum at the center of the nozzle tip base and decreases to a minimum away from the center in radial direction. It is this pressure gradient and the recirculation gas flow that cause the radial flow of liquid metal, forming a thin liquid film and then a thin liquid sheet prior to primary breakup. Therefore, the strong recirculation flow and the adjacent free shear flow are responsible for the liquid film-sheet breakup mechanism, and critically affect the resultant droplet properties, as will be discussed in the next subsection.

The recirculation region has been experimentally observed in the gas flow fields generated using this type of atomizer configuration[160] or similar atomizer configurations.[169] Mi et al.'s numerical modeling of high pressure gas flows near a close-coupled atomizer[161] also demonstrated that downstream of the delivery nozzle there exist a strong recirculation region and an adjacent mixing layer at all gas stagnation pressures, and normal shock structures at some high stagnation pressures. The intensity of the recirculation flow and the mixing layer increases with increasing stagnation pressure.

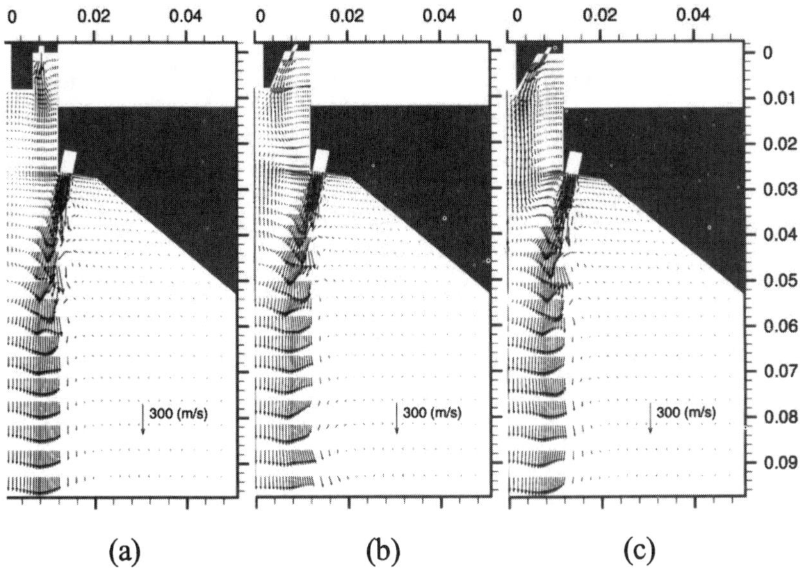

Figure 5.2. Calculated gas flow fields in the near-nozzle region of free-fall atomizers. Primary gas pressure: 0.140 MPa; secondary gas pressure: 0.189 MPa; angle of secondary gas nozzle relative to the spray centerline: 10°; angle of primary gas nozzle relative to the spray centerline: *(a)* 0°, *(b)* 22.5°, and *(c)* 30° designed for minimizing recirculation gas flow. *(Reprinted from Ref. 612.)*

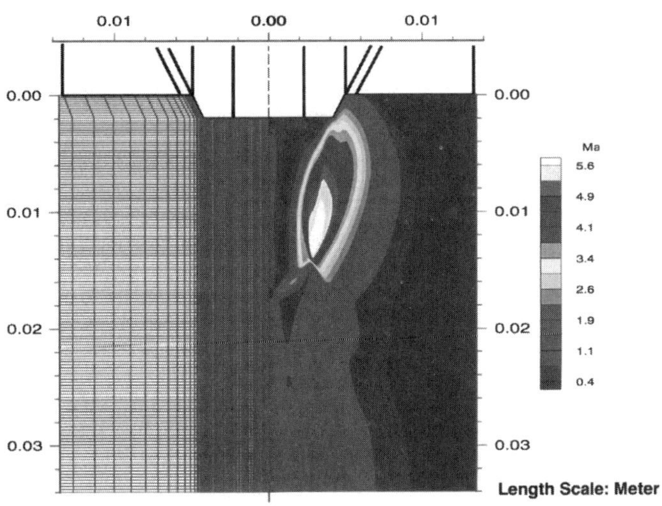

Figure 5.3. Atomizer geometry (*top*),computational grid mesh (*left*) and calculated Mach number contour (*right*) near a close-coupled atomizer (Atomization gas: Ar, Ma = 1 at nozzle exit). *(Reprinted with permission from Ref. 325.)*

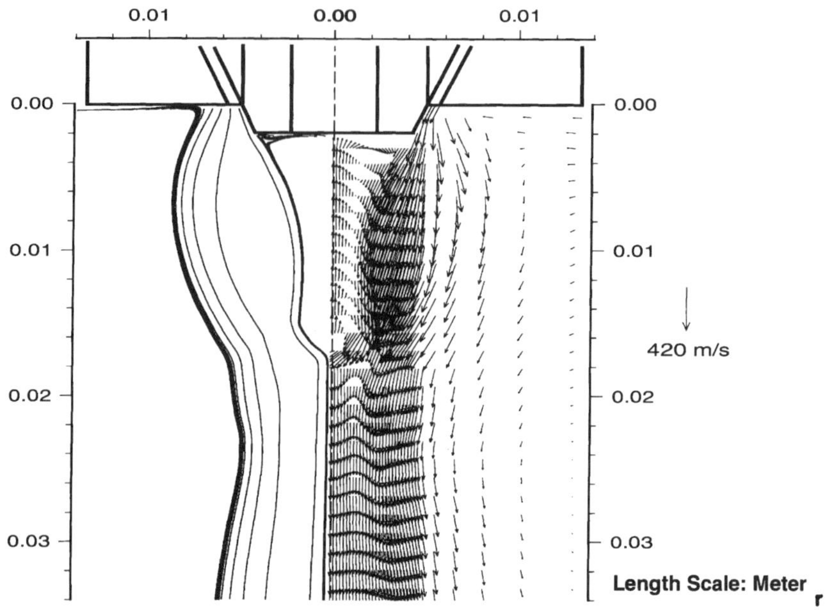

Figure 5.4. Calculated gas velocity field near a close-coupled atomizer: velocity vectors (*right*) and stream lines (*left*) (Atomization gas: Ar, Ma = 1 at nozzle exit). *(Reprinted with permission from Ref. 325.)*

5.3.3 Modeling of Droplet Formation and Breakup in Sprays

Gas atomization of liquid metals involves complex physical phenomena, such as high-speed flow of atomization gas, primary and secondary breakup of liquid metals, and droplet dynamics and heat transfer. Atomization is a key stage in spray atomization and spray forming processes. It determines initial conditions of droplets. It also influences droplet velocity, temperature, cooling rate, and eventually the microstructure and mechanical properties of the sprayed products. Numerical modeling proves to be an effective means for understanding atomization mechanisms and studying effects of process parameters and atomizer designs on resultant droplet properties. By optimizing process parameters and atomizer designs, the ultimate objective of the numerical modeling of atomization processes is to provide guidelines for controlling droplet size distribution, increasing atomization efficiency, and reducing gas consumption.

Although spray forming has been extensively developed for producing preforms from diverse alloy compositions, the difference between atomization processes using different gases (for example, nitrogen and argon) is not well understood, nor has a correlation between process parameters and final part quality been established for spray forming using argon. For example, to convert a spray forming process from using nitrogen to argon, the ratio of mass flow rate of gas to metal has to be doubled according to a rule of thumb based on the differences in the density, heat capacity and thermal conductivity between the gases. However, the spray enthalpy or solid fraction of the spray developed when using the rule of thumb is not comparable, leading to different levels of microstructural characteristics. In order to produce the same microstructure using argon as using nitrogen, to achieve a narrow spray distribution and to reduce total gas consumption for cost-effective manufacturing, it is necessary to investigate atomization mechanisms and ascertain the effects of thermophysical properties and process parameters on atomization processes. Zeng et al.[173][500] and Liu et al.[154][500] studied the influence of gas chemistry on atomization processes of liquid metals. The results showed that gas chemistry critically affects both the size distribution and the thermal history of the atomized droplets.

Over the past decade, a number of numerical studies have been performed to model the liquid metal and gas delivery, spray, deposition, and consolidation stages in spray forming.[18][51][52][154][156][174][338][339][378][379][388][389][495][496][606][607][614]-[616] However, only limited modeling studies have been conducted for the atomization stage.[161][324][325]

On the basis of the experimental observations,[160][169][327] Liu[325] conceived a liquid film-sheet breakup model for atomization of liquid metals with close-coupled atomizers. In this atomization model, it was postulated that atomization of a liquid metal with a close-coupled atomizer may occur in the following sequence: *(1)* formation of a liquid film, *(2)* conversion of the liquid film into a liquid sheet, *(3)* primary breakup of the liquid sheet into droplets, *(4)* droplet deformation and cooling, and *(5)* secondary breakup and/or solidification of the droplets. Computational methods and a computer code were developed to model liquid metal breakup, droplet dynamics and heat transfer during gas atomization. Empirical and experimental correlations were used along with basic conservation equations to predict the droplet size distribution and eventually the properties of the metal powder produced by gas atomization.

In this numerical model,[325] the flow characteristics of the liquid metal film were calculated on the basis of a force balance under the assumption of a steady-state Couette flow. These flow characteristics along with the empirical correlation suggested by Eroglu et al.[262] were used to estimate the mean sheet length, assuming that the sheet trajectory follows the gas stream line prior to primary breakup. The Sauter mean diameter of the droplets after primary breakup was calculated using Knoll and Sojka's semi-empirical correlation.[263] The droplet size distribution after primary breakup and prior to any secondary breakup was described by Simmons' universal root-normal distribution function. In this distribution, a straight line can be generated by plotting $(D/\mathrm{MMD})^{0.5}$ *vs.* cumulative volume of droplets on a normal-probability scale. The slope of the straight line can be specified by the ratio MMD/SMD that may take a value of 1.1, 1.2 or 1.5.[249] After primary breakup, the droplets cool while they accelerate in the gas flow field. The droplet trajectories were calculated using Thomas' 2-D equation of motion for a sphere[579] with some simplifications.[154][156] Henderson's formulations[575] were used for the calculation of drag coefficients of a sphere in subsonic and supersonic gas flows. The effect of droplet distortion on the drag coefficients was estimated with the experimental results in Ref. 285. The heat transfer between the droplets and atomization gas leads to rapid cooling and eventual solidification of the droplets, terminating the deformation and further breakup of the droplets. Due to the small Biot number, a uniform droplet temperature was assumed and the lumped parameter model[156] was used. Convection and radiation were considered in the heat transfer model. The heat transfer coefficient was calculated using the Ranz-Marshall correlation[505] after appropriate modifications.[154][156]

Following primary breakup, the droplets deform and may further break up, depending on their interactions with the atomization gas. To determine if a droplet will undergo secondary breakup in addition to deformation, the breakup regime map developed by Hsiang and Faeth[285] was used. The time scale and degree of deformation prior to secondary breakup as well as the Sauter mean diameter of droplets after secondary breakup were calculated using the correlations suggested by Hsiang and Faeth.[285] The droplet size distribution after secondary breakup also follows the Simmons' root-normal distribution pattern with MMD/SMD taking a value of 1.1, 1.2 or 1.5.[285] Similarly to primary breakup, once SMD was calculated, the entire droplet size distribution after secondary breakup could be determined. Following secondary breakup, no further breakup was considered since no evidence of subsequent breakup was found in the experiments.[285]

The solution of the gas flow and temperature fields in the near-nozzle region (as described in the previous subsection), along with process parameters, thermophysical properties, and atomizer geometry parameters, were used as inputs for this liquid metal breakup model to calculate the liquid film and sheet characteristics, primary and secondary breakup, as well as droplet dynamics and cooling. The trajectories and temperatures of droplets were calculated until the onset of secondary breakup, the onset of solidification, or the attainment of the computational domain boundary. This procedure was repeated for all droplet size classes. Finally, the droplets were numerically sieved and the droplet size distribution was determined.

The modeling results suggested that gas atomization of a liquid metal with a close-coupled atomizer may occur in the following sequence, as illustrated by the calculated trajectories of the liquid metal sheet and droplets in Fig. 5.5 (a). Initially, the liquid metal flows downwards through the delivery nozzle. At the exit of the nozzle, it changes its flow direction to radially outwards along the nozzle base plane, forming a thin liquid film. The formation of the liquid film is caused by the interaction between the liquid metal and the gas flow in the recirculation region. It is also this interaction that leads to the turning of the liquid film at the edge of the nozzle base plane into a liquid sheet which extends downwards following gas stream lines. This is followed by the primary breakup of the sheet into droplets. The droplets accelerate, deform, and cool in the gas flow field. The trajectories of small droplets also follow the gas stream lines closely. Secondary breakup of the droplets may occur, depending on their deformation and cooling/solidification rates. This liquid film-sheet breakup mechanism may be considered as one of the operating atomization mechanisms in gas atomization processes with close-coupled atomizers.

Using this model and computer code, the calculated sheet and droplet trajectories showed a good qualitative agreement with high-speed video imaging of actual gas atomization processes,[160][327] as evident in Fig. 5.5. The code-predicted droplet size distribution is also in a fairly good agreement with the experimentally measured distribution,[163][164] with the calculated droplet size distribution narrower than the measured data. Droplet coalescence caused by collision and aggregation, particularly in the vicinity of the spray centerline, may explain this discrepancy.

Droplet collision is a phenomenon inherent in the dense region of a spray. Droplet collisions may lead to local agglomeration that affects the droplet size distribution. There have been considerable efforts in modeling droplet-droplet collisions and coalescence,[229] but the models are still not generally applicable.[576] Moreover, the calculations in the dense region of a metal spray is much more complicated than in a diesel spray because the physical phenomena and mechanisms in the dense region are not well understood.

Regarding the effects of process parameters on gas atomization of melts, modeling and experimental studies[163][164][324] revealed that the mass median droplet diameter decreases with increasing atomization gas pressure and gas to metal mass flow rate ratio. The standard deviation decreases with increasing gas to metal mass flow rate ratio. As the melt superheat increases, both the mass median droplet diameter and the standard deviation decrease.

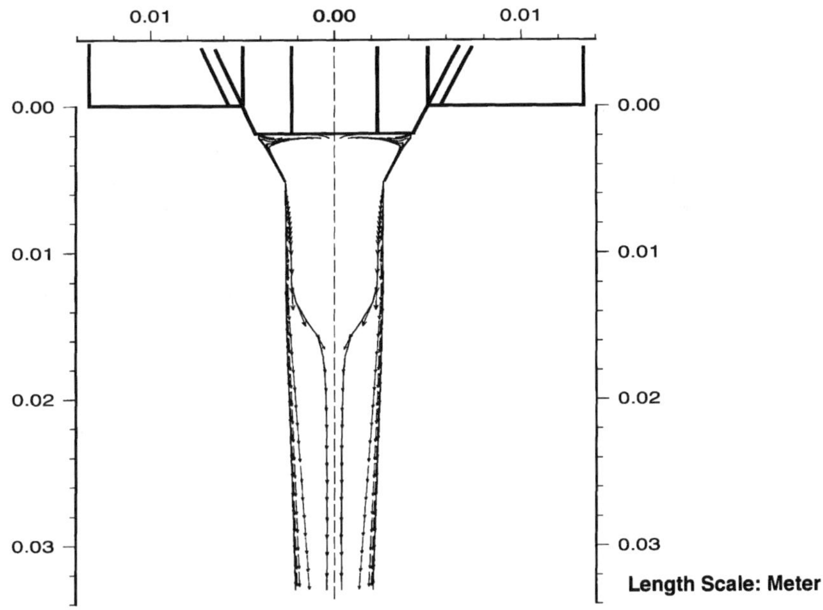

(a)

Figure 5.5. *(a)* Calculated trajectories of liquid metal sheet and droplets near a close-coupled atomizer (Atomization gas: Ar, Ma = 1 at nozzle exit, Metal: Ni, T^0 = 1877 K, $\dot{m}_G/\dot{m}_L$ = 3.74). *(Reprinted with permission from Ref. 325.) (b)* High-speed video imaging of actual gas atomization process with a close-coupled atomizer. *(Courtesy of Mr. Paul Martiniano and Dr. Paul Follansbee, General Electric Corporate R & D, Schenectady, NY, USA.)*

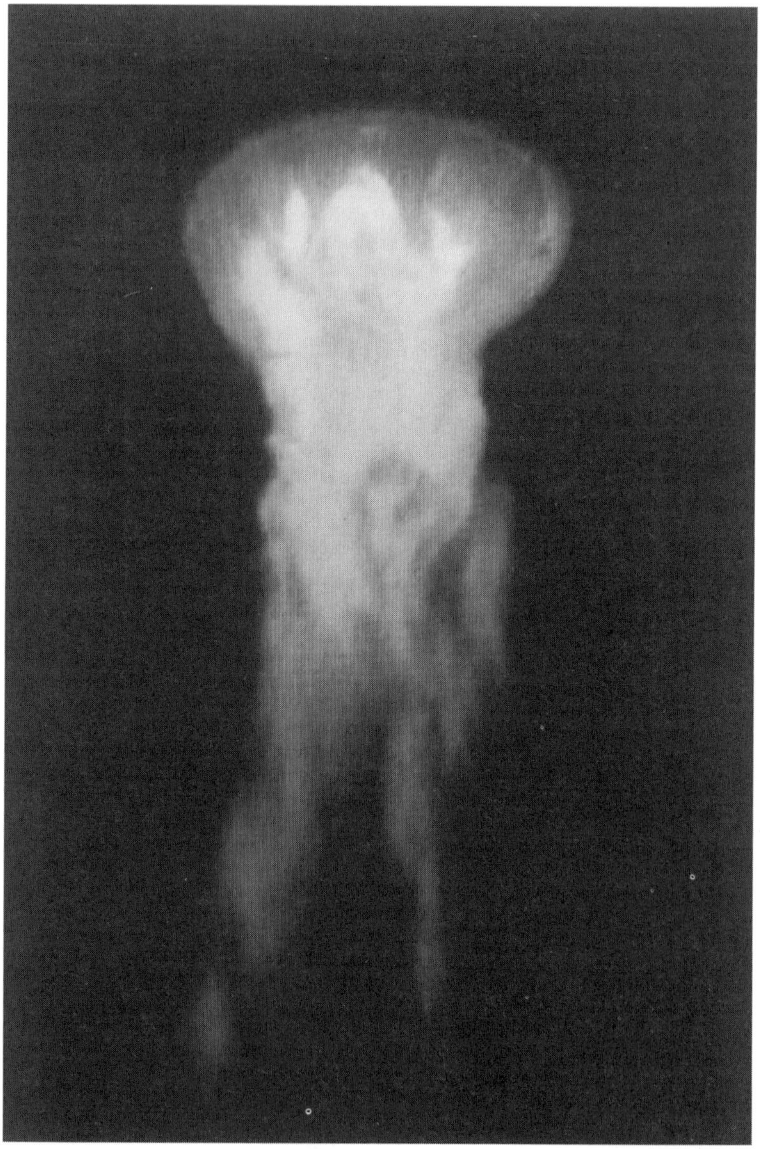

(b)

Figure 5.5. (*Cont'd.*)

5.3.4 Modeling of Multiphase Flows and Heat Transfer in Sprays

A metal spray is comprised of a dispersion of metal droplets of different sizes in atomization gas. Droplet-gas interactions in the spray determine droplet velocity, temperature, and solidification history, and consequently affect the spray enthalpy and radial mass distribution. These factors in-turn critically influence the microstructural evolution and the ultimate microstructural characteristics, such as average grain size and porosity, in the sprayed materials. Variations in microstructure at different locations within a sprayed material have been observed in experiments, suggesting the importance of the spatial distributions of thermal and solidification histories of the droplets in the spray. To elucidate the microstructural evolution of the droplets, it is necessary to systematically analyze the multidimensional flow, momentum and heat transfer, as well as rapid solidification phenomena, which occur simultaneously during droplet-gas interactions. Accordingly, the modeling of droplet-gas interactions and rapid solidification kinetics in the spray is aimed at controlling the spray mass distribution to achieve the required spray pattern, and controlling the thermal and solidification histories to obtain the desired spray enthalpy and microstructural characteristics.

Generally, 3-D models are essential for calculating the radial distributions of spray mass, spray enthalpy, and microstructural characteristics. In some applications, axisymmetry conditions may be assumed, so that 2-D models are adequate. Similarly to normal liquid sprays, the momentum, heat and mass transfer processes between atomization gas and metal droplets may be treated using either an Eulerian or a Lagrangian approach.

Crowe[576] classified the numerical models for dilute sprays as two-fluid model and discrete element model. The *two-fluid model* treats the dispersed phase as another fluid with appropriate constitutive relationships for effective viscosity and thermal conductivity. The advantage of the two-fluid model is that the same algorithm used for the gas phase can be applied directly to the droplet phase. The drawback is the lack of information on the dispersion coefficient and the effective viscosity of droplets that must be estimated empirically. The *discrete element model* is similar to the *stochastic particle method* suggested by Dukowicz.[549] In this model, droplets are simulated by the motion of discrete droplet parcels. The parcels of identical droplets are released in each time step. It is assumed that a parcel moves through computational cells in the flow field as the motion of an individual droplet and the temperature history of the droplets in the parcel is the same as that of an individual droplet. The model accounts for the full two-way coupling between droplets and gas due to mass, momentum, and energy exchanges. The dispersion of droplets due to turbulence is accomplished by adding a fluctuation velocity component to the gas phase velocity. There is no need to select an effective droplet viscosity. The approach is suitable for dilute, unsteady flows because it captures the essence of the Lagrangian nature of the dispersed phase. It can also be used for dense flows where collisions dominate the particle motion. The disadvantage of this approach is the number of droplets needed to represent the droplet field. The viability of the predictions by both the two-fluid model and discrete element model depends on the reliability of initial conditions, i.e., the flow properties at the end of the dense flow region or after the completion of secondary breakup.

A simpler version of the discrete element model is the so-called *trajectory model*. In the trajectory approach, droplet field is modeled as a series of trajectories that emanate from the atomizer or a starting point. The coupling effects are included by summing the heat release to and the drag force on the gas phase. This approach can be used for steady dilute flows.

Numerical modeling of multiphase flows, heat transfer and droplet-gas interactions in metal sprays has been reported by a number of investigators.[154][495]-[497][576][617]-[620] In limited studies,[617][621][622] 2-D experimental measurements of atomization gas and droplet velocities were conducted. A few complex 2-D Eulerian models[615][618] were used to calculate the droplet-gas interactions in metal sprays. Most of the previous studies, however, were confined to 1-D, and only the droplet trajectory and thermal history along the spray centerline were addressed using uncoupled trajectory approach.

The spatial motion of a droplet during spray forming, not only determines its residence time in the spray, but also critically influences its cooling and solidification by changing the boundary conditions around the droplet surface. Therefore, the knowledge of the velocity fields of both atomization gas and droplets is a precondition for numerical calculations of droplet temperature and solidification histories. The momentum and energy exchanges between the gas phase and the discrete droplet phase not only determine the droplet size, velocity, cooling rate, and solidification time, but also affect the gas flow. Thus, the two- (or three-) phase flows in the spray during spray atomization or spray forming should be solved in a fully coupled way. However, some experimental studies[169] showed that the droplets have little effect on the gas flow, as supported by the nearly identical shock structures and shear layer shapes in a gas-only flow field and a droplet-gas two-phase flow field under the same operation conditions. Hence, reasonably accurate results may be obtained from an uncoupled solution. This simplification not only makes the problem tractable but also leads to much less computational cost and time relative to a fully coupled solution. In the uncoupled approach, the gas flow field is calculated separately and its solution is used for modeling liquid metal breakup and droplet dynamics during atomization and spray. There is evidence, however, that a relatively large discrepancy may occur between experimental data and the predictions by the uncoupled solution approach. The spray predicted by the uncoupled approach was narrower than that measured experimentally, while reasonably good agreement was obtained between experimental data and the prediction by the coupled solution approach.[576]

In the Lagrangian frame, droplet trajectories in the spray may be calculated using Thomas' 2-D equations of motion for a sphere[579] or the simplified forms.[154][156] The gas velocity distribution in the spray can be determined by either numerical modeling or direct experimental measurements. Using the uncoupled solution approach, many CFD software packages or Navier-Stokes solvers can be used to calculate the gas velocity distribution for various process parameters and atomizer geometries/configurations. On the other hand, some simple expressions for the gas velocity distribution can be derived from experimental measurements. For example, on the basis of the experimental results of Bewlay et al.,[609][617] the decay in the mean axial gas velocity at the spray centerline has been formulated using an exponential function,[156] and the radial distributions of the axial and radial gas velocities specified with power functions.[156] Although these experimental correlations are very simple and easy to use, some parameters in the correlations are specific to the atomizer geometry and operation conditions under which they are derived, and therefore must be determined by experimental measurements.

Grant et al.[619] presented a modeling study of droplet dynamic and thermal histories during spray forming. In spite of the 1-D nature of this modeling, a major advance in the formulation of atomization gas velocity was made by analyzing previous experimental and theoretical results and correlating them into a simple, non-apparatus-specific form. In addition, a detailed solidification model was proposed that involves three thermal regions (i.e., nucleation and recalescence, segregated solidification, and eutectic solidification) and hence is applicable to rapid solidification processes of eutectic alloys.

Using the Lagrangian approach, Liu et al. developed both 1-D[154] and 2-D[156] flow and heat transfer models and rapid solidification kinetics models, including a reaction kinetics model for reactive atomization and deposition, to study droplet-gas interactions in the spray. The formulations are suitable for both Newtonian and non-Newtonian conditions. The models and the related numerical methods as well as the computer codes developed[154][156] were tested and validated for Ni_3Al intermetallic compound and Ta-W alloy droplets sprayed using nitrogen-oxygen mixture gases, argon and nitrogen, respectively. These models and codes can be used to calculate the momentum, heat and mass transfer in the spray and to predict the spatial distributions of droplet velocity, temperature, cooling rate, solid fraction, and spray enthalpy, prior to deposition. Some microstructural characteristics in semi-solid or fully solidified droplets/particles (for example, secondary dendrite arm spacing, SDAS) can be determined on the basis of the thermal histories predicted by the codes.

In some spray forming processes, the atomization gas exits the atomizer nozzle(s) at sonic or supersonic velocities. The gas may expand

and accelerate within a short distance and then decay through shocks and diffusion to subsonic speed in a non-uniform "roller-coaster" fashion. The compressibility of the gas may influence the transfer processes between the gas and droplets, particularly in the near-nozzle region. Usually, Henderson's formulations[575] may be used for calculating the drag coefficients of droplets in subsonic and supersonic gas flows. However, within certain Reynolds number range, the dependence of drag coefficients on the Mach number seems to be negligibly small. Thus, the approximation[615] of the standard drag curve may be used in the numerical calculations of droplet trajectories in the spray. In addition, other effects on the drag coefficients need to be considered in the numerical calculations, such as the effects of local void (air/gas) fraction,[550] droplet oscillation and distortion,[315][572] droplet vaporization,[573] and turbulence.[577][578]

For a given system of metal/alloy and atomization gas, the 2-D velocity distributions of the gas and droplets in the spray can be then calculated using the above-described models, once the initial droplet sizes and velocities are known from the modeling of the atomization stage, as described in the previous subsection. With the uncoupled solution of the gas velocity field in the spray, the simplified Thomas' 2-D nonlinear differential equations for droplet trajectories may be solved simultaneously using a 4th-order Runge-Kutta algorithm, as detailed in Refs. 154 and 156.

During the flight of droplets in the spray, the forced convective and radiative heat exchanges with the atomization gas lead to a rapid heat extraction from the droplets. A droplet undergoing cooling and phase change may experience three states: *(a)* fully liquid, *(b)* semi-solid, and *(c)* fully solid. If the Biot number of a droplet in all three states is smaller than 0.1, the lumped parameter model[156] can be used for the calculation of droplet temperature. Otherwise, the distributed parameter model[154] should be used.

In the *lumped parameter model*, the transient temperature of a single droplet during flight in a high speed atomization gas is calculated using the modified Newton's law of cooling,[156] considering the frictional heat produced by the violent gas-droplet interactions due to the compressibility of the gas. In the *distributed parameter model*, the Fourier equation is used to describe the transient temperature variation within a droplet. Thus, boundary conditions at the droplet surface and center as well as moving interfaces within the droplet are to be specified in addition to the initial droplet temperature.

In a supersonic gas flow, the convective heat transfer coefficient is not only a function of the Reynolds and Prandtl numbers, but also depends on the droplet surface temperature and the Mach number (compressibility of gas).[154][156] However, the effects of the surface temperature and the Mach number may be substantially eliminated if all properties are evaluated at a film temperature defined in Ref. 623. Thus, the convective heat transfer coefficient may still be estimated using the experimental correlation proposed by Ranz and Marshall[505] with appropriate modifications to account for various effects such as turbulence,[587] droplet oscillation and distortion,[585] and droplet vaporization and mass transfer.[555] It has been demonstrated[156] that using the modified Newton's law of cooling and evaluating the heat transfer coefficient at the film temperature allow numerical calculations of droplet cooling and solidification histories in both subsonic and supersonic gas flows in the spray.

For the purpose of analysis, the thermal history of a metal droplet in the spray can be divided into six regions:

(1) rapid cooling in fully liquid state,
(2) undercooling,
(3) nucleation,
(4) recalescence,
(5) post-recalescence solidification,
(6) cooling in fully solid state.

For an alloy droplet, the post-recalescence solidification involves segregated solidification and eutectic solidification.[619] Droplet cooling in the region *(1)*, *(2)* and *(6)* can be calculated directly with the above-described heat transfer model. The nucleation temperature (the achievable undercooling) and the solid fraction evolution during recalescence and post-recalescence solidification need to be determined additionally on the basis of the rapid solidification kinetics.[154][156]

In spray forming, a large undercooling may generally be attained prior to the onset of nucleation. The most likely solidification mechanism is dendrite growth. The upper bound of the degree of undercooling may be estimated using Hirth's formulation[624] for homogeneous nucleation within a droplet, coupled with the expression for the homogeneous nucleation rate[156] derived on the basis of the classical nucleation theory.[625] In view of the high probability of heterogeneous nucleation associated with the conditions common in spray forming processes, the undercooling calculated assuming homogeneous nucleation needs to be modified properly by means of theoretical analyses or experimental measurements.[156][620]

Following nucleation, the release of the latent heat of fusion generally occurs at a rate that is substantially higher than that of convective and radiative heat dissipation at the droplet surface, leading to an overall temperature rise in the droplet (lumped parameter model) or a temperature rise at the solid-liquid interface (distributed parameter model). This phenomenon is termed recalescence. The extent of the temperature rise during recalescence depends primarily on the undercooling and cooling rate. The solid-liquid interface velocity during recalescence is also a function of the undercooling and cooling rate. Different expressions for the interface velocity have been used in previous studies, including:

(1) Linear and exponential laws for planar continuous growth at small and large undercoolings, respectively,[155][619][626]

(2) Ivantsov equation for growth of paraboloid dendrites,[497]

(3) Power law for dendrite growth.[627]

In view of the flexibility of power laws with different power indices to describe different growth mechanisms, the power law may be used to estimate the growth velocity of dendrites into the undercooled droplet and the solid fraction evolution during recalescence.[155][626] The limit of the droplet temperature rise during recalescence is the liquidus temperature. Satisfying this criterion, the maximum solid fraction during recalescence may be calculated on the basis of an energy balance.[156] As recalescence proceeds, the interface velocity and hence the release rate of the latent heat of fusion decrease. When the latter equals the rate of heat extraction at the droplet surface, or the solid fraction during recalescence attains the maximum, recalescence terminates. Following recalescence, the interface velocity decreases to a relatively steady value. Further solidification of the droplet in the mushy (semi-solid) zone, referred to as post-recalescence solidification, is dictated by the Scheil equation.[619][628] After all the latent heat of fusion is removed, the droplet cooling enters into the region (6).

With the above-described heat transfer model and rapid solidification kinetic model, along with the related process parameters and thermophysical properties of atomization gases (Tables 2.6 and 2.7) and metals/alloys (Tables 2.8, 2.9, 2.10 and 2.11), the 2-D distributions of transient droplet temperatures, cooling rates, achievable undercoolings, and solid fractions in the spray can be calculated, once the initial droplet sizes, temperatures, and velocities are established by the modeling of the atomization stage, as discussed in the previous subsection. For the implementation of the heat transfer model and the rapid solidification kinetic model, finite difference methods or finite element methods may be used. To characterize the entire size distribution of droplets, some specific droplet sizes (for example, $D_{0.16}$, $D_{0.5}$, and $D_{0.84}$) are to be considered in the calculations of the 2-D motion, cooling and solidification histories.

For the numerical calculations, the gas temperature may be assumed to be constant due to its large mass flow rates under some operation conditions. However, modeling studies using coupled solution approach[576][620] have showed that thermal coupling may lead to a hot core of the spray, a reduced gas density in the core, and thermal expansion of the spray to satisfy continuity. The gas temperature exhibits a gradient across the spray that affects the momentum and heat transfer between the gas and droplets. Due partly to this gas temperature distribution, the droplets in the center region of the spray may still remain in fully liquid state while those at the edge of the spray may be already completely solidified at certain axial distance. The gas temperature distribution may also influence the radial distributions of droplet sizes, velocities, volume fractions, and mass fluxes in the spray.

The numerical modeling of the spray stage[154][156] revealed some features of gas flows that are in qualitative agreement with the LDA-measured gas velocities.[609][617] Fig. 5.6 shows the calculated velocities of gas and droplets in the spray. In this figure and the following figures, u is the axial velocity, u_{max} is the axial gas velocity at the starting point of the calculations, r is the radial distance from the spray centerline, z is the axial distance from the starting point, and z_{max} is the maximum axial distance from the starting point in the calculations. The axial gas velocity decays exponentially along the spray axis as the spray spreads. The radial profiles of axial gas velocities exhibit a shape akin to a Gaussian probability distribution. The axial droplet velocities increase initially along the axial direction and attain rapidly their maximum values. With increasing axial distance, the radial profiles of the axial droplet velocities become wider and approach the profiles of gas velocities. The calculated droplet trajectories show a good qualitative agreement with the high-speed video imaging of the actual spray, as illustrated in Fig. 5.7.

The calculated droplet temperatures and solid fractions in the spray are depicted in Fig. 5.8 and Fig. 5.9, respectively. In these figures, T is the droplet temperature, T^0 is the initial droplet temperature, and f_s is the solid fraction of droplet. At any axial distance, the droplet velocity, temperature, cooling rate and solidification rate all exhibit a maximum at the spray axis, and decrease to a minimum at the periphery of the spray, except for the radial locations where solidification occurs. For a given droplet diameter, the achievable undercooling is smaller, the flight time required to reach a given axial distance is longer, and the secondary dendrite arm spacing formed during post-recalescence solidification is larger in the periphery region of the spray than elsewhere in the radial direction. Hence, the droplets in the periphery region of the spray solidify within a shorter axial distance relative to those at the spray axis due partly to the longer flight time in the periphery. These modeling results are consistent with those obtained from microstructural characterization studies[629] in which thick prior-droplet-boundaries (more porosity and less droplet deformation) were observed in the fringe region of spray-deposited materials. In addition, the calculated radial distributions of droplet velocity, cooling rate and SDAS are in qualitative agreement with the findings of the microstructural characterization studies of spray-deposited materials,[629][630] in which a distribution of the microstructural characteristics in the radial direction was found and finer grain sizes were observed in the edges of the deposits.

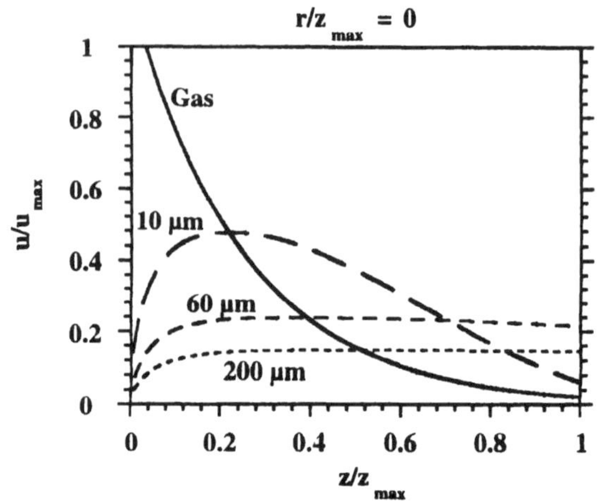

(a)

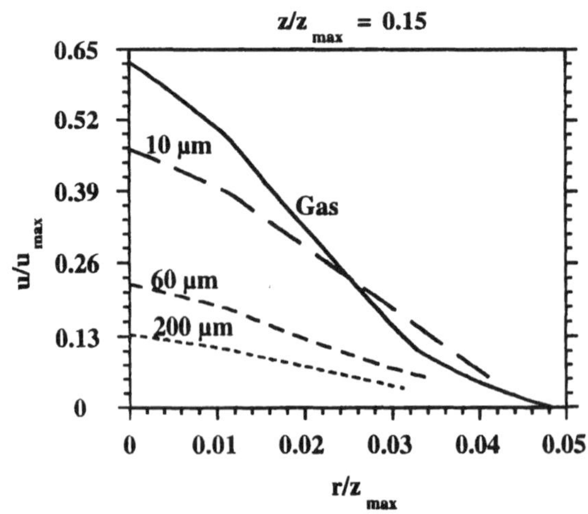

(b)

Figure 5.6. Calculated axial velocity evolution along the spray centerline *(a)* and distribution in the radial direction of the spray *(b)*.

Figure 5.7. Comparison of numerical modeling to high-speed video imaging of droplet trajectories in the spray during spray forming. *Left*: calculated droplet trajectories; *Right*: high-speed video imaging of an actual spray *(Courtesy of Prof. Dr.-Ing. Klaus Bauckhage at University of Bremen, Germany.)*

The numerical modeling of the spray stage[154][156] also provided insight into the effect of droplet size on droplet trajectory and thermal history in the spray (Figs. 5.6, 5.8, and 5.9). With decreasing droplet size, the flight distance required to attain the maximum droplet velocities decreases and the flight time required to reach any given flight distance also decreases. Compared to a large droplet, a small droplet exhibits a large axial velocity, a low temperature, a high cooling rate, and a large undercooling and solid fraction at any radial distance, except for the locations where solidification takes place and the fringe region where the relative velocity between the gas and small droplets is low. In addition, a small droplet has a wider radial distribution at any axial distance. Hence, it may be postulated that coarse droplets comprise the core of the spray whereas the periphery of the spray is populated by fine droplets. These findings are supported by the experimental analyses of particle size distributions[497][630][631] which demonstrated that the oversprayed particles collected during atomization and deposition exhibit a smaller mean size than the average size of powders generated and acquired during atomization,[630] and the mass median droplet diameter decreases along the radial direction from the spray axis.[497][631]

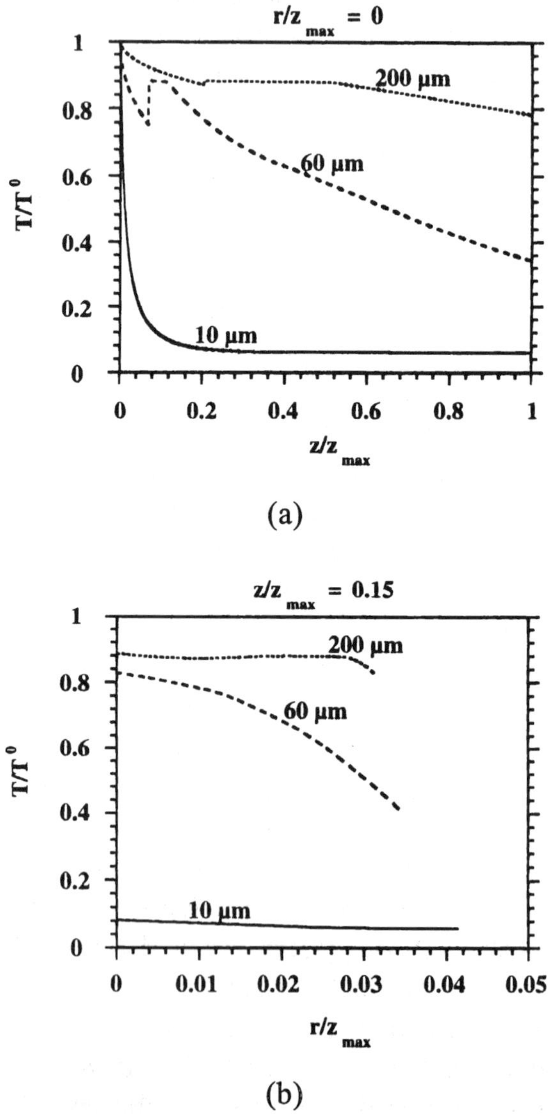

Figure 5.8. Calculated droplet temperature evolution along the spray centerline *(a)* and distribution in the radial direction of the spray *(b)*.

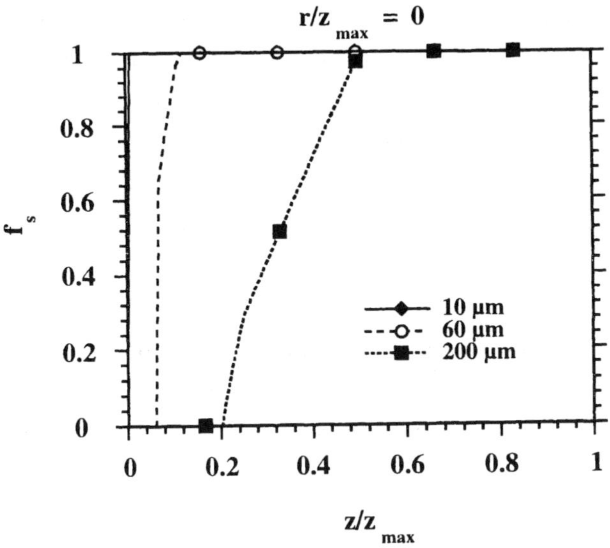

(a)

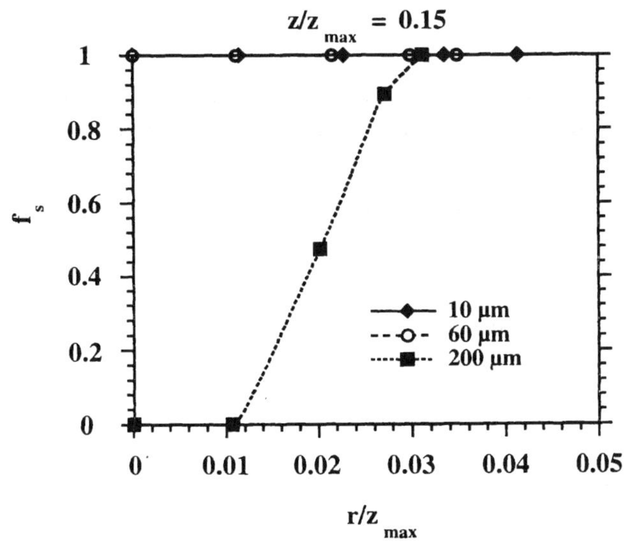

(b)

Figure 5.9. Calculated droplet solid fraction evolution along the spray centerline (*a*) and distribution in the radial direction of the spray (*b*).

It should be noted, however, that some of the tendencies described above may become invalid for very small droplets (for example, smaller than 10 μm under conditions in Ref. 156). Such small droplets may require a longer flight time to a given axial distance far from the atomizer due to the high deceleration, and their cooling rates may decrease as a result of the reduced relative velocity and temperature. In addition, the two-way coupling[576] may affect the momentum and heat transfer between atomization gas and droplets so that the droplet behavior may be different from that discussed above, particularly the radial distributions of droplet sizes and velocities.

For a stationary spray without scanning, a Gaussian shaped mass distribution typically develops with an annular-jet or discrete-jet atomizer. The radial mass distribution in the spray can be formulated in terms of mass flux:[632]

Eq. (12) $$\dot{M} = \dot{M}_0 \exp(-\frac{r}{r_c})^m$$

where $\dot{M}$ is the mass flux in the spray, $\dot{M}_0$ is the maximum mass flux at the spray axis, r is the radial coordinate, r_c is the characteristic radius that controls the spread of the spray, and m is the sensitivity index that influences the shape of mass flux profile ($m = 1.2 \sim 2.0$). More delicate formulation for the mass flux has been developed recently by Uhlenwinkel et al.[631]

The mass flux in the spray scales with liquid metal flow rate. Gas pressure tends to narrow the spray whereas melt superheat tends to flatten the spray.[3] By changing the process parameters and/or manipulating the configuration and/or motion of the spray, the mass distribution profile can be tailored to the desired shape. For example, a linear atomizer produces a relatively uniform mass distribution in the spray. The mass flux distribution in the spray generated with a linear atomizer has been proposed to follow the elliptical form of the Gaussian distribution:[178]

Eq. (13) $$\dot{M} = \dot{M}_0 \exp\left\{-(\frac{x}{B_x})^2 - (\frac{y}{B_y})^2\right\}$$

where B_x and B_y are the characteristic width along the x and y axis, respectively.

The modeling results of the spray stage provide input data and initial conditions for the modeling of droplet-substrate interactions in the ensuing deposition stage of the spray forming process.

5.4.0 MODELING OF DROPLET DEFORMATION ON A SURFACE

Several research groups have conducted numerical analyses and simulations of the transient flow of a single liquid droplet impinging onto a flat surface, into a shallow or deep pool,[397] and related phenomena such as liquid breakup and dispersion,[422] heat transfer,[334][368] and solidification.[50][157][371][401][515] Recently, important progress in the modeling of droplet impingement processes has been accomplished by Liu et al.[18][51][52][144][338][339][378][379][388][389] The modeling studies and their applications will be reviewed in this section.

5.4.1 Modeling of Deformation of a Single Droplet on a Flat Surface

Numerous 2-D models have been developed to simulate droplet deformation processes during impact on a smooth surface. Most of these models assumed axisymmetric deformation of a spherical or cylindrical droplet. The models may be conveniently divided into two groups, i.e., compressible and incompressible.

The *compressible models* resolve a very early stage of droplet deformation when the compressible wave generated by impact has not yet traveled throughout the droplet. In this stage, the shock wave separates the compressed liquid from the undisturbed liquid that is above the compressed liquid and at the initial impact velocity. The intersection of the compressed and undisturbed liquids constitutes a contact ring. As long as the velocity of the contact ring is greater than the sonic speed in the liquid, the shock wave will be attached to the contact ring and the shape of the droplet is a truncated sphere. This stage is referred to as *subcritical flow period* and the stage following it is termed *postcritical flow period*. Radial spreading of the droplet starts after the subcritical flow period and is driven by the high pressures in the compressed liquid. A simple scaling argument shows that the time scale for this early stage is on the order of:

Eq. (14) $$t_{com} = \frac{D_0}{U_s}$$

where U_s is the speed of sound in the liquid. Due to the complexity, the compressible models have only been used to study the initial stage of droplet impact on a solid substrate to obtain the peak pressure. No results have been available for final splat size.

The *incompressible models* offer a simplified examination of the radial flow during droplet impact. The time scale of the impact process may be estimated using the following equation:[515]

Eq. (15) $$t_{incom} = \frac{2D_0}{3u_0}\mathrm{Re}^{0.2}$$

which is at least an order of magnitude longer than the time scale for the compressible flow. Thus, the incompressible models should yield reasonably accurate results. It should be indicated, however, that the extension of the numerical results from 2-D models to actual spherical droplets is not readily achievable. Further, instabilities that may cause breakup upon impact (for example, the classical crown splash geometries presented in Ref. 337) cannot be resolved in any 2-D models.

The first numerical study on the transient flow of a single liquid droplet impinging onto a flat surface, into a shallow or deep pool was performed by Harlow and Shannon.[397] In their work, the full Navier-Stokes equations were solved numerically in cylindrical coordinates using the Marker-and-Cell (MAC) technique.[633] The axisymmetric calculations showed that the liquid sheet jet created on a flat surface by an impinging droplet spreads radially at a velocity 1.6 times the droplet impact velocity, independent of any other parameters. Splashing was resolved for the cases where a droplet impinges into a shallow or deep pool. However, the effects of the surface tension and viscosity, that are important to the deformation of molten metal droplets on a surface, were not taken into consideration. In addition, the numerical simulations did not yield a final splat diameter because all the mechanisms that would limit droplet spreading were ignored. Tsurutani et al.[334] extended the method of Harlow and Shannon by considering the effects of the surface tension and viscosity, and applied the method to a single droplet impinging onto a hot flat surface. These pioneering studies provided valuable insight into the problem of droplet impingement. In both studies, however, interactions between multiple droplets during spreading were not taken into account. In addition, the impact speed, density, viscosity and surface tension of the liquid droplets considered in these studies were significantly lower than those present normally in spray forming and thermal spray processes.

Montavon et al.[634] investigated the transient contact pressure at the interface between an impinging droplet and a flat substrate surface under thermal spray conditions. These researchers divided a droplet impingement process into two stages: *(a)* impact and *(b)* flattening and solidification. To address the transient dynamic phenomena in the impact stage, they developed a finite element model to solve the transient nonlinear Navier-Stokes equations in primitive variables on an Eulerian rectangular mesh with the Euler implicit and Newton-Raphson methods. The numerical results showed that the impact of a droplet on a substrate surface may induce a high pressure at the droplet-substrate interface, and this contact pressure is strongly influenced by spray process parameters, especially droplet diameter, density and impact velocity. The contact pressure increases almost linearly with increasing Reynolds number of droplet at impact. As the interaction time increases, the contact pressure diminishes.

Samsonov and Shcherbakov[635] used the Gibbs energy approach based on non-equilibrium thermodynamics to study the evolution of a spherical droplet after point contact with a solid surface. The study was restricted to the isothermal spreading of an incompressible and non-volatile liquid droplet on a flat, chemically inert, rigid solid surface in an atmosphere of inert gas that is not soluble in the liquid. A two-cell model for the droplet bulk volume and the region of wetting perimeter was used to derive an integral free-energy balance equation for the droplet-substrate-gas system and to determine the thermodynamic forces that govern the spreading kinetics. They also analyzed the role of the linear free energy of the wetting perimeter and the energy of point contact in the primary wetting action.

For droplet impingement processes at very low or zero impact velocities, the driving force for droplet spreading is the difference between two competing factors, i.e., the attractive substrate potential that promotes the spreading, and the surface tension of the liquid that attempts to keep the droplet in spherical shape in order to minimize its surface area.[342] Direct numerical solution of the Langevin equation of motion, Monte Carlo method, and molecular dynamics method have been used to study the spreading dynamics.[48][342][636] Heiniö et al.[48] investigated the spreading of a viscous fluid droplet on a solid surface using a horizontal solid-on-solid model with Langevin theory for 2-D case and the Monte Carlo method for both 2-D and 3-D cases. The droplet-substrate interaction was described first with one-monolayer-range chemical potential and second with a long-range van der Waals type potential. Dynamic droplet profiles, spreading rates, contact angles, and their characteristic time behaviors were examined for both partial and complete wetting regimes. The transition from partial to complete wetting occurs when the total droplet-substrate interaction exceeds the interatomic interaction between fluid particles. Thus, both the thickness and speed of the liquid film are determined by the value and range of the substrate potential. Nieminen et al.[636] conducted molecular dynamic simulations of Lennard-Jones systems to analyze the spreading of a microscopic nonvolatile liquid droplet on a homogeneous solid substrate. The substrate was assumed to be continuous without atomic structure and exerting a long-range Lennard-Jones potential on liquid particles in the droplet. The molecular film was observed to spread initially with nearly constant speed, crossing over later to diffusive spreading. Cheng and Ebner[342] used the Monte Carlo method to implement a 2-D non-volatile Ising-lattice-gas model with particle-number conservation. The dynamics of droplet spreading on a solid surface was simulated with different substrate-droplet long-range attractive-potential strengths and under both wetting and non-wetting conditions. This model was deemed to be an improvement over the previous horizontal solid-on-solid model.

5.4.2 Modeling of Droplet Deformation and Solidification on a Cold Surface

Deformation and solidification of a single molten metal droplet on a cold surface are of particular importance in thermal spray processes. Madejski[401] developed a simplified flow model, including inertial, viscous and surface tension effects, to predict the final splat diameter and height. Solidification during droplet spreading was determined by incorporating the 1-D Stefan solution into the macroscopic flow model. A recent analytical study[157] modified some of the assumptions of Madejski's model and addressed the effects of different solid fractions of a droplet prior to impact. Using Madejski's model, however, the details of the transient deformation process during droplet spreading, especially during the interaction of multiple droplets, cannot be determined.

Trapaga and Szekely[515] conducted a mathematical modeling study of the isothermal impingement of liquid droplets in spray processes using a commercial CFD code called FLOW-3D. Their model is similar to that of Harlow and Shannon[397] except that viscosity and surface tension were included and wetting was simulated with a contact angle of 10°. In a subsequent study,[371] heat transfer and solidification phenomena were also addressed. These studies provided detailed information of the spreading process of a single droplet on a flat surface and related the final splat diameter to operation parameters such as initial droplet velocity, diameter and material properties. The numerical results revealed that the liquid film created on the substrate surface by an impinging droplet spreads radially at a velocity about three times the droplet impact velocity, in agreement with the experimental data of Shi and Dear.[637] The final splat size was shown not sensitive to the liquid surface tension under the operation conditions typical of plasma spray applications. This is especially true for cases where a liquid wets the target substrate. Breakup of a splat of a non-wetting liquid was also reported, although the results were dependent on the mesh size used. Generally, the numerical results showed the same trends as those found by Madejski[401] except for a slightly smaller splat diameter. Trapaga and Szekely also compared their modeling results to their own experimental data for millimeter-sized droplets and indicated that the existence of a contact resistance does not greatly influence the final splat size. However, Dykhuizen[390] argued that this conclusion would not likely be valid for thermal spray applications. In addition, only preliminary numerical results were presented in these studies regarding the interactions of two droplets on a flat surface, a single droplet into a liquid pool and onto a fiber.

Watanabe et al.[517] conducted numerical modeling of deformation and solidification of a molten droplet onto a cold substrate using the Simplified Marker and Cell (SMAC) algorithm and obtained results very similar to those of Trapaga and Szekely.[515] The numerical simulation showed that solidification occurs immediately after deformation, and the solidification time is two orders of magnitude longer than the deformation time. The normalized deformation time (time corresponding to 90% deformation), $t_{\max,d}$, and solidification time, $t_{\max,s}$, are both proportional to the 0.2 power of the Reynolds number:

Eq. (16) $$t_{\max,d} u_0 / D_0 = 0.31 \mathrm{Re}^{0.2}$$

Eq. (17) $$t_{\max,s}u_0 / D_0 = 0.492\,\mathrm{Re}^{0.2}\,\mathrm{Pr}\,\mathrm{St}^{-1}(k_L / k_S)$$

Using a finite element method, Bertagnolli et al.[518] simulated the simultaneous spreading and solidification of a ceramic droplet on a rigid substrate under plasma spray conditions. Numerical simulations of similar phenomena have also been conducted by Pasandideh-Fard et al.[366][367][638] using an enthalpy approach and the SOLA-VOF method.[639] A comparison of the computer generated images with photographs of a spreading droplet showed that the numerical model could predict, with reasonable accuracy, the droplet shape and contact diameter during spreading and at equilibrium,[366] as well as during simultaneous deformation and solidification.[367] The model, however, overpredicted the droplet contact diameter during recoiling.[366]

A series of theoretical and numerical studies of droplet spreading on a surface have been performed by Fukai et al.,[341][362] Kang et al.,[381] Zhao et al.,[368] and Waldvogel and Poulikakos.[50] Zhao et al.[368] investigated the fluid dynamics and heat transfer phenomena during the impingement of a single liquid metal droplet on a cold surface and a single water droplet on a hot substrate. In their modeling study, the basic conservation equations in the Lagrangian formulation were solved numerically by extending the model and associated numerical methodologies of Fukai et al.[341] based on a finite element technique. A deforming mesh was utilized to accurately simulate the large deformations and to accommodate the domain non-uniformity characteristic of droplet spreading processes. The temperature variations in both the droplet and substrate were calculated during droplet deformation. The effects of thermophysical properties of the droplet and substrate, droplet diameter and impact velocity on the fluid dynamics and heat transfer during droplet spreading were examined. The results showed that the impact dynamics is characterized by droplet spreading, recoiling and oscillations. Droplet spreading is retarded and finally halted by the action of the surface tension and viscosity forces. Subsequently, the liquid flow reversal (recoiling) sets in and the main flow is directed toward the axis of symmetry. The fluid motion in the first spreading-recoiling cycle is substantially stronger than that in subsequent oscillations, suggesting that within the impact velocity range considered, most of the droplet impact energy is dissipated within the first cycle. The duration of the spreading motion in the first cycle is considerably shorter than the duration of the recoiling motion. The spreading and recoiling motions depend upon the initial droplet velocity and properties. The occurrence of the recoiling and mass accumulation around the splat periphery were deemed to be standout features of this type of numerical simulations under the impact conditions considered. The recoiling leads to a non-monotonic dependence of the maximum splat diameter on time. Accordingly, the maximum splat thickness does not always occur at the axis of symmetry, but instead at the splat periphery in the late stage of spreading. The predicted temperature distribution in a molten metal droplet during quenching exhibits strong 2-D features. Convection effects on the temperature evolution were found to be important for the entire history of spreading. These effects sometimes result in a practically radial temperature variation in the late stage of spreading, particularly in the cases of high impact velocities, although the splat diameter is many times larger than the splat thickness. The fluid temperature at the spreading front is substantially lower than the temperature at the splat center. Thus, solidification may be expected to initiate around the periphery of the flattening droplet in the neighborhood of the contact line and advance towards the splat center. Under the impact conditions considered, the heat transfer time scales were found to be comparable to the deformation time scales.

Waldvogel and Poulikakos[50] extended the model and numerical techniques of Zhao et al.[368] by incorporating solidification and droplet-substrate contact resistance in the heat transfer model. They conducted both theoretical and experimental studies on the impact and solidification of molten solder droplets on a multilayer substrate. The theoretical model was based on the Lagrangian formulation, and accounted for a host of thermal-fluid phenomena, including surface tension and heat transfer with solidification. Deforming finite elements with integrated automatic mesh generation were utilized to accommodate the large deformations that develop during the computations. The influences of initial droplet temperature, impact velocity, thermal contact resistance and initial substrate temperature on droplet spreading, final deposit shape, and onset and completion times of solidification were examined in the simulations. Many non-intuitive results were obtained, such as the non-monotonic dependence of solidification time on variations of many parameters considered. The modeling results also suggested strong coupling between the droplet dynamics and solidification behavior.

It should be noted that it is difficult to obtain models that can accurately predict thermal contact resistance and rapid solidification parameters, in addition to the difficulties in obtaining thermophysical properties of liquid metals/alloys, especially refractory metals/alloys. These make the precise numerical modeling of flattening processes of molten metal droplets extremely difficult. Therefore, experimental studies are required. However, the scaling of the experimental results for millimeter-sized droplets to micrometer-sized droplets under rapid solidification conditions seems to be questionable if not impossible,[390] while experimental studies of micrometer-sized droplets under rapid solidification conditions are very difficult, and only inconclusive, sparse and scattered data are available.

5.4.3 Modeling of Interactions of Multiple Droplets on Flat and Non-Flat Surfaces

Most previous modeling studies focused on the spreading process of a single droplet on a flat surface, not taking into account the influence of surface roughness and interactions of multiple droplets. Under the practical conditions in spray deposition and thermal spray processes, the physical phenomena of droplet spreading and consolidation take place at a microscopically rough and non-flat surface on a target substrate. It is widely accepted that the most critical factors governing the quality of as-sprayed materials are bonding and porosity which are directly related to the interactions of droplets on the target substrate surface. Therefore, it is necessary to investigate the transient deformation behavior of molten droplets during impingement onto a rough and non-flat surface, including the interactions of multiple droplets. It is also desired to ascertain the effects of process parameters, such as substrate surface condition, initial droplet velocity and diameter, as well as material properties, on the deformation and interactions. In addition, the mechanisms of the formation and evolution of various types of porosity need to be addressed.

Recently, Liu et al.[18][51][52][144][338][339][378][379][388][389] performed extensive modeling studies on droplet impingement processes. In their work, the deformation and interactions of multiple droplets on both flat and non-flat surfaces were numerically investigated. The governing equations describing the fluid flow and interactions during substrate impact of droplets were composed of the continuity equation, the complete Navier-Stokes equations for transient, axisymmetric, viscous, incompressible fluid flow, including surface tension effects, and the equations for the VOF function. The full Navier-Stokes equations in primitive variables were solved on an Eulerian, rectangular, staggered mesh in cylindrical geometry using the extended SOLA-VOF method.[530][639] Free surfaces were represented with data of the VOF function on the numerical mesh. Surface tension was modeled as a volume force derived from the continuum surface force model.[530] The numerical calculations were conducted by modifying a computer code called RIPPLE developed by Kothe et al.[530] at Los Alamos National Laboratory. A two-domain method was used to simulate the thermal field and solidification in the flattening droplets and to track the moving solid-liquid interfaces. The Stefan solution of solidification problem was incorporated into a two-phase flow continuum model to simulate the liquid flow with a growing solid layer. The numerical results were used to determine the exact motion, solidification and interactions of droplets with arbitrary free surfaces and solid-liquid interfaces on the deposition substrates of various surface morphologies. The micro-porosity was quantitatively calculated from the VOF function data based on an algorithm developed by Liu et al.[52]

The modeling studies by Liu et al.[18][51][52][144][338][339][378][379][388][389] not only provided detailed information on spreading and interactions of multiple droplets on flat and non-flat surfaces, but also improved the understanding of the effects of important process parameters on the spreading and interactions of multiple droplets. In addition, the numerical results provided insight into the formation and evolution of vortices and the breakup and ejection of liquid from deposition surface during impingement of multiple droplets. Therefore, these studies formed a useful basis for exploring the mechanisms that govern the micro-porosity evolution and for correlating the sticking efficiency of droplets on the deposition surface to thermal, dynamic and geometry conditions during spray forming. Moreover, the work established a good starting point for modeling droplet spreading with simultaneous solidification.[390] As solidification models are improved, incorporating thermal contact resistance and nucleation delay, more accurate predictions of droplet impact dynamics may become possible.

The numerical simulations[388][389] predicted that the liquid film created on a substrate surface by an impinging droplet spreads radially at a velocity twice the droplet impact velocity. A fully liquid droplet impinging onto a solid particle on a flat surface leads to the formation of macro-pores between the particle boundary and the surface (Fig. 5.10). Multiple fully liquid droplets striking onto a solid particle on a flat surface substantially eliminate the macro-pores, but simultaneously produce vortices within the liquid, as well as breakup and ejection of the liquid. Multiple fully liquid droplets striking onto multiple solid particles on a flat surface not only lead to the vortices and ejection of the liquid, but also to the formation of the macro-pores due to the large solid fraction. Therefore, increasing the liquid fraction in the spray may reduce the macro-porosity significantly. However, an excessive liquid fraction in the spray may decrease the sticking efficiency as a result of the increased ejection of the liquid from the deposition surface. Detailed descriptions of various mechanisms of porosity formation and effects of thermal, dynamic and geometry conditions on the sticking efficiency have been given in Ref. 3 for spray forming processes. Further discussions on the interactions of multiple droplets on flat and non-flat surfaces can be found in Sec. 3.2.5.

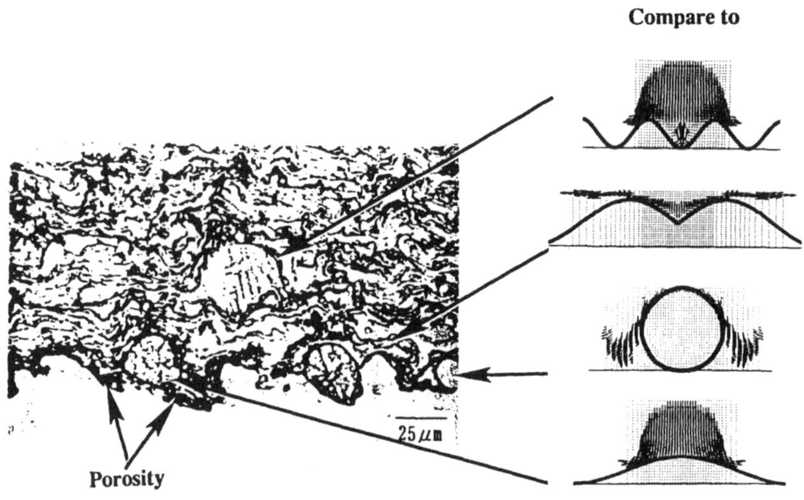

Figure 5.10. Droplet deformation and splashing on deposition surfaces of various shapes/morphologies. *(Micrograph reprinted with permission from Re. 640; Plots reprinted from Ref. 389, © 1995, with kind permission from Elsevier Science Ltd., The Boulevard, Lamgford Lane, Kidlington 0X5 1GB, UK.)*

The numerical calculations[389] also revealed that a large droplet diameter (Fig. 5.11), a high impact velocity (Fig. 5.12), and/or a high liquid density (Fig. 5.13) leads to a large final splat diameter. A high impact velocity, a large roughness height, a small roughness spacing, and/or a low liquid viscosity (high liquid temperature) results in an early occurrence of a violent breakup of the liquid, reducing the sticking efficiency. Decreasing the roughness height and/or increasing the roughness spacing of the deposition surface may improve the extent of droplet flattening, reduce the liquid breakup and porosity, and enhance the sticking efficiency.

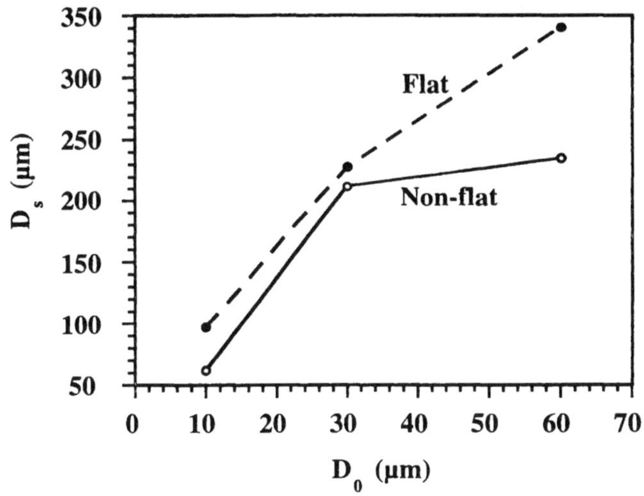

Figure 5.11. Final splat diameter as a function of initial droplet diameter: Re = 3177, 9530, 19060; We = 177, 530, 1059; $\varepsilon/D_0 = 0$ (*flat*); $\varepsilon/D_0 = 0.33$ and $\lambda/D_0 = 2.8$ (*non-flat*). *(Reprinted from Ref. 389. © 1995, with kind permission from Elsevier Science Ltd., The Boulevard, Langford Lane, Kidlington 0X5 1GB, UK.)*

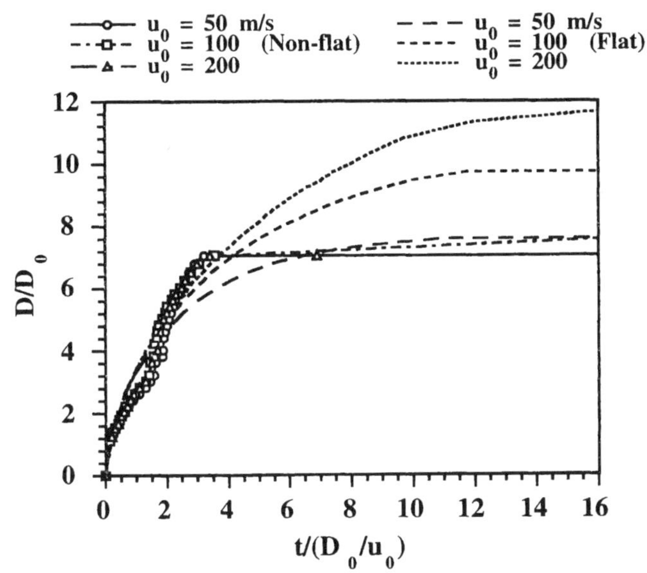

Figure 5.12. Comparison of spreading behavior of a single droplet at different initial impact velocities: Re = 9530, 19060, 38115; We = 530, 2119, 8474; $\varepsilon/D_0 = 0$ (*flat*), $\varepsilon/D_0 = 0.33$ and $\lambda/D_0 = 2.8$ (*non-flat*).*(Reprinted from Ref. 389. © 1995, with kind permission from Elsevier Science Ltd., The Boulevard, Langford Lane, Kidlington 0X5 1GB, UK.)*

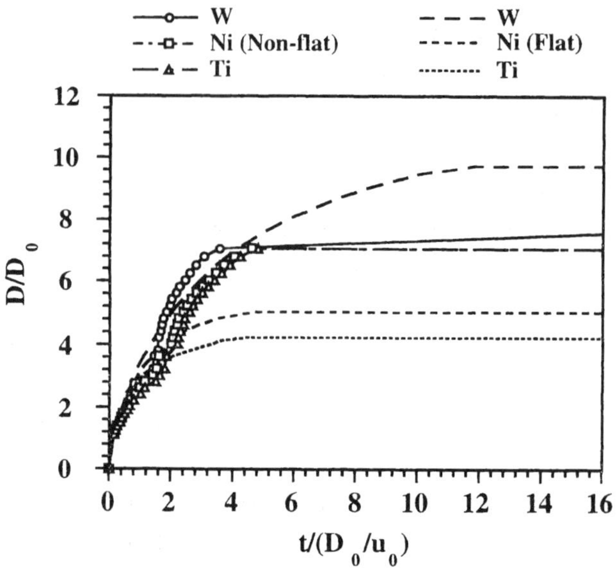

Figure 5.13. Comparison of spreading behavior of a single droplet of different materials: Re = 19060, 4323, 2372; We = 2119, 1334, 747; $\varepsilon/D_0 = 0$ (*flat*); $\varepsilon/D_0 = 0.33$ and $\lambda/D_0 = 2.8$ (*non-flat*). *(Reprinted from Ref. 389, © 1995, with kind permission from Elsevier Science Ltd., The Boulevard, Langford Lane, Kidlington 0X5 1GB, UK.)*

5.4.4 Modeling of Droplet Deformation and Evaporation on a Hot Surface

Deformation, heat transfer and evaporation of liquid droplets impinging onto a hot surface have been studied by many researchers in combustion community.[422][599][641]-[643] It was found that the droplet Weber number is an important parameter for the impingement process. At low Weber numbers, a droplet rebounds elastically from a hot surface. A cushion of vapor forms under the droplet during the interaction, preventing it from wetting the surface. With increasing droplet Weber number, the normal velocity component of the rebounding droplet decreases and the droplet may actually break up into smaller droplets. For droplet Weber numbers higher than about

40, the droplet spreads out into a thin liquid film on the surface with little or no normal velocity component.[641] This criterion, as suggested by Wachters and Westerling,[641] has been used by other researchers in subsequent modeling studies.[574][599] The degree of droplet-surface wetting depends on many factors, such as surface temperature, material and roughness, and droplet impingement angle, etc.[641] At high temperatures, film boiling may set in.

Naber and Reitz[599] proposed three impingement models:

(a) Stick model: Droplets adhere to wall surface at the impingement location and continue to vaporize.

(b) Reflect model: Droplets that impinge the wall rebound with their tangential and normal velocity components unchanged but an opposite sign for the normal velocity component.

(c) Jet model: Droplets glide along the wall surface.

Computed spray structure showed that a wall-jet emanates radially outward from the spray impingement site. The leading edge of the jet contains a large vortex. Small droplets are deflected away from the wall by the gas wall-jet flow, forming a wall spray with the small droplets entrained in the gas wall-jet vortex. Large droplets that impinge on the wall with high momentum move along the wall. The spray-wall interaction is weak for large impingement angles (i.e., angle between spray jet centerline and wall surface normal). Comparison of the model predictions with experimental data[644] suggested that the jet model agrees best with the experiments.

Gonzalez et al.[574] modified these models for cold-starting diesel engine applications. Naber and Farrell[645] studied hydrodynamics of droplet impingement on a heated surface under the diesel engine conditions where the combustion chamber pressures are higher than the critical pressure of the impinging fuel. They showed that the formation of a liquid film on the wall surface influences heat transfer and vaporization rates. For surface temperatures below the critical temperature, wall-wetting takes place even at supercritical pressures. For surface temperatures above the critical temperature, non-wetting

(Leidenfrost) phenomenon occurs. For diesel engines, the wall-wetting regime is the appropriate hydrodynamic regime. Eckhause and Reitz[646] formulated a spray-wall heat transfer model based on these findings and simulated the breakup of droplets during impingement on a wall using the KIVA code.[555] In the flooded regime, heat transfer was modeled by means of the boundary layer correlations. In the non-flooded regime, heat transfer was modeled with the correlations for individual droplets impinging on a surface.

In a series of numerical modeling studies of diesel spray impingement on a flat wall, Senda et al.[422] fine-tuned the impingement models developed by Naber and Reitz.[599] In the modeling, Senda et al. considered the following phenomena: *(a)* formation of fuel film on the wall, *(b)* breakup of the film and impinging droplets, *(c)* heat transfer from the wall to the droplets, *(d)* droplet-droplet and droplet-film interactions, and *(e)* dispersion of secondary droplets. They proposed new sub-models for droplet deformation, heat transfer, fuel film formation and dispersion process of an impinging spray near a flat wall, and incorporated the sub-models into the KIVA code[555] to simulate the spray impingement on both low-temperature and high-temperature walls. The numerical results showed both upward and radial dispersions of secondary droplets due to the breakup and splashing of impacting droplets during the spray impingement.

6

Measurement Techniques for Droplet Properties and Intelligent Control of Droplet Processes

Measurements of droplet properties are required in many areas of scientific research and engineering applications. For example, accurate and detailed measurements of distributions of droplet properties in sprays are needed for understanding spray processes and validating spray modeling. The droplet properties to be measured include droplet size, velocity, mass flow rate (flux), temperature, deformation history, and final splat dimensions, among others. Accurate spray diagnostics require developments in instrumentation to increase the span of droplet sizes down to submicron scale and to measure distributions and flux quantities in sprays with significantly higher number densities than current normal levels. It is also desired to measure droplet temperature, composition (for multi-component fuels), and non-sphericity (for liquid breakup), as well as other properties, jointly with the measurements of droplet size and velocity in high number-density, highly turbulent sprays.

Various measurement techniques have been developed and applied with different degrees of success. Some techniques, that are well applicable to the measurement of single droplet properties, find difficulties in the application to measurements of droplet properties in practical sprays. The difficulties may be caused by the large number of droplets, wide range of droplet properties (such as wide droplet size range, high and varying droplet velocities and temperatures), and/or simultaneous phase changes that lead to variations of droplet properties with time through evaporation, solidification, chemical reactions and/or coalescence. In many spray combustion processes, for example, dramatic variations in droplet properties may occur due to the high pressures and temperatures in the sprays. Such severe conditions further increase the difficulties in accurate and reliable spray diagnostics. These are among the factors to be considered when selecting a measurement technique to meet the requirements of a specific application.

For the measurement of droplet properties in a spray, no intrusive techniques are often desired. The measurement should not create disturbance to the spray pattern. An ideal measurement technique should have a large range of capability to measure both spatial and temporal distributions. Such technique should tolerate wide variations in droplet properties at some extreme conditions present in sprays in various engineering applications. An appropriate technique should also be able to acquire sufficient representative samples to ensure a reasonable measurement accuracy. Rapid sampling and data processing means are hence needed for the analysis of measurement results. The sampling, data acquisition and processing system must be fast enough to record every droplet passing through a measurement volume when measuring the number density of droplets in a spray. The system should be able to reproduce in real time the mean droplet properties, their distributions and standard deviations in histogram form. Although these features are often required in practical applications, it is virtually impossible for a single measurement system to provide all these functions.

Many laser-based droplet diagnostic techniques have evolved from the fields such as spray combustion and spray drying. Phase-Doppler particle analyzer is now recognized as the most successful and advanced diagnostic instrument for spray characterization. Other proven diagnostic techniques include laser velocimetry and interferometry for measurements of droplet velocity and size, various forms of Rayleigh, Raman and fluorescence spectroscopies for measurements of temperature and species concentration.[647] The use of the laser-based diagnostic techniques is considered as the state-of-the-art in spray characterization.

In this chapter, the primary measurement techniques for droplet properties are outlined, and their capabilities and limitations are discussed. The approaches to intelligent process control as related to droplet processes are presented along with descriptions of recent developments.

6.1.0 MEASUREMENT TECHNIQUES FOR DROPLET SIZE

Many useful measurement techniques for droplet size have been developed. Each of these methods has its advantages and limitations, and none of the methods is fully satisfactory. In addition, no single method/instrument can cover the entire range of droplet sizes or other properties that need to be characterized. The measurement techniques for droplet sizing may be grouped conveniently into four primary categories (Table 6.1): *(a)* mechanical methods, *(b)* electrical methods, *(c)* optical methods, and *(d)* acoustical method. *Mechanical methods* are relatively simple and of low cost. Droplets are collected in either liquid or frozen state, followed by microscopic or sieving analysis. *Electrical methods* involve the detection and analysis of electronic pulses generated by droplets in a measurement volume or on a wire. The electronic signals are then converted into digital data and calibrated to produce information on droplet size distribution. *Optical methods* have been developed in recent years and are finding an increasing range of applications. Some of the optical methods are capable of simultaneously measuring droplet size and velocity,[648]-[650] as well as velocity and number density.[651] However, many important optical measurement techniques provide only the information of spatial distributions of droplet properties in a spray, whereas the information of temporal distributions is generally more desirable. An *acoustical method*[6] has been evaluated for the measurements of fine droplets. Probe techniques, such as hot wire technique, have a size range from 1 to 600 μm, while imaging methods offer a size range above 5 μm depending on the system used.

Table 6.1. Measurement Techniques for Droplet Size

Categories	Methods	Size Range (μm)
Mechanical	Collection of droplets on slides or in cells	≥~3
	Cascade impactor	≥~3
	Molten-wax and frozen-drop techniques	
	Mechanical sieving of metal powder	
Electrical	Pulse counting technique	
	Charged wire technique	
	Hot wire technique	1–600
Optical	*Imaging* Photography	>~5
	Videography	
	Holography	5–1000
	Non-imaging Light-scattering interferometry	5–3000
	Phase-Doppler anemometry	0.5–3000
	Light intensity deconvolution technique	0.2–200
	Light scattering technique	10–250
	Malvern particle analyzer	1–500
	Polarization ratio particle sizer	
	Intensity ratio method	
	Phase optical-microwave method	
	Dual-cylindrical wave laser technique	
Acoustical	—	5–30

In some droplet sizing techniques, such as the TSI aerodynamic particle sizer,[652] both droplet shape and density affect the response of the sizer. A higher-density droplet has a smaller drag, leading to a longer flight time through the sensing zone and a larger measurement size. On the contrary, a non-spherical droplet has a larger drag than a spherical droplet with the same diameter and density, giving rise to a shorter flight time through the sensing zone and a smaller measurement size. Depending on the surface tension, viscosity, and original size of a droplet, the degree of droplet shape deformation varies, and thus the measurement size may change with all these parameters.

Extensive reviews of droplet size measurement techniques have been given by Chigier,[647][653] Bachalo,[654][655] Hirleman,[656] Jones,[657] and Lefebvre.[1] A comparison of performance of seventeen droplet-sizing instruments from six different types has been made by Dodge.[658]

6.1.1 Mechanical Methods

Mechanical methods are fairly simple. The methods may be divided into two major groups. One involves the capture of a sample of droplets on a solid surface or in a cell containing a special liquid. The captured droplets are then observed or photographed by means of a microscope, generating information on droplet size. The other involves freeze-up of droplets into solid particles and subsequent sieving to generate droplet size distribution. The major problem is associated with the extraction and collection of representative spray samples.

Collection Techniques. The simplest mechanical method for normal liquid droplets is the *slide collection* (*slide sampling*) or *impression method*. This method was extensively used three decades ago, and has been rarely employed since then. In this method, when the slide is exposed to a spray, droplets impinging on it make impressions. The impressions are then observed and measured usually using a Quantimet image analyzer, although a microscope fitted with a traversing scale may be used for the measurement. The measured data are subsequently converted to actual droplet sizes based on a correction factor proposed by May.[659]

The slide used in this method is usually made of glass and coated with a material of very fine grain to generate clear impressions of droplets on it. For example, a thin soot layer with a very fine surface can be deposited on a glass slide by burning a kerosene-soaked wick under the glass,[438] and a thin film of magnesium oxide can be produced by burning a magnesium ribbon.[659] The soot layer thickness may be determined from the relation between the droplet Weber number and impression diameter, and thus varies for different liquids.[438] Droplets as small as 3 μm in diameter can be observed and measured using slides with these coatings. In addition, spatial and overall droplet size distributions may be measured with this method by properly manipulating the slide.[459]

The fraction of slide surface to be covered by collected droplets is an important factor influencing overall measurement accuracy and time. If the slide is covered by too many droplets, the measurement error and time will increase due to droplet overlap and tedious counting. If too few droplets are collected, the sample may not be large enough to generate statistically representative data. For the measurement accuracy and ease, a coverage of 0.2% has been found to be sufficient and satisfactory with an upper limit of 1.0%.[1]

Size bias associated with this measurement method can be caused by many phenomena, such as droplet evaporation, selective collection, and correction factor. In a spray of a normal liquid, small droplets have short life time due to evaporation. This effect is especially significant in the measurements of fine sprays. The selective collection may occur in the measurements of the sprays generated with air-blast atomizers. Since small droplets tend to follow streamlines of the air (gas) flow field around the collection slide surface while large droplets impinge onto the slide surface due to large inertia, more large droplets are collected. Thus, the measured droplet sizes tend to be larger than the actual sizes. In addition, the diameter of a deformed (flattened) droplet on the collection surface needs to be corrected to the diameter of the original spherical droplet by multiplying a correction factor. The determination of the value of this correction factor is not trivial because it varies with droplet properties (notably surface tension) and velocity as well as coating material properties. It may change from about 0.5 to 0.86 for oil droplets on a clean glass and a magnesium-coated slide, respectively.

To reduce the bias caused by droplet deformation during impingement, droplets can be collected on slides[660] or in cells[661] containing an appropriate immersion liquid in which the droplets do not dissolve but remain stable and suspended during counting and measuring.[317] This is the so-called *liquid immersion sampling technique*. In this technique, an appropriate immersion liquid must be selected for each given liquid. For example, Stoddard solvent and white kerosene may be the suitable immersion liquids for the droplets of water and water-alcohol mixtures.[661] For an immersion liquid with a density slightly smaller than that of droplets, the shape of the droplets can remain almost unchanged if the evaporation and disintegration of the droplets are prevented during impinging the immersion liquid. Thus, the droplet sizes can be measured directly without need for correction. However, this method may generate large measurement errors when used for coarse sprays because large droplets tend to split into smaller ones upon impact with the immersion liquid. In addition, droplets may coalesce during collection. To avoid or reduce the breakup and coalescence of droplets, immersion liquids of low viscosity and surface tension may be used to facilitate the penetration of liquid surface by the droplets and to reduce shear stresses acting on the droplets during the penetration. Thus, the droplets may remain stable and suspended near the bottom of the collection cell for some time allowing photographing at high magnification. The magnified images of the droplets can be processed by means of a computer to generate data for droplet size distribution.

Zhang and Ishii[662] used an isokinetic sampling probe and image processing system for droplet size measurements in two-phase flows. Droplet samples were extracted from flow field by the sampling probe and collected in an immiscible liquid. Droplet images were grabbed and digitized by the image processing system. Image processing software was developed for droplet counting and sizing, with the capability of distinguishing overlapping droplets.

Possible measurement bias factors such as droplet deposition in the probe, droplet breakup and coalescence were studied. A simple criterion for minimizing measurement bias was proposed. The system can be used for both water and liquid-metal droplets.

Cascade Impactor method is conceptually a slide collection method. Theoretically, any number of stages may be used to make the desired droplet size groupings. In single-stage impactor method, a slide (i.e., an impaction surface) is placed in a spray below a nozzle. The slide is usually coated with a mixture of carbon and magnesium oxide to retain the droplets impinging on it. A large droplet traveling at high velocity most likely impacts the slide placed in its path due to its large momentum, while a small droplet traveling at low velocity tends to follow air flow around the slide. With increasing air flow velocity, an increased number of small droplets will impact the slide due to increased momentum. In multi-stage (cascade) impactor method, air flow velocity is increased gradually, thus both the velocity and impaction efficiency of droplets increase from slide to slide. For the improvement of impaction efficiency, impactors of large dimensions should be used. The advantages of the cascade impactor method include *(a)* isokinetic sampling that can overcome discrimination against very large or very small droplets in a spray; *(b)* easy assessment of the quantity of liquid for each droplet size class collected in each stage of a cascade impactor by gravimetric or chemical methods; *(c)* satisfactory operation under arduous conditions; *(d)* capability to withstand the rigors of routine handling; and *(e)* ease and simplicity in design and operation relative to optical methods.

When a droplet falling at a certain velocity impacts on an object, it will wet and spread through the surface if the contact angle is less than 90°. If the object is very thin, the droplet may all soak into it and the liquid may reach the other side that can be seen if the droplet contains a tracer dye. In a thick and porous Kromekote® card,[507] a droplet spreads to form a circular stain with simultaneous absorption and penetration into the card. Kromekote® cards have been used in forestry spray trials to collect pesticide droplets for assessing target coverage (number of droplets per unit area) and size spectra of the spray cloud reaching the treatment site.

Molten Wax and Frozen-Drop Techniques. *Molten wax technique* is a mechanical method developed in the early 1940's.[663] In this method, paraffin wax is heated to an appropriate temperature above its melting point so that its properties approach those of aviation kerosene. The molten wax is then injected into a large pressure chamber, in which the droplets of the molten wax rapidly cool and solidify. The wax particles are subsequently collected and separated into size groups using a sieve. Each size group is weighted to obtain the volume or mass fraction of the corresponding size range, generating directly the cumulative volume or mass distribution and mass median size. Thus, it alleviates large expenditure of time and efforts for sizing and counting of numerous individual droplets. In addition, this technique offers better statistical accuracy due to the huge number of droplets/particles in a sample.

The *molten wax method* requires that the properties of a simulant are very close to those of the liquid of interest. Thus, the choice of suitable materials is limited. The method also suffers from some practical problems in preheating the wax and errors incurred by changes in physical properties of the wax during cooling after leaving the injector. Since the properties of the wax (notably surface tension and viscosity) critically influence the process of droplet formation, it may not be accurately reproduced due to the changes in these properties with temperature. Therefore, it may be required that the air in the near-nozzle region, where the key process of droplet formation occurs, be heated to the same temperature as that of the molten wax.

To get rid of the drawbacks associated with the molten wax technique, an alternative is to solidify the droplets of the liquid of interest as soon as they are formed after leaving the nozzle. This is the basic concept of the *frozen-drop technique*, a natural extension of the molten wax technique. The freeze-up and collection of droplets may be carried out in many different ways. For example, liquid droplets from a fuel spray can be collected into a stream of fluid at room temperature. The fluid carrying the droplets is then introduced into an alcohol bath at the temperature of dry ice where the droplets freeze into solid particles. The solid particles are subsequently sieved to obtain the size distribution. Water droplets can be collected in a hexane pan,[664] wrapped around with dry ice and kept at –20 °C. The droplets freeze up during falling through the hexane and settle on a shutter. The shutter is then opened, allowing the frozen particles to fall 30 cm and come to rest on a scale pan. The time needed for the traveling and the variation of the scale pan weight with time are measured. The traveling time is dependent on the relative density of the droplets and the carrying liquid, the liquid viscosity, the falling distance, and the droplet diameter. Thus, the droplet size can be calculated on the basis of the measured time data. The entire measurement takes very long time to complete, particularly for small droplets.

Droplets can also be collected and frozen in a liquid nitrogen bath placed below an atomizer.[665] After sufficient droplets are collected, the frozen particles are passed through a series of screens of different sizes to determine the size distribution. The density of the liquid being atomized is required to exceed 1200 kg/m^3 to avoid any agglomeration of droplets on the surface of the liquid nitrogen bath. Thus, the method is not applicable to many normal liquids such as water, fuel oils, and kerosene. In a modified version of the frozen-drop method, frozen particles are collected in a nitrogen-cooled perspex pot and photographed through a microscope. The final prints are enlarged and analyzed to obtain the droplet size distribution. This method is simple and convenient. However, similarly to the original version, the change in droplet size due to freezing requires a correction factor for the determination of the actual droplet size.

The frozen-drop technique was naturally adopted in measuring molten metal droplet size before any other methods became available. Similarly to the methods for normal liquids, the freeze-up and collection of molten metal droplets may be carried out in many different ways. For example, metal droplets can solidify during flight in gaseous or liquid medium in a spray chamber.[3]-[5] The solidified particles are subsequently sieved to obtain the size distribution.

6.1.2 Electrical Methods

Electrical methods involve the detection and analysis of electronic pulses generated by droplets in a measurement volume or on a wire. The electronic signals are then converted into digital data and calibrated to produce information on droplet size distribution. A detailed review of electrical methods for droplet size measurements has been made by Jones.[657]

In one of the early electrical methods,[666] droplet size is measured by counting pulses that are generated by droplets momentarily bridging a gap between two sharp needles across which a potential difference is applied. The width of the gap can be adjusted and measured by means of frequency counts that in turn can be converted into the droplet size. The method is easy to implement, and therefore may be used in certain types of applications. However, the method has some inherent limitations. For example, the apparent resistance of a droplet bridging the gap is a function of its velocity and the depth of immersion of the needles into it. Thus, a large proportion of the pulses may be below the threshold preset to avoid counting noise, leading to errors in droplet size. In addition, changes in droplet velocity exert the same influence on counting rate as changes in droplet size, so that a change in measured droplet size may occur when merely the droplet velocity is varied. To quantify the sources of errors and identify the optimum operation range, it may be necessary to calibrate the method by comparing to a proven droplet size measurement technique.

Another early electrical method is the *charged wire technique*.[667] In this method, a droplet impinges on an electrically charged wire and removes a certain amount of charge that is a function of the droplet size. This charge amount is converted into a measurable voltage pulse and the information on the droplet size is subsequently extracted. The electrical conductivity and the flux of droplets all influence the measurement results. Liquids of low conductivity generate pulses of long duration, increasing the probability of counting errors due to the superimposition of successively arriving droplets on each other. Thus, the effective application of this method may be restricted to dilute sprays of high-conductivity liquids.

The third primary electrical method is the *hot wire technique.*[668] The associated device is termed *hot wire anemometer.* Similarly to other electrical methods, it is an intrusive technique. However, the wire causes very little disturbance to the flow in a spray. The hot wire anemometer operates on the principle that when a droplet impinges on a heated wire it causes local cooling of the wire as it evaporates. The cooling process depends on the size and number of droplets deposited on the wire so that the information on the droplet size and concentration in a spray can be obtained. For example, for a constant-temperature hot wire anemometer, the electrical resistance of the wire is high and sensibly uniform along its length if no droplets are deposited. When a droplet impinges on the wire, the local cooling by the droplet reduces the resistance in proportion to the droplet size. This reduction in the resistance is manifested as a voltage drop across the wire supports. The constant-current electrical energy supplied to the wire then evaporates the droplet, leaving the wire ready to accept other droplets. Depending upon droplet sizes, such a process may take a few milliseconds. The principle and calibration procedures of a commercial hot wire instrument have been described by Mahler and Magnus.[668] When appropriately used, a hot wire instrument can yield accurate measurements of Sauter mean diameter, mass median diameter, volume flow rate, and concentration of droplets. Typically, a platinum wire is used with a diameter of 5 μm and a length of 1 mm. For large droplets, the droplet velocity range that can be measured may be limited to below about 10 m/s due to droplet shattering on the wire. This method may not be applicable to liquids that leave a residue on the wire because it may influence the calibration.

The hot wire technique appears to be the most useful one among the various electrical methods. However, the hot wire anemometer can operate only at relatively low droplet velocities.

6.1.3 Optical Methods

Various optical methods have been developed and used in spray diagnostics. Although each technique has its own advantages and limitations, all optical methods have an important, common attribute of allowing droplet size measurements to be made without disturbing the flow field being measured. *Optical methods* for droplet size measurements include two primary types, i.e., imaging and non-imaging techniques.

In *imaging methods*, such as flash photography, high-speed videography/cinematography, and holography, droplets are visualized at the time when they pass through the measurement point. Therefore, the imaging methods can eliminate the errors associated with evaporation and/or coalescence of droplets after sampling. The imaging methods are easy to understand and independent of droplet composition. They can record droplet distribution in space, identify ligaments and other non-spherical particles, and measure droplet shape. However, the methods are practically limited to droplet sizes larger than about 5 μm. The data reduction is slow; the number of samples is limited by the recording volume and droplet number density; and the working distance is limited as well. In addition, the sample volume is a function of droplet diameter, and high quality optics and windows are required. Extensive descriptions have been given by Chigier[669] on various optical imaging methods, data processing, and image analysis.

Non-imaging methods may be further divided into two groups: *(a)* single-particle counters that count and size individual droplets one at a time (such as light-scattering interferometry, Phase-Doppler anemometry, and intensity deconvolution technique, etc.), and *(b)* ensemble methods that measure a large number of droplets simultaneously (such as Fraunhofer diffraction methods and Malvern particle analyzer). *Single-particle counters* have the capability to perform measurements non-intrusively with high spatial resolutions over a large size range. They are extremely fast, measuring samples of 10000 or more particles in a second. They also have the potential of producing both droplet size and velocity distributions directly. These distributions are based on the measurements of individual droplets and do not require the use of distribution functions. *Ensemble methods* are effective in measuring small particles that do not scatter much light and in particle fields with high number densities. In low number density environments, however, they lose sensitivity and suffer from inadequate signals. They also require deconvolution schemes to extract the size information. Single-particle counters measure individual droplets or particles that pass through a focused laser beam, only one droplet at a time, so that the size of the measurement volume (or probe volume) sets the limitation on the droplet number density that can be handled accurately by the counters. In practice, the measurement volume is controlled by the diameter of the focused beam, the *f* number of the receiver and its angle to the transmitted beam, as well as the aperture of the photodetector.[654] Originally, single-particle counters were limited to dilute sprays. Recently, the use of large off-axis angles of light scatter detection has made these counters applicable to relatively dense sprays.[670] The restriction in the number density becomes similar to that of Fraunhofer diffraction methods. However, single-particle counters are limited by high particle number densities and beam extinction, while diffraction instruments are limited by low number densities and beam extinction in high number-density environment. Since beam extinction depends on the optical path length and number density, both single-particle counters and Fraunhofer diffraction methods are limited by beam extinction for particle fields larger than about 100 mm.

In the following, various optical methods will be described along with discussions of their capabilities and limitations.

Photography and Videography. Photography is probably one of the most accurate and least expensive techniques for measuring droplet size and velocity.[653][671][672] It has the potential of measuring droplets in dense, fast-moving sprays[657] that are of particular interest in power generation. In a *photography method*, a photograph of droplets is taken with a light pulse of sufficient intensity and sufficiently short duration to yield a sharp image of the droplets. The images on a processed film are then analyzed to count and size the droplets. High-intensity light sources of short duration used in photography include mercury vapor lamps, electrical sparks, flash lights, and laser pulses. Flash light sources have a time duration on the order of microseconds, whereas laser sources have a pulse duration on the order of nanoseconds. Flow visualization using laser light sheet or other suitable illumination to scatter light from droplets can be used to study droplet interactions with large eddies in sprays.

Some problems associated with photography methods include, for example, difficulties and inaccuracy in analyses of the photographic images. Manual analyses are tedious and subject to operator fatigue and bias, while automatic image analyzers also entail certain degree of human judgement. These data processing problems may be largely alleviated by using the TV analyzer developed by the Parker-Hannifin Corporation.[673] In highly dense sprays, droplet images may become very closelyspaced or even overlapped on films. Thus, the accuracy of image analyses may be detracted from to varying degrees. An approach to alleviate this problem is to increase the magnification at the imaging stage. This also allows to reduce the minimum droplet size that can be reliably measured, at the cost of forfeiting the depth of field.[657] The effective depth of field may be determined using the calibration procedure described by Chigier.[671]

Holography. Holography has much in common with high-speed photography in terms of droplet size measurements.[650][653][657][674][675] It produces a recorded image that can be studied later at leisure. However, holographic techniques have the advantage of allowing a sizable region to be captured instead of the limited depth of field afforded by photographic methods. In addition, the holographic techniques require, in principle, no calibration. The primary drawback of the holographic techniques is that their applications are limited to dilute sprays.[657]

A *holographic method* is basically a two-step imaging process that captures in permanent form the sizes and locations of moving droplets and then produces a stationary 3-D image of all the droplets contained within a sample volume. In a holographic method (Fig. 6.1), a sample volume of moving droplets in a spray is typically illuminated with a coherent beam of light (laser) in the form of a short pulse. The measurement volume is a cylinder with a length equal to the total width of the spray at the specific axial station and with a diameter equal to that of the laser beam. The droplets contained within the measurement volume may be deemed as effectively "frozen" since the duration of the laser pulse is extremely short (about 20 ns). The resultant hologram provides a complete three-dimensional image of the droplets with clear visibility. The hologram is then illuminated with a coherent beam of light to produce a stationary image of all the droplets contained within the sample volume at their correct relative locations in space. The accuracy of a holographic method is essentially set by the wavelength of the light used. Theoretically, a resolution of about 2 μm may be obtained.

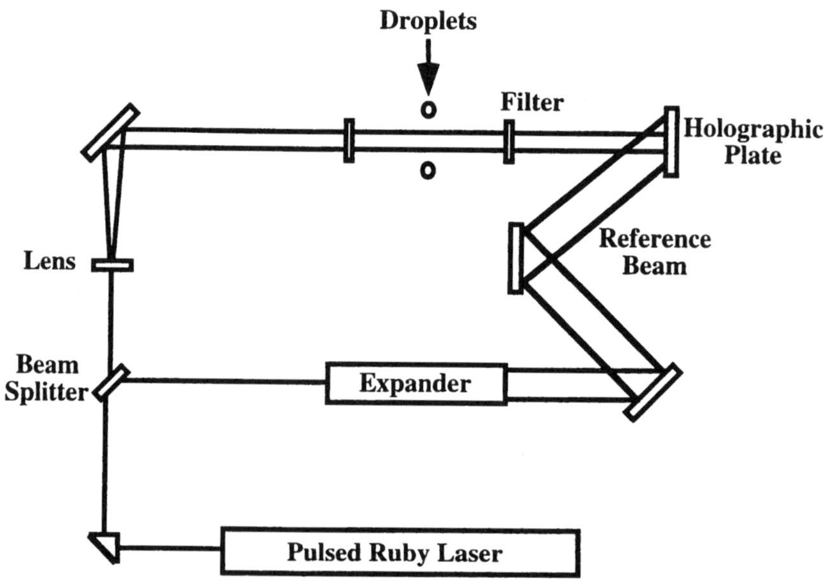

Figure 6.1. Schematic of a typical laser holographic measurement method.

Wu et al.[249] used a single-pulse holography system to measure droplet size distributions and a flash photography to observe the overall appearance of a spray. In their experiments, the Spectron Development Laboratories Model HTRC-5000 system was used in an off-axis arrangement with an angle of 28° between the object and reference beams. Adequate spatial resolution and sufficient light intensity to penetrate the spray were produced by reducing the diameter of the object beam through the spray and then expanding it back to the same diameter as the reference beam when the two signals were optically mixed to form holograms. The holograms were generated by a ruby laser that deposited 50 mJ in roughly 20 ns. This effectively "froze" the spray so that droplets as small as 5 μm in diameter could be measured. The holograms were obtained in a darkened room using AGFA 8E75HD-NAH unbacked holographic film plates in 100 × 125 mm format. The holograms were reconstructed using a 15-mW He-Ne laser beam that was expanded to a diameter of 50 mm to provide real images of the spray in front of the holograms. The reconstructed holograms were observed using a MTI Model 65 video camera with optics to provide a field of view of 1.4 × 1.6 mm. Computer controlled x-y traversing of the holograms (1 μm resolution) and z traversing of the video camera (5 μm resolution) allowed observations of the entire object field. The video images were analyzed using a Gould FD5000 image processing system. Reference pins of known sizes and locations near the edge of the object field provided direct calibration of distances on the reconstruction.

It is also feasible to measure both droplet size and velocity simultaneously with the holographic method. Such a measurement system has been developed by Murakami and Ishikawa.[650] They used two pulse lasers of different wavelengths to record two holograms on two different films, separated by an appropriate time interval. The displacements of moving particles were measured on a superimposed picture of the two films to determine the velocities of the particles.

Ridder et al.[676][677] applied laser holography along with high-speed cinematography to record the disintegration of liquid metals and study the breakup mechanisms of molten metals in gas atomization. They used an illuminating laser beam of short pulse duration (20 ns) and high light energy (10 mJ) that produced superior freeze-frame images of the atomization phenomena. Regeneration of three-dimensional images allowed analysis and photography of selected two-dimensional slices.

Light-Scattering Interferometry. The principle and practice of laser anemometry have been outlined by Durst et al.[678] Farmer[679][680] investigated the possibility of particle sizing using the visibility of Doppler burst signals. The greatest limitation of the method was the requirement of using on-axis forward- or back-scattered light detection. This resulted in an excessively large sample volume and the detection of particles that passed outside the region of complete overlap of two laser beams. Moreover, the size range was also limited to a factor of 10 or less. Yule et al.[649] sought to combine the measurement of scattered light intensities with the laser Doppler velocimeter to obtain the data of particle size and velocity simultaneously. Instead of measuring signal modulation, the peak mean value of the Doppler signal for droplets larger than the fringe spacing was measured. A linear relationship between pedestal peak amplitude and droplet diameter was shown to exist for droplets of 30–300 μm. Due to the use of near-forward light scatter detection, this method was also limited to very dilute sprays. It was not until Bachalo[670] conducted a theoretical analysis of dual-beam light scattering at large off-axis angles that light-scattering interferometry became a viable method for spray diagnostics.

In the *light-scattering interferometry*, the measurement of droplet size and velocity is based on the observation of the light scattered by droplets passing through the crossover region of two intersecting laser beams. In practice, a single laser beam is split into two coherent beams of equal intensity and parallel polarization, intersecting to build a measurement volume (Fig. 6.2). Droplets passing through the measurement volume scatter light, producing signals from which the information on droplet velocity and size can be extracted. The droplet velocity is calculated from the following equation:

Eq. (1) $$u = \frac{\lambda f_D}{2\sin(\theta/2)}$$

where u is the droplet velocity, λ is the wavelength of laser light, f_D is the Doppler frequency, and θ is the laser beam intersection angle (Fig. 6.2). The droplet size information can be extracted from the relative modulation or visibility of the scattered signals on the basis of a fringe model. The term *visibility* was introduced by Michelson as:

Eq. (2) $$\text{Visibility} = \frac{I_{max} - I_{min}}{I_{max} + I_{min}}$$

where I_{max} and I_{min} are illustrated in Fig. 6.3. On the basis of the scalar diffraction theory and Farmer's derivation, the droplet size can be related to the visibility as:

Eq. (3) $$\text{Visibility} = \frac{2J_1(\pi D/\delta)}{\pi D/\delta}$$

where J_1 is a first-order Bessel function of the first kind, D is the droplet diameter, and δ is the interference fringe spacing, $\delta = \lambda/(2\sin(\theta/2))$. The number of fringes equals d_w/δ, where d_w is the distance as depicted in Fig. 6.3. Based on the relationship between the dimensionless parameter D/δ and the signal visibility, the maximum measurement range is about 10:1. With increasing droplet size relative to the fringe spacing, a droplet scatters light from more than one fringe at a time, reducing the signal visibility (Fig. 6.3). Larger particles scatter light that is more narrowly distributed in the forward direction, leading to less intensity overlap for a given beam intersection angle.

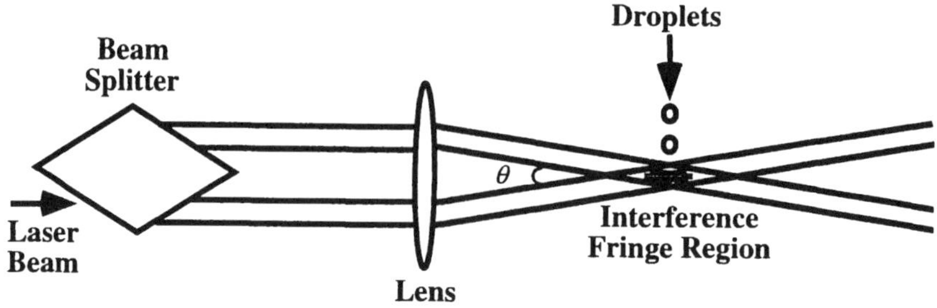

Figure 6.2. Schematic of light-scattering interferometry method.

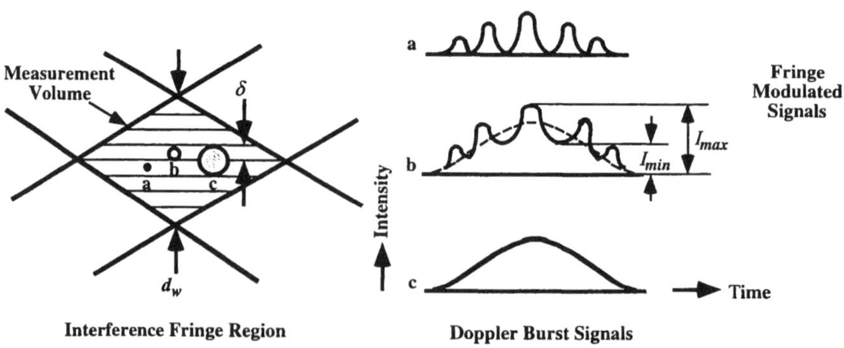

Figure 6.3. Schematic of interference fringe region, measurement volume, Doppler burst signals, and effects of particle size on signal modulation.

In a more rigorous description, the superposition of the light scattered separately from each beam that then interferes is considered. In the near-forward direction, the scattered light may be described by the *Fraunhofer diffraction theory* for particles much larger than the light wavelength, or more accurately, by the *Lorenz-Mie theory* for spherical particles of arbitrary size. The Mie theory dictates a correlation between the measured droplet size and the amount of scattered light. The relative visibility of scattered interference fringe pattern is determined by the relative overlap of forward-scattered light intensities.

In the *on-axis light scatter detection mode*, an interferometer based on the visibility sizing method can be used only for droplets smaller than the fringe spacing. Thus, large beam diameters must be used. For example, to measure droplets of about 200 μm in diameter, the beam diameter must be increased to approximately 10 times the droplet diameter and the beam intersection angle must be made very small.[1] This leads to an excessively large sample volume so that it is unlikely that merely one droplet would be in the sample volume at one time. Thus, for sprays with a wide range of droplet sizes, droplet coincidence errors will occur.

The need for sizing large droplets on the order of 100 μm in diameter stimulated the development of more reliable methods on the basis of the light scattered by reflection and refraction[670] and the ensuing use of large off-axis detection as required by these light-scattering components. The large off-axis detection offers advantageous features desired for sizing droplets in high number-density sprays. The required beam diameter and length of the sample volume are significantly reduced and more precisely defined by the off-axis detection. For example, the sample volume can be reduced by as much as two to three orders of magnitude in size.

In the *large off-axis light scatter detection method*, the droplet size can be related, for example, to the phase shift of the light scattered by refraction in the off-axis forward direction as:

Eq. (4) $$D = \frac{\Phi_B \lambda}{2\pi(\sin\tau - n_c \sin\tau')}$$

where Φ_B is the phase shift of the light scattered by refraction, τ is the incidence angle of the ray with the surface tangent, τ' is the refracted angle of the transmitted ray given by Snell's law, and n_c is the index of refraction of droplets. Since the parameters λ and n_c are constant for a given measurement system and droplet material, and τ and τ' are fixed by the receiver aperture, the droplet diameter is directly proportional to the phase shift. In the case of dual-beam light scattering, the phase shift manifests as an interference fringe pattern with a spatial frequency that is inversely proportional to the droplet diameter. An angle of 30° off the forward optical axis and orthogonal to the plane of the two beams is typically used to detect the light scattered predominantly by refraction without the detection of the light scattered by diffraction. On the basis of the refraction or reflection mechanism, good sensitivity in the droplet size range of 5–3000 μm can be obtained using optical configurations allowing discrete measurement steps. The dynamic size range for any one optical configuration is, however, still limited to less than 10:1.[1]

A typical example of large off-axis light scatter detection method is the droplet sizing interferometer that incorporates a laser velocimeter, as described by Bachalo et al.[681] The system consists of two basic components, i.e., a transmitter and a receiver. The transmitter optics determine the fringe spacing and the position of sample volume by defining the intersection angle of laser beams. The receiver optics detect and collect the light scattered by droplets passing through the sample volume. A collecting lens is used to image the sample volume on the aperture of a photomultiplier with a collection angle of 30° off the forward optical axis. The Doppler signals superimposed on Gaussian pedestals are registered at the photomultiplier and processed by a visibility processor. The electronic processor separates the Doppler component from the pedestal component, integrates the areas under the respective signals, and divides the results to produce the signal visibility. The droplet diameter and velocity are then determined from the signal visibility and the Doppler frequency, respectively.

Spectron Development Labs at Costa Mesa, CA, has commercialized this droplet sizing interferometer system and extended its capability to collect the information of amplitude of the scattered light in addition to the visibility.[682] On the basis of the Mie theory, the correlation between these two parameters that are obtained from each signal for each droplet can be used to select right signals. Moreover, in this system, the voltage is automatically set to the photomultiplier so that a direct calibration of intensity is not required. This advantage of the measurement system has led to significant improvement in instrument accuracy, particularly for dense sprays.[654] However, the lack of a very large number of samples in this approach significantly increases the sampling time.

The conditions necessary for accurate measurements of droplet size include highly coherent beams, equal beam intensities, complete beam overlap at the intersection, and prefect fringe visibility. The visibility method can provide accurate measurements of droplet size only if the fringe visibility is precisely determined. In dense sprays, droplets may interact with the transmitted laser beam upstream of the sample volume, and thus disrupt randomly the conditions required for ideal fringe visibility in the sample volume. This reduction in the fringe visibility may result in a measured droplet size that is larger than the actual size. On the other hand, the fringe visibility may also be increased as a result of masking of signals from out-of-focus droplets by the aperture or by the visibility processor's response to a signal with a high rise.[683]

Droplets with erroneous visibility scatter light with intensities lower than those corresponding to their actual sizes, while droplets passing the sample volume at the point of less intensity have the correct visibility but scatter light with different intensities. These are the two major problems associated with the droplet sizing interferometry method. An intensity validation technique has been developed and used to resolve these two problems. Signals yielding wrong droplet sizes are eliminated by setting limits on the intensity of the pedestal based on the Mie theory.

Glover et al.[684] reported a full-field, time-resolved interferometric laser imaging method for the characterization of sparse, polydisperse spray systems. The method makes use of the angular intensity oscillations in the wide-angle forward-scatter region. A pulsed laser is used to illuminate a planar sheet through a spray, which is imaged, out of focus, from the 45° direction. The image consists of a set of out-of-focus spots, each of which represents an individual droplet, and superimposed on which is a set of fringes corresponding to the angular intensity oscillations of that droplet. Macrophotographic recording with high-resolution digitization for image analysis provides a full-field capability. The spatial frequency of fringes on each spot in the image plane is dependent on the diameter of the corresponding droplet in the object plane, and a simple geometric analysis is shown to be appropriate for the calculation of the spatial frequency of fringes as a function of droplet size. Images are analyzed automatically by a software suite that uses Gaussian blur, Canny edge detection, and Hough transforms to locate individual droplets in the image field. Fringe spatial frequency is then determined by least-squares fitting to a Chirp function. This method is applicable to droplets with diameters in the range of several micrometers to several hundred micrometers and number densities of up to 10^3–10^4. The accuracy of this method for droplet size determination was evaluated by measurements of monodisperse aerosols of known droplet size, and the method was used to measure the droplet size distribution in a polydisperse aerosol produced by a gasoline fuel injector. The method may be extended by using high-speed photography to measure two velocity components in addition to the size and position of droplets. It may also be feasible to measure the concentration of model fuel additives in droplets using a two-wavelength approach.

Phase-Doppler Anemometry. Theoretical analyses on dual-beam light scattering with off-axis detection[670] showed that the spatial frequency of the scattered interference fringe pattern is inversely proportional to the droplet diameter, and the temporal frequency is related to the beam intersection angle, the light wavelength and the droplet velocity. Bachalo[655] invented a method for simultaneous measurement of both spatial and temporal frequencies using pairs of detectors at fixed spacings in the image of the interference fringe pattern. In this method, measurements are relatively unaffected by the random beam attenuation if the signal-to-noise ratio is sufficiently large. The instrument response is linear over the entire operation range. In addition, the dynamic size range is potentially large, and only limited by the detector response and signal-to-noise ratio.

On the basis of this method, a *phase-Doppler anemometer* has been developed by Aerometrics, Inc. at Mountain View, CA, and tested extensively.[685] Phase-Doppler anemometry (PDA) is a laser-Doppler velocimeter (LDV)-based method for non-intrusive, simultaneous measurements of size and velocity of spherical particles.[686] The phase-Doppler anemometer differs from the conventional dual-beam laser Doppler velocimeter only in the receiver where two or three photodetectors are used. As schematically shown in Fig. 6.4, the anemometer consists of a laser light source, a transmitter and a receiver. Three photodetectors are used to eliminate measurement ambiguity, improve reliability and provide high sensitivity over a large size range. Doppler burst signals are produced by each detector with a relative phase shift that is linearly related to droplet size. Signals from the photodetectors are amplified and transferred to a signal processor from that histograms of droplet size and velocity distributions are derived and displayed on a video monitor in virtually real time. Mean droplet sizes and velocities can be determined at the completion of data acquisition. The overall measurement size range for spherical droplets is 0.5–3000 μm. A single optical setting may cover a size range of 105, and the dynamic range is a factor of 35, limited by signal level and signal-to-noise ratio.

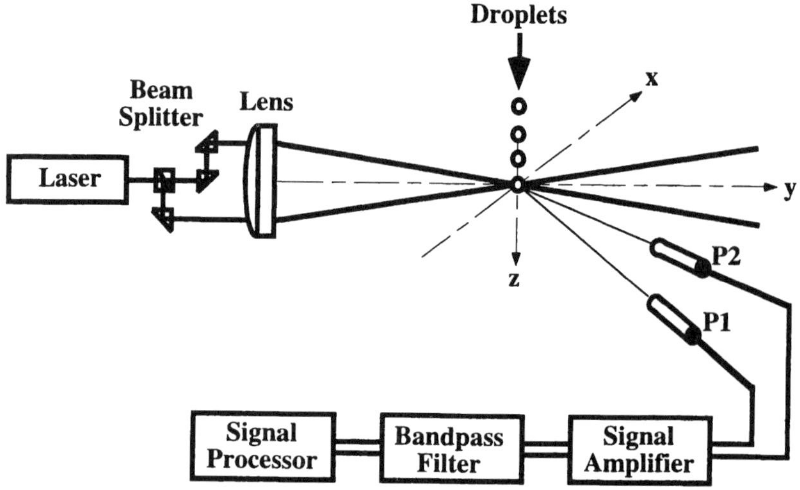

Figure 6.4. Schematic of optical arrangement in a phase-Doppler anemometer *(Reprinted from Ref. 615.)*

The phase-Doppler anemometry has been used successfully in measurements of various particles,[686] droplets of both normal liquids[685][687] and molten metals,[614][621][622] and bubbles.[688] The particles/droplets may be homogeneous and either transparent or opaque. Recently, Manasse et al.[689][690] extended the method to simultaneous size and velocity measurements of real fluid droplets that are optically absorbent and/or inhomogeneous. They used semiconductor devices with a larger wavelength (830 nm) as compared to Ar-ion laser (488 nm) so that the ambiguity in phase difference-particle diameter relationships due to absorption is effectively avoided. For optically absorbent homogeneous liquid droplets, the Mie theory proves to be an indispensable means for selecting the appropriate PDA set-up parameters. For inhomogeneous liquid droplets, the Mie theory and Mie scattering computer programs can be used to describe the scattering behavior of the droplets. Thus, the computer aided PDA technique allows *in-situ* measurements of real liquid sprays.

Light Intensity Deconvolution Technique. *Light intensity deconvolution technique* is a single-particle counting method that measures the absolute intensity of the light scattered in the near-forward direction by particles/droplets traversing a sample volume defined by the intersection of a single focused laser beam and detector optics. The sample volume has a non-uniform light intensity distribution. Thus, the intensity of the scattered light signals detected by a photomultiplier in the receiver depends on both droplet size and trajectory. To avoid the potential ambiguity in droplet size and concentration measurements, a deconvolution algorithm has been used that requires the measurement of the absolute intensity of the light scattered by a large number of droplets for a known sample volume intensity distribution. The count spectrum of the scattered light intensities is then deconvoluted to yield the absolute droplet concentration as a function of droplet size.

An example of this technique is the *particle counter-sizer velocimeter* (PCSV) developed by Insitec at San Ramon, CA, that can measure particle size, velocity and concentration.[691]-[693] The technique is capable of measuring droplets, slurries and solid particles. The droplets/particles may be spherical or of irregular shapes. The overall measurement size range is 0.2–200 μm, and particle velocities up to 200 m/s can be measured using a transit timing technique.[693] The typical measurement size range is 0.3–100 μm for an optical configuration of a typical instrument. Measurements in combustion conditions can be made by using an interference filter. On-line data analysis is possible using rapid signal processing.

Light Scattering Technique. Properties of the light scattered by a large number of droplets can be used to determine droplet size distribution. Dobbins et al.[694] first derived the theoretical formulation of scattering properties of particles of arbitrary sizes and refractive indices in polydispersions of finite optical depth. Based on the relationship between the angular distribution of scattered light and particle size distribution, the experimentally determined angular distribution of scattered light can be used to evaluate the Sauter mean diameter of droplets.

Extensive measurements using the *light scattering technique* have been conducted by Lefebvre and co-workers.[80][695] In the optical system, a 5-mW He-Ne laser at a wavelength of 632.8 nm was used as light source. The laser beam was spatially filtered to pass only the fundamental laser mode. An expanding telescope was used to produce a highly collimated beam that was then chopped by a rotating perforated disk to reduce the stray light and increase the sensitivity of the system. The parallel light beam was diffracted through the spray being measured and focused by a receiver of 600 mm focal length onto a circular aperture of 20 μm in diameter. The light was detected by a photomultiplier located at the focal plane of the receiving lens and was arranged to traverse horizontally in a direction at right angles to the optical axis. Signals from both the photomultiplier and a position transducer were fed simultaneously into an *x-y* plotter to obtain the light intensity profile. The Sauter mean diameter of droplets was then determined by measuring the radial distance from the optical axis to a point on the curve at which the light intensity is one-tenth of its maximum value.

This light scattering technique has the advantages of simplicity and low cost. By recording light intensity profiles, spurious data can be discarded. However, the technique is practically limited to the size range of 10–250 μm. For small particles, the intensity variation is weak. For large particles, the light intensity is concentrated near the axis and accuracy is difficult to obtain. In addition, there exists ambiguity in the maximum light intensity at the optical axis due to the strong on-axis peak from the unscattered beam. It also lacks the speed, convenience and data-storing capacity of the Malvern particle sizer.

The original version of the light scattering technique has been limited primarily to the measurement of Sauter mean diameter. Rizk and Lefebvre[695] developed a method based on the scattered energy distribution rather than the intensity distribution. The plots of light intensity distributions for the determination of the Sauter mean diameter are converted into the distributions of light energy to estimate the complete droplet size distribution.

Benayahu et al.[10] proposed a method for calculating droplet size distribution in atmospheric clouds, based on the measurement of laser back scattering and multiple scattering from water clouds. The lidar used a Nd:YAG laser that emits short pulses at a moderate repetition rate. The back scattering, that is composed mainly of single scattering, was measured with a detector pointing along the laser beam. The multiple scattering, that is mainly double scattering, was measured with a second detector, pointing at a specified angle to the laser beam. The domain of scattering angles that contribute to the double-scattering signal increased monotonically as the pulse penetrated the cloud. The water droplets within the probed volume were assumed to have a constant size distribution. Thus, from the double-scattering signal as a function of penetration depth within the cloud, the double-scattering phase function of the scattering volume was derived. Inverting the phase function resulted in the cloud droplet size distribution in the form of a log-normal function.

Huckaby et al.[696] measured the intensities of the light scattered in the planes parallel and perpendicular to the polarization plane of an incident light. The size, refractive index, and dispersion of a single droplet suspended in an electrodynamic balance were determined from wavelength-dependent scattering spectra. Wavelengths of TE- and TM-mode resonances were determined independently with high precision when a ring dye laser was scanned. Resonating wavelengths were matched with theoretical intensity peaks to determine the constants of a dispersion formula and the size that minimizes the difference between the observed and calculated wavelengths. This method permits the determination of the size and refractive index with relative errors of 3×10^{-5} and the dispersion with an absolute error of 2×10^{-5} over the experimental spectral range.

Malvern Particle Analyzer. *Malvern particle sizer* is a non-intrusive technique for the measurement of droplet size distribution. It was developed by Swithenbank et al.[697] manufactured and commercialized by Malvern Instruments Ltd. in Malvern, England. The solution strategy of the Malvern particle sizer is analogous to that of the intensity deconvolution technique, although the physics of the two measurement methods is completely different. The Malvern particle sizer is based on the Fraunhofer diffraction of a parallel beam of monochromatic light by a moving droplet. As schematically depicted in Fig. 6.5, when a droplet interacts with a parallel beam of light, a diffraction pattern is formed. For a monodisperse spray, the diffraction pattern is of the Fraunhofer form. It comprises a series of alternate light and dark concentric rings whose spacing is related to the droplet size. For a polydisperse spray, the diffraction pattern comprises a number of the Fraunhofer patterns with a series of overlapping diffraction rings, each of which is produced by a different group of droplet sizes. The Fourier transform receiver lens focuses the diffraction patterns onto a multi-element photodetector that measures the light energy distribution. The photodetector consists of 31 semicircular photosensitive rings surrounding a central circle. Each ring is most sensitive to a particular small range of droplet sizes. The output of the photodetector is multiplexed through an analog-digital converter. The measured light energy distribution is then converted to the droplet size distribution by a computer, allowing instant display of the size distribution. The measured data may be either analyzed in terms of a histogram with 15 size classes or presented in the format of normal, log-normal, Rosin-Rammler, or other modes[683] that are independent of the known models. Information on the droplet size distribution, such as the Sauter mean diameter and the parameter in the size distribution, can be obtained from the printouts of the computer in the Malvern measurement system.

The measurement size range of the Malvern particle sizer is dependent on the focal length of the Fourier transform receiver lens. Each detector-lens combination corresponds to approximately a 100 to 1 droplet size range. For example, a 100-mm lens gives a droplet size range of 2–197 μm, whereas a 300-mm lens corresponds to a droplet size range of 5.8–564 μm. Lenses of 600, 800 or 1000 mm in focal length may be used for coarser sprays. With a proper choice of lens, most practical sprays can be readily measured.

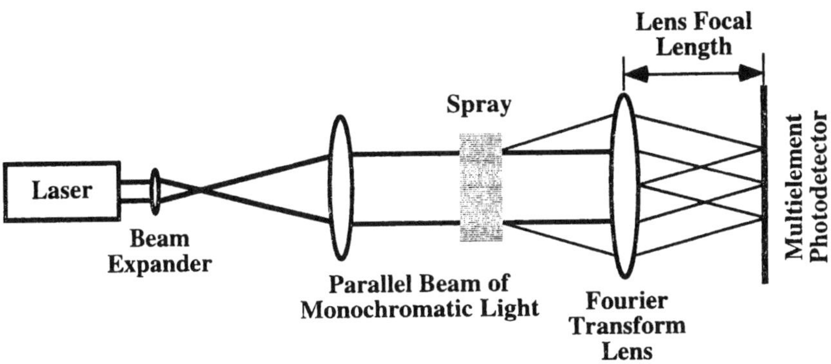

Figure 6.5. Schematic of optical arrangement in a Malvern particle sizer.

The Malvern particle sizer is one of the most widely used, most effective, simple, and reliable methods commercially available for rapid measurements of ensemble characteristics of a spray. It is able to handle high droplet concentrations. It is easy to use and does not require comprehensive knowledge of its basic principles for operation. The primary advantage of the system is the speed of data acquisition and analysis. In addition, measurements of droplet size distributions can be made at any droplet velocities due to the fact that the diffraction patterns generated by droplets are independent of the positions of the droplets in the light beam. However, the Malvern particle sizer is capable of measuring droplet sizes only within a certain range, and inaccuracies may be incurred when the droplet sizes in the spray being measured are beyond the size range corresponding to the receiver lens used. Other potential problems, that may influence the data accuracy and repeatability of the Malvern particle sizer, may include variations in detector sensitivity/responsivity,[698] beam steering,[699] multiple scattering,[700] and vignetting,[698] as reviewed by Lefebvre.[1]

Ridder et al.[676] used a laser diffraction instrument, operating on the principle of the Fraunhofer diffraction, to measure the size distribution of molten metal droplets in gas atomization. The instrument provided adequate time resolution for high gas flow rates (~0.18 kg/s or 230 scfm), particle concentrations (~10^6 particles/cm^3), and short run times (~2–5 min) typical of the supersonic inert gas metal atomization process. The instrument has a data-acquisition repetition rate of 30 scans/s and a resolution of 10 ms/reading. Mathematical modeling of the Fraunhofer diffraction was conducted, addressing the light-scattering phenomena and relating the particle size distribution vector to the light energy distribution vector due to the diffraction from the particles. An inverse transformation method capable of converting the measured light intensity distribution into the particle size distribution was developed on the basis of pattern recognition using the approach of feature map and content-addressable memories. Experiments were conducted to evaluate this novel pattern recognition scheme, and its performance was found to be fairly high for real-time particle size analysis.

6.1.4 Other Methods

In addition to the techniques outlined above, many other methods have been developed for droplet sizing. These include, for example, intensity ratio method, phase optical-microwave method, and *dual-cylindrical wave laser technique*,[701] etc.

Intensity ratio method for particle sizing is based on the ratio of measured light intensities at different angles and the dependence of the ratio on particle size.[702][703] Droplets in a spray scatter light in all directions when a light beam is passed through the spray. The relative intensity of the light scattered by droplets in the forward direction is a function of angle and may be calculated on the basis of the diffraction theory. The method is fairly insensitive to refractive index. However, the measurement size range is limited. An intensity ratio system for particle sizing has been developed by Spectron.

Zemlianskii and Yanovsky[704] suggested a strategy for sizing aerosols and water droplets in air using multiwavelength and multisensor technique. They conducted analyses on the basis of the Mie and Rayleigh theories. The *phase optical-microwave system* with fixed differential frequency shift between two beams of different wavelengths consists of two microwave detectors and two detector receiving optics. The proposed optical-microwave measurement method, theory and numerical treatment are based upon the assumptions that the particles to be sized are spherical and traverse a uniform field of plane waves.

An *acoustical particle counter* for counting and sizing fog droplets has been evaluated by Singh and Reist.[6] Fog droplets, mostly in the size range of 5–30 μm, were measured by the acoustical particle counter as well as an optical and an electron microscope for comparison. The mean droplet diameters estimated from the acoustical particle counter were in agreement with the microscope values. A Rich 100 condensation nuclei monitor was also operated simultaneously during the fog droplet counting to monitor condensation nuclei counts.

Korobochka and Pavlenko[705] proposed a simple model and nozzle design for the determination of exact droplet size distribution generated by an air-assist nozzle. The approach enables the direct measurements of droplet size and allows generation of a very narrow range of droplet size distribution.

6.2.0 MEASUREMENT TECHNIQUES FOR DROPLET VELOCITY

In sprays, droplet trajectories in the Lagrangian reference frame are probably one of the most interesting quantities to be measured. To optimize atomizer design, it is necessary to obtain spray characteristics such as droplet size and velocity. The primary methods for simultaneous measurements of droplet size and velocity may include high-speed photography,[706] cinematography, holography,[650] and phase-Doppler particle analyzer.[686][689][690] Bachalo[655] has presented a brief description of a variety of optical measurement techniques, and reviewed the capabilities and limitations of these methods along with discussions on the range of reliability, accuracy, and droplet number density considerations.

High-speed photography techniques can be used to measure droplet velocity. During a measurement, two light pulses are generated in rapid succession, yielding a double image of a single droplet on the photographic plate. The droplet velocity can be then determined from the image by measuring the distance traveled by the droplet and dividing it by the time interval between the two pulses. The direction of the velocity vector can also be directly determined from the image as an angle of flight with respect to the central axis of spray. From a series of such instantaneous measurements of individual droplets, time-averaged and space-averaged velocities as well as standard deviations can be obtained.[457][671][672] High-magnification double-pulse laser photography techniques can be used to measure both droplet size and velocity simultaneously. However, the measurement range may be limited.[706]

To extract spatial and temporal information on 3-D trajectories of droplets, high-speed photography, cinematography and/or holography should be used together with a digital image analyzer. Ante et al.[651] combined a double pulse *holographic system* with digital image processing to measure droplet velocities and number densities in the injection spray of a model diesel engine. The model engine consisted of a conventional injection pump and a rest chamber that could be operated at high pressures and temperatures to simulate the conditions in an operating diesel engine. The holographic images were reconstructed and fed into a PC by means of a CCD camera. Up to 30 image sections were necessary to store the information of a whole hologram. A software package was developed for the segmentation of droplets and the determination of droplet velocities.

In imaging methods, droplet velocity is generally determined by measuring the displacement of droplet images over a known time interval. This family of techniques is commonly referred to as *particle image velocimetry* (PIV), although a number of other terms are also often used to describe various subsets of the wide range of similar techniques.[707] The PIV can generate information of instantaneous velocity vectors across planes, from which vortex and cluster structures can be discerned. In a measurement with the PIV, droplets in a spray are illuminated with a sheet of light, normally provided by a pulsed laser, and scatter light to an imaging device. The droplet images may be recorded photographically on films or electronically using a CCD or CID camera. High resolution (>25 million pixels per image mark) CCD and CID cameras eliminate the need for both the chemical processing of films and the subsequent digitization of images. Alternatively to the 2-D recording with CCDs, holographic recording may be used to obtain 3-D components of droplet velocities over an entire flow volume. To obtain reliable velocity information using the PIV, the time between exposures must be selected so as to obtain a sufficient displacement for an acceptable velocity resolution, but not so large that the droplet moves out of the field of view or the plane of illumination. A variety of techniques have been used for the analysis of images from the PIV.[655]

The phase-Doppler method is capable of accurately measuring particle size distribution and velocity.[655] The most recent models of *phase-Doppler particle analyzer* (PDPA) can generate data of droplet size and velocity simultaneously as a function of time, from that droplet drag can be calculated and clustering phenomenon can be studied. PDPA is now recognized as the most successful and advanced instrument for simultaneous measurements of droplet size and velocity.[686][689][690][708] A *phase-Doppler anemometer* (PDA) and associated instrument for simultaneous measurements of droplet size and velocity are shown in Fig. 6.6. The mathematical formulations for droplet size and velocity have been discussed in detail by Manasse et al.[689] and the measurement results have been compared to the Mie theory in Ref. 690.

Figure 6.6. Experimental setup of a PDA and associated instrument for simultaneous measurements of droplet size and velocity *(Courtesy of Prof. Dr.-Ing. Klaus Bauckhage at University of Bremen, Germany.)*

An experimental method based on the theories for *rainbow* phenomena has been applied to the measurement of droplet size and velocity and to the detection of non-sphericity.[7] In this method, a comparison between two droplet diameters is deduced from two different optical interference patterns observed in a rainbow that is created by a droplet scattering laser light. Once a rainbow pattern is identified as coming from a spherical droplet, reliable droplet velocity and diameter can be derived from the same interference patterns, using the theories for rainbow that are valid only for spherical droplets. Preliminary experiments and validation have been carried out by means of a laser beam along with a photomultiplier and a CCD camera, respectively.

6.3.0 MEASUREMENT TECHNIQUES FOR DROPLET NUMBER DENSITY

Droplet number density and volume flux are important quantities in studying droplet dispersion and mixing processes in multiphase flows and sprays. The *phase-Doppler particle analyzer* has the potential of measuring local droplet number density and volume flux with sufficient accuracy.[655] In addition, it can measure the time varying nature of these quantities that may result from the interaction with large-scale structures in the flows/sprays. The measurement techniques for droplet number density and volume flux have been discussed by Bachalo.[655]

In the phase-Doppler method, the number density, N_d, is determined from:

Eq. (5)
$$N_d = \frac{1}{t}\sum_i \frac{\sum_j t_{i,j}(D_i)}{\Omega_i(D_i)}$$

where t is the total sampling time, $t_{i,j}$ is the transit time of the ith droplet of size class j, D_i is the droplet diameter in each size class i, and $\Omega(D_i)$ is the sample volume that depends on the probe cross sectional area and the length along the beam axis. The reliable measurement of the number density does not depend on the accurate sizing of droplets but is very dependent on the reliable determination of the measurement volume and the counting of droplets. It requires reliable detection of the smallest, most numerous droplets and accurate detection of the transit time of each droplet as it passes the sample volume relative to the total time for accumulating a sample. Such a measurement can be evaluated by comparing to beam extinction measurements and using Beer's law. In this method, the measured intensity of a beam passed through a spray is compared to the intensity without the spray present that is a measure of the light extinction produced by the spray. The mean droplet size D_{20} and the number density at points along the extinction beam path are measured using the PDPA. These results are used in the expression of Beer's law to calculate the effective beam extinction that is then compared to the measured value.

In the phase-Doppler method, the volume flux, V_{fx}, is obtained from:

Eq. (6) $$V_{fx} = \frac{\pi}{6} \frac{D_{30}^3}{A} \frac{N_t}{t}$$

where A is the probe area normalized to that for the maximum particle diameter and N_t is the total number of particles counted, corrected for probe area variations. The volume flux is not very dependent on the counting of small droplets since they have a minimal effect on D_{30}. However, it is obvious from Eq. (6) that the volume flux is most dependent on the accurate measurement of the size of large droplets in the size distribution. Since the relative population of the large droplets is small in a size distribution, it is important to collect a sufficiently large total number of samples to ensure that the large droplets are statistically representative. Moreover, the sampling cross section must be accurately determined as in the measurement of droplet number density.

Liu et al.[622] used a *laser Doppler velocity and size* (LDVS) measurement technique to determine the local size, velocity, and number flow density of droplets in the spray cone during spray deposition of a liquid steel. The experimental setup is schematically depicted in Fig. 6.7.[615] The measured results showed that smaller particles may be underrepresented due to the rapid solidification typical of the spray deposition processes of molten metals/alloys.

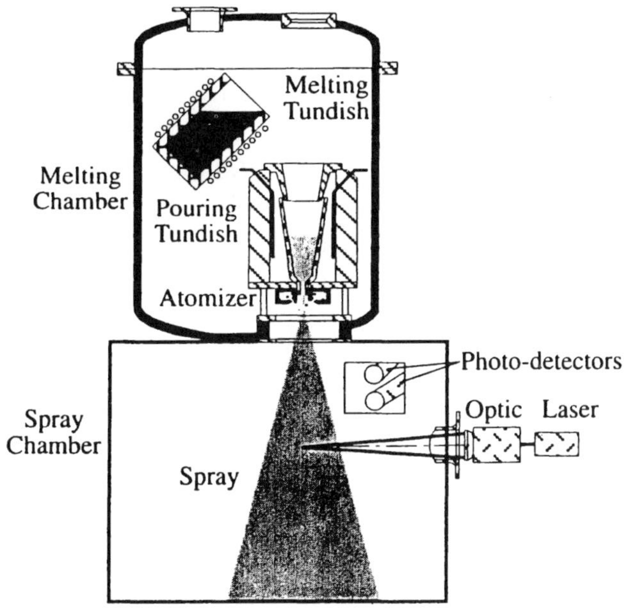

Figure 6.7. Schematic of experimental setup of a LDVS measurement system for local size, velocity, and number flow density of droplets in the spray cone during spray deposition of a liquid steel *(Reprinted from Ref. 615.)*

The technique of *stimulated Raman scattering* (SRS) has been demonstrated as a practical method for the simultaneous measurement of diameter, number density and constituent material of micrometer-sized droplets.[709] The SRS method is applicable to all Raman active materials and to droplets larger than 8 μm in diameter. Experimental studies were conducted for water and ethanol monodisperse droplets in the diameter range of 40–90 μm. Results with a single laser pulse and multiple pulses showed that the SRS method can be used to diagnose droplets of mixed liquids and ensembles of polydisperse droplets.

6.4.0 MEASUREMENT TECHNIQUES FOR DROPLET TEMPERATURE

Droplet temperature is of interest in practical spray processes since it influences the associated heat and mass transfer, chemical reactions, and phase changes such as evaporation or solidification. Various forms of Rayleigh, Raman and fluorescence spectroscopies have been developed for measurements of droplet temperature and species concentration in sprays.[647] Rainbow refractometry (thermometry), polarization ratioing thermometry, and exciplex method are some examples of the droplet temperature measurement techniques.

In the *rainbow thermometry*, a linear CCD array is used to determine the location of the primary rainbow that is formed when a droplet is illuminated by a laser beam. The refractive index of the droplet is obtained by measuring the rainbow location assuming that the droplet is spherical. Since the refractive index varies with temperature, the droplet temperature is then inferred from a relationship between the refractive index and temperature if the relationship is known a priori. For accurate measurements of droplet temperatures in spray flames, the information of droplet sizes is required because rainbow angles also depend on droplet sizes. Accordingly, a rainbow thermometer should be integrated with an independent particle sizing instrument, such as the well established and widely used phase-Doppler interferometer, for simultaneous, accurate measurements of individual droplet size, velocity and temperature (refractive index) in complex sprays. Such an integrated system has been developed by Aerometrics, Inc. at Mountain View, CA.[655] The accuracy of the rainbow method may be limited by non-uniform heating of droplets and when measuring temperatures of multi-component fuel droplets. However, the rainbow method appears to have the most potential for accurate measurements of droplet temperatures *via* the refractive index in a combustion environment as well as in a wide range of applications.

Several theories have been developed to explain the rainbow phenomena, including the Lorenz-Mie theory, Airy's theory, the complex angular momentum theory that provides an approximation to the Lorenz-Mie theory, and the theory based on Huygen's principle. Among these theories, only the Lorenz-Mie theory provides an exact solution for the scattering of electromagnetic waves by a spherical particle. The implementation of the rainbow thermometry for droplet temperature measurement necessitates two functional relationships. One relates the rainbow angle to the droplet refractive index and size, and the other describes the dependence of the refractive index on temperature of the liquid of interest. The former can be calculated on the basis of the Lorenz-Mie theory, whereas the latter may be either found in reference handbooks/literature or calibrated in laboratory.

A critical review of the rainbow method for droplet temperature measurement has been presented by Massoli et al.[710] They conducted a light-scattering analysis based on the Lorenz-Mie theory and showed that the size and refractive index of transparent droplets can be determined by measuring the polarized components of the scattered light at two angles in the forward direction. The horizontally polarized cross section $C_{HH}(33°)$ depends exclusively on the droplet diameter, whereas the ratio $C_{HH}(33°)/C_{HH}(60°)$ is a sensitive function of the refractive index and hence of the droplet temperature. On the basis of the light scattering in the forward and rainbow regions, these researchers developed a new optical system for measuring the temperature, size, and velocity of individual transparent droplets. This system is capable of determining droplet temperatures with a resolution of a few degrees Centigrade. The techniques have been applied to vaporizing tetradecane droplets (D_0 = 72 μm), which were heated up in a tube furnace with a temperature range of 20–200 °C.

Polarization ratioing thermometry has been proposed as a means of measuring the refractive index of a droplet and relating it to the droplet temperature. However, this approach does not have the potential resolution of the rainbow method whereas it has the similar limitations.

Exciplex method has also been proposed for droplet temperature measurement. In an oxygen environment, however, the fluorescence from the exciplex is quenched by the oxygen. In addition, fuel droplets may contain aromatic hydrocarbons that can produce fluorescence emissions, masking the fluorescence spectrum of the dopants used for the temperature determination.

A *two-color pyrometer* has been used along with the phase-Doppler anemometer to simultaneously measure the local velocity and size of kerosene droplets and the temperature of burning soot mantle in a swirl burner.[648] The measurements were conducted within the flame brush that develops in the shear layer of a swirl-stabilized, gas-supported kerosene flame with a swirl number of about 0.19 and potential heat releases of 10.6 and 15.5 kW, respectively. The results showed that the maximum burning fraction of the droplets occurs adjacent to the region denoted as gas flame but the value ranges from 20±5 to 40±5% depending on the axial station, and decreases sharply across the shear layer. The flame mantle temperature was found to be independent of droplet diameter, which agrees with previous results in the literature.

A *non-invasive infrared* (IR) *method* has been developed for the measurement of temperatures of small moving fuel droplets in combustion chambers.[711] The IR system is composed of two coupled off-axis parabolic mirrors and a MCT LWIR detector. The system was used to measure the temperature variations in a chain of mono-sized droplets generated with equal spacing and diameter (200 μm), moving at a velocity of >5 m/s and evaporating in ambient air. The system was also evaluated for droplet temperature measurements in flames under combustion conditions.

6.5.0 MEASUREMENT TECHNIQUES FOR DROPLET DEFORMATION ON A SURFACE

Experimental observations of the spreading of a liquid droplet impinging on a flat surface were first reported by Worthington[337] in the late nineteenth century. In his experiments, Worthington designed an ingenious inductor circuit to produce an electric spark for illumination at different stages of droplet spreading and to sketch artistically the various forms assumed by a droplet during its impact on a surface.

Recently, experimental investigations of droplet spreading processes have received increasing attention. A variety of techniques have been used to record and measure the dynamic shapes of a droplet during its interaction with a surface. These include, for example, *(a)* high-speed film camera,[334][357][395][411] high-speed digital video system,[371][400][409] electronically controlled short-duration flash light photography, and holography[368] for recording droplet deformation details, and *(b)* high-speed two-color pyrometers[406] and high-speed thermal imaging systems[410] for recording and measuring both spreading and solidification. The clarity of pictures has been greatly improved by using single-shot flash illumination.[357][411] In the single-shot method, only one image of a droplet is recorded at one instant of the droplet impact process. By recording successive stages of the impact of several different droplets, the entire impact process can be pieced together from individual images.[411] Key parameters describing a droplet spreading process, such as splat-substrate contact area, splat diameter, and splat thickness, may be measured subsequently from the recorded pictures.

Generally, *visualization methods* record deforming droplet shapes at discrete stages of a continuous deformation process. The splat size is usually measured from visualization images with an accuracy on the order of 10 microns.[411] The visualization methods, while providing pictures of droplet spreading events at different stages, have certain limitations[368] For example, high-speed cinematography is expensive to implement and records multiple images of a single deforming droplet at the cost of reduction of image clarity so that it is difficult to make quantitative measurements. Single-shot flash photography can produce high clarity images of deforming droplets, but only one image can be rendered for each impinging droplet. Some details of the entire droplet spreading and possibly recoiling process may be skipped due to insufficient information at certain stages.

Efforts have been made to search for alternative experimental methods that do not involve direct visualization. Senda et al.[335] presented a simple method for the measurement of heat flux and heat transfer coefficient between a droplet and a hot surface on which it impinges. Shi et al.[336] developed simple experimental devices and measurement methods for the instantaneous diameter of a spreading droplet and the transient surface temperature of a solid substrate to assess cooling rates during droplet impingement without resort to photographic interpretation. In the work of Shi et al.,[336] the instantaneous diameter of a spreading water droplet was measured using an *electrical resistance probe method*. In their experimental set-up, a glass plate was coated with a 10-μm thick aluminum layer. A gap of 0.2 mm in width and 0.01 mm in depth was scribed across the center, separating the conducting surface into two halves. An electrical DC voltage was applied to each half of the conducting surface. As a water droplet spreads at the center of the gap, the gap is bridged and the voltage across the gap decreases. By monitoring the voltage change across the gap, the temporal change of the splat radius may be obtained. The calibration was performed by measuring the final splat area vs. the output signal voltage for an impacting water droplet of known initial radius. While this method is of considerable merit, it has certain limitations. For example, different droplet sizes require different calibration curves. The recoiling of the splat may result in loss of data in high-speed impact calibration. The gap is intrusive and may affect the spreading and recoiling dynamics, particularly for small droplets. In addition, an off-center droplet spreading may cause measurement errors.

In an effort to invent fast, reliable and affordable measurement techniques, Zhao et al.[368] developed a simple and accurate non-intrusive photoelectric method to measure the transient splat diameter and the entire evolution of the maximum splat diameter during the collision of a droplet on a substrate. In this measurement system, a collimated laser beam and a linear response photodiode are used, as schematically depicted in Fig. 6.8. As a droplet spreads out on a horizontal quartz plate, the splat area increases and blocks more laser light. Thus, the light intensity collected by the photodiode decreases as the spreading proceeds. By recording the output signals from the photodiode, the transient splat diameter can be measured. In the experiments of Zhao et al.,[368] the probe beam was provided by a 15-mW He-Ne laser. The laser beam was focused and passed through a 20-μm diameter spatial filter in order to produce a smooth irradiance across the expanded and collimated probe beam. An iris was used to improve the system sensitivity for small droplets. After intercepting the splat, the partially blocked beam was collected onto the sensor area of the photodiode by a focusing lens. A Si-photodiode (E2RUV passive silicon photo detector of type S 1226-5BQ, Spindler and Hoyer Inc.) was chosen to detect the laser light due to its fast response and high linearity. The photo detector rise time is 0.5 μs and the sensitive area is 2.4×2.4 mm^2. A 100-MHz oscilloscope (HP 54601 A Oscilloscope, Hewlett-Packard Co.) was used to record the output voltage signals. The angle between the probe beam axis and the vertical direction was 8.5°. The measurement system was calibrated by placing successively opaque circular disks of known area on the clear quartz to block the probe beam. The sample disks were precisely manufactured from a black cardboard using a computer-operated CO_2 laser (C-80 Laser System, Laser Machining, Inc.). The system output signals are a linear function of the sample areas. The measurement uncertainty of this technique is directly related to the sensitivity of the oscilloscope and the measurement system calibration. The uncertainty of splat area measurements was estimated to be less than 6.6×10^{-3} mm^2, corresponding to uncertainties less than 2 μm and 1 μm for splat diameters larger than 2.1 mm and 4.2 mm, respectively. The time and length scales in this technique were on the order of microseconds and micrometers, respectively. This method requires that the droplet to be studied must be opaque to the probe beam. To alleviate this limitation, a special probe beam may be selected such that the droplet fluid is opaque to it. The preferred probe beam wavelength is generally in the range from visible to ultraviolet, and a special wavelength window may be needed to reduce the environmental radiation that could influence the photoelectric measurements.

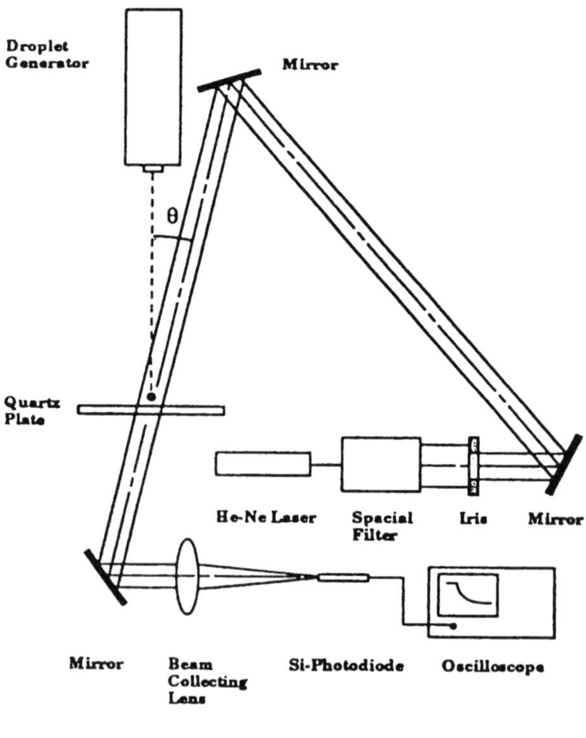

Figure 6.8. Schematic showing photoelectric method for measurement of splat diameter *(Reprinted with permission from Ref. 368.)*

Zhao et al.[368] also designed a *two-reference-beam double-pulse holography* system to visualize the deformation process of a droplet during impingement on a substrate. The system can record and reconstruct two holographic images of a droplet from one hologram at two instances of a deformation process. In this experimental method, the field distribution and phase information from the light scattered by a deforming droplet is recorded on a hologram. A wealth of information about the object can be then extracted from the hologram. For example, one hologram can readily generate different real images of a deforming splat that correspond to different viewing angles. The axial symmetry of a droplet can be examined for each exposure. As schematically depicted in Fig. 6.9 (*top*), the key components in this optical arrangement are the Pockels cell and the polarizing beam splitter. During each experiment, the ruby laser fires two pulses of laser light. The Pockels cell, driven by a high-voltage pulse generator, is electronically switched such that the polarization direction of the second pulse is rotated by 90°. This enables the second beam to pass through the polarizing beam splitter and follow the second reference path. The images of a deforming droplet from both the first and the second pulse are accordingly recorded on the same holographic plate, but with spatially different reference beams. The Pockels cell modulates both the object beam and the reference beam. This optical arrangement enables the polarization directions for the object beam and the reference beam to be used on the hologram plane. Therefore, a high degree of optical visibility is ensured for the holographic images recorded for both pulses. One reference beam at a time is used for reconstruction. Thus, two images of a splat at different stages of spreading can be separately reconstructed with exact information about the order. The separation time between the first pulse image and the second pulse image is the same as the electronically set laser pulse separation time. For the HLS-2 Ruby Laser (Lumonics, Inc.) used in this study, the pulse separation time varied in the range of 1–800 μs with a pulse duration of 20 ns. The maximum laser beam power was 3 joules. Figure 6.9 (*bottom*) shows the overall experimental setup of the holography system for

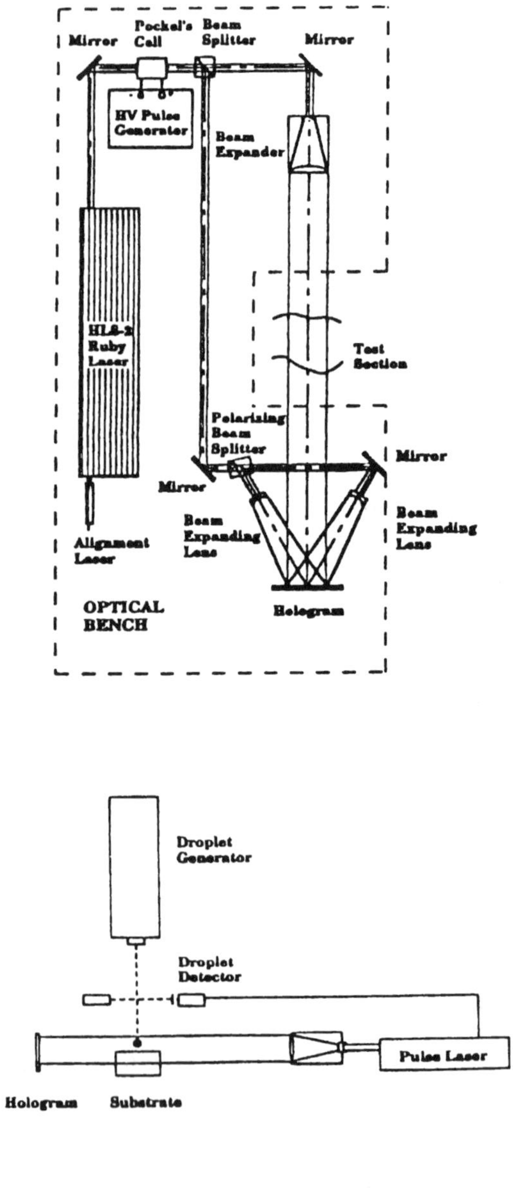

Figure 6.9. Schematic of optical arrangement *(top)* and experimental setup *(bottom)* of two reference beam holography for recording images of a deforming droplet at different stages of spreading. (*Reprinted from Ref. 368.*)

recording images of a deforming droplet at different stages of spreading. In this setup, a droplet is released from the droplet generator and detected by a detector during its free-fall. After a preset time delay, the droplet detector sends a signal to trigger the pulse laser that in turn fires two pulses consecutively with a preset separation time. Two images of a deforming droplet are recorded on the same holographic plate. The images are then reconstructed from the holograms and digitized and stored in a personal computer for image analysis (IMAGE ANALYST® version 8.0, 1992, Automatix, Inc.).

Experimental observations of droplet impact under conditions typical of thermal spray applications are difficult due partly to the difficulties in accurate determination of droplet size, velocity, and energy state prior to impact. In addition, high-temperature properties may not be available, limiting the usefulness of flattening data in determining the proper scaling.[390] Vardelle et al.[712] used two *high-speed two-color pyrometers* to monitor the flattening, cooling and solidification of droplets on a substrate during plasma spray deposition. In this measurement system, the light emitted by a flattening droplet was collected by one of the pyrometers that was focused on the substrate. The optical signal was then imaged on the entrance slit of a monochromator, and two output signals filtered at 632.8 and 832.8 nm were transmitted to two photomultipliers. After amplification, the signals were recorded by a numerical oscilloscope. Another identical pyrometer was focused on a point 1.8 mm above the substrate to determine the droplet size, velocity and surface temperature prior to impact. A coincidence sensor was focused on the center of the field of view of this pyrometer to ensure that the radiation detected by both pyrometers came from the same droplet in-flight and at impact. Thus, the measurement system can provide data on the deformation and solidification of a droplet during impingement on a substrate surface, along with data on its size, velocity and molten state at impact. The spatial resolution for the in-flight measurement was about 0.015 mm^3 and the monitored area on the substrate surface was 0.5 mm^2.

Hofmeister et al.[410] used two *high-speed thermal imaging* systems to record spatial and temporal temperature distributions at the splat-substrate interface and to observe the droplet spreading during impact and solidification on a quartz plate. In their experimental setup, a surface-aluminized mirror was positioned below the quartz plate to allow viewing of the splat-substrate interface through an 8" view port at the end of a chamber. A Kodak Ektapro scanning array camera was operated at 40500 frames/s and a HSDA96 parallel tapped thermal imaging system was operated at 250000 frames/s. The two thermal imaging systems were positioned to simultaneously view the splat-substrate interface *via* a beam splitting cube. The HSDA96 was fitted with a narrow bandpass optical filter centered at 950 nm, and both instruments were calibrated for temperature measurement using a standard tungsten strip lamp. The Ektapro has a higher spatial resolution, 4096 pixels as compared to 96 active pixels in the HSDA96. However, the response time of the HSDA96 (less than one μs) is two orders of magnitude shorter than the Ektapro (about 75 μs). Thus, the two systems complement each other and together provide the necessary bandwidth and spatial resolution to capture the details of droplet spreading and solidification on a substrate surface. The imaging experiments provided detailed information on the deformation, heat transfer and solidification during droplet impingement on a surface.

6.6.0 INTELLIGENT CONTROL OF DROPLET PROCESSES

Over the past decade, instrumentation has undergone substantial developments in the areas of optics, electronics and data processing software. These have led to improvements in the reliability, accuracy and ease of use of associated instruments. Particularly, the introduction of fiber optics has allowed a greater degree of flexibility in application areas. These include, for example, measurement of oil droplet dispersion in the ocean, on-line monitoring of polymer bead formation, in-flight measurement of cloud droplets pertinent to aircraft icing, measurements of droplets in rocket and automotive injectors, investigation of gas turbine fuel spray interactions with turbulent flows, paint sprays, and molten metal sprays. The optical probes employed in these applications have been designed in a variety of configurations. For example, very small probes of only 15 mm in diameter have been developed for industrial applications, and very large probes with optics of 500 mm in diameter for large-scale wind tunnel testing.

Techniques are being developed to directly measure single droplet trajectories in the Lagrangian reference frame and to provide information on their temporal variations. These have stimulated new developments in instrumentation, including instantaneous planar imaging velocimetry, and digital image analysis of high-speed photography, cinematography, videography and holography. As new experimental data are becoming available for droplet trajectories and their temporal variations in the Lagrangian reference frame, droplet processes will benefit from improving spray diagnostic techniques that will allow measurements of the joint distribution functions of droplet sizes, 3-D components of velocities, temperatures and evaporation rates.

In droplet processes, particularly spray atomization and spray forming processes, a major obstacle to achieving consistently high-quality spray products has been the limited ability to accurately monitor and effectively control various operation parameters *on line* and in real time. For example, in monitoring and controlling spray processes, sensors used to monitor spray angle, penetration, shape and optical density may relay signals to control systems for actuation of changes in liquid and gas supply pressures and flow rates. In an effort to improve the quality of final products, various diagnostic techniques have been evaluated for monitoring spray process parameters, such as droplet size, velocity, concentration, temperature, and size distribution. Sensing techniques and instruments evaluated[713][714] include high-speed cinematography, off-axis holographic cinematography, infrared thermal imaging, PCSV-P probe, phase-Doppler particle analyzer, EPMP monitor, and other intelligent sensors.[327][715] Some methods have capabilities to provide *on-line* feedback of the spray parameters. Generally, laser diffraction, phase-Doppler and digital imaging instruments may be simplified for *on-line* monitoring of droplet size and velocity histograms and distributions. Sensors and techniques used to monitor the process parameters during spray forming have been reviewed and discerned for their usefulness in making spray atomization and deposition a more controllable process.[713]

Methods of *near-field, midfield and ensemble (global) imaging and real-time visualization* have been developed for monitoring gas atomization of liquid metals.[327] The primary process sensors and monitors used include high-speed video and infrared imaging systems. The process monitors allowed continuous and detailed observations of the atomization process and enabled measurements of the key parameters necessary for adequate control and optimization of the process. The sensors provided the operators with real-time information on the temperature of nozzle tip, visual characteristics of atomization plume, and gas and metal flow rates. The images can be displayed in real time, offering the potential for more responsive process control.

A number of intelligent sensors have been used for *on-line* monitoring of gas atomization of liquid metals.[715] These *on-line* sensing techniques enabled measurements of liquid metal flow rate, *on-line* droplet size and size distribution, as well as video imaging of spray plume. The near-field measurements provided detailed information on atomization characteristics for better understanding of the atomization process. The downstream measurements along with the near-field measurements demonstrated the applicability of *on-line* process monitoring and control to spray atomization and spray deposition. It was also showed that, with the diagnostic tools currently available for use in industrial gas atomization, significantly improved control over the atomization operation is achievable by implementing *on-line* measurements and process-product know-how.

An *intelligent process diagnostics and control system* has been integrated for supersonic inert gas metal atomization (SIGMA) at the National Institute of Standards and Technology (NIST) in the

US.[163][676][677] An expert system based on the principle of hierarchical control was developed that features a generic structure allowing easy adaption to other processes. As schematically depicted in Fig. 6.10,[163] the diagnostics and control system consists of a product sensor (particle sizer), actuators, process sensors (temperature, pressure, etc.), a low-level (local) controller, a high-level (learning) controller, and an operator interface. The high-level controller refers to a physically remote microprocessor with which an operator can interact *via* the keyboard and graphics monitor of the operator interface. This controller has access to a "world model," i.e., a database describing the performance characteristics of the atomizer as provided by knowledgeable experts and by continuous updates during atomization operations. The local low-level controller is a "stand-alone" device capable of maintaining status quo during an atomization operation and providing necessary process feedback. The product sensor is an *in-situ* particle sizer that is connected to the high-level controller directly, bypassing the local low-level controller, so as to improve response of the control system to changes in particle size distribution. Some fundamental issues in gas atomization that relate to process control have been studied, including fluid dynamics, process parameterization and particle size sensing. The high-level controller incorporated the results of these studies into a multiple level I/O command interface for intelligent decision-making and *on-line* control.

In summary, it has become increasingly evident that further technological developments, particularly in such challenging areas as real-time, *on-line* control, require a detailed understanding and effective control of the fundamental phenomena associated with droplet processes. The ultimate realization of the real-time process control relies on the development of robust process models and fast solution techniques and the implementation of the models on a computer coupled with sensing and control systems. Faster, more accurate, reliable and affordable systems integrating process models, monitoring sensors and control units are required to achieve final products of consistently high qualities and to optimize droplet processes.

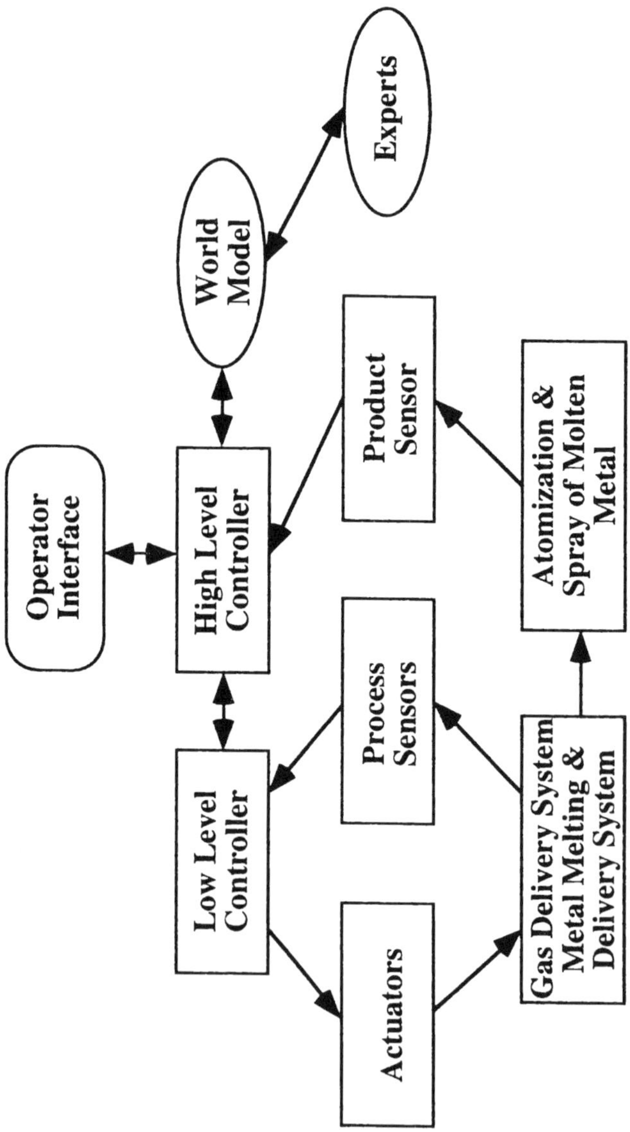

Figure 6.10. Schematic of the diagnostics and control system for gas atomization developed at NIST.

References

1. Lefebvre, A. H., *Atomization and Sprays*, Hemisphere Publishing Corporation, New York, NY, USA (1989)
2. Masters, K., *Spray Drying*, Second Edition, John Wiley and Sons, New York, NY, USA (1976)
3. Liu, H., *Spray Forming*, in: *Modeling and Simulation for Casting and Solidification: Theory and Applications* (Yu, K.O., ed.), Marcel Dekker, Inc., New York, NY, USA (1999)
4. Lawley, A., *Atomization - The Production of Metal Powders*, MPIF, Princeton, NJ, USA (1992)
5. Yule, A. J., and Dunkley, J. J., *Atomization of Melts for Powder Production and Spray Deposition*, Clarendon Press, Oxford, UK (1994)
6. Singh, D. P., and Reist, P. C., Evaluation of Acoustical Particle Counter for the Sizing of Fog Droplets, *Environmental Monitoring and Assessment,* 24(1):35–44 (1993)
7. Van Beeck, J. P. A. J., and Riethmuller, M. L., Rainbow Phenomena Applied to the Measurement of Droplet Size and Velocity and to the Detection of Nonsphericity, *Appl. Opt.,* 35(13):2259–2266 (1996)
8. Nye, J. F., Rainbows from Ellipsoidal Water Drops, *Proc. R. Soc. Lond. A,* 438(1903):397–417 (1992)
9. Chylek, P., and Hallett, J., Enhanced Absorption of Solar Radiation by Cloud Droplets Containing Soot Particles in Their Surface, *Q.J.R. Meteorol. Soc.,* 118(503):167–172 (1992)

10. Benayahu, Y., Ben-David, A., Fastig, S., and Cohen, A., Cloud-Droplet-Size Distribution from Lidar Multiple-Scattering Measurements, *Appl. Opt.,* 34(9):1569–1578 (1995)

11. Ghan, S. J., Chuang, C. C., and Penner, J. E., A Parameterization of Cloud Droplet Nucleation. I. Single Aerosol Type, *Atmospheric Res.,* 30(4):197–221 (1993)

12. Edson, J. B., and Fairall, C. W., Spray Droplet Modeling. 1. Lagrangian Model Simulation of the Turbulent Transport of Evaporating Droplets, *J. Geophys. Res.,* 99(C12):25295–25311 (1994); Edson, J. B., Anquetin, S., Mestayer, P. G., and Sini, J. F., Spray Droplet Modeling. 2. An Interactive Eulerian-Lagrangian Model of Evaporating Spray Droplets, *J. Geophys. Res.,* 101(C1):1279–1293 (1996)

13. Zhao, F. Q., Lai, M. C., and Harrington, D. L., The Spray Characteristics of Automotive Port Fuel Injection – A Critical Review, *SAE Paper 950506* (1995)

14. Sirignano, W. A., Fuel Droplet Vaporization and Spray Combustion Theory, *Prog. Energy Combust. Sci.*, 9:291–322 (1983)

15. Sirignano, W. A., Spray Combustion Review, *Trans. ASME J. Fluids Eng.*, 115:345–378 (1993)

16. Bär, P., *Über die physikalischen Grundlagen der Zerstäubungstrocknung*, Ph.D. Thesis, TH Karlsruhe, Germany, February (1935)

17. Chandra, S., and Aziz, S. D., Leidenfrost Evaporation of Liquid Nitrogen Droplets, *Trans. ASME J. Heat Transfer,* 116(11):999–1006 (1994); Qiao, Y. M., and Chandra, S., Boiling of Droplets on a Hot Surface in Low Gravity, *Int. J. Heat Mass Transfer,* 39(7):1379–1393 (1996); Qiao, Y. M., and Chandra, S., Experiments on Adding a Surfactant to Water Drops Boiling on a Hot Surface, *Proc. R. Soc. Lond. A,* 453:673–689 (1997); Qiao, Y. M., and Chandra, S., Spray Cooling Enhancement by Addition of a Surfactant, *Trans. ASME J. Heat Transfer,* 120(2):92–98 (1998)

18. Liu, H., Lavernia, E. J., and Rangel, R. H., Numerical Simulation of Impingement of Molten Ti, Ni and W Droplets on a Flat Substrate in Plasma Spray Processes, *J. Thermal Spray Technol.*, 2(4):369–378 (1993)

19. Cabler, P., Geddes, L. A., and Rosborough, J., The Use of Ultrasonic Energy to Vaporize Anaesthetic Liquids, *Br. J. Anaesth.*, 47:541–545 (1975)

20. Sumiyoshitani, S., Effects of an Applied Electric Field on Collection Efficiency by a Charged Droplet for Dust Particles in Charged Droplet Scrubbers, *Aerosol Sci. Technol.,* 20(1):71–82 (1994)

21. Abu-Zaid, M., and Atreya, A., Transient Cooling of Hot Porous and Nonporous Ceramic Solids by Droplet Evaporation, *J. Heat Transfer,* 116(3):694–701 (1994)

22. Ono, H., and Kawatsuki, N., New Fabrication Method for Poly (Vinyl Alcohol)/Liquid Crystal Composite Films with Controlled Droplet Size, *Jpn. J. Appl. Phys., Pt. 2,* 34(1A):L54–56 (1995)

23. Nazarenko, V. G., Sarala, S., and Madhusudana, N. V., Kinetics of Droplet Formation in UV-Cured Polymer-Dispersed Liquid Crystal Films, *Jpn. J. Appl. Phys., Pt. 1,* 33(5A):2641–2647 (1994)

24. Teixeira, P. I. C., and Mulder, B. M., Cell Dynamics Model of Droplet Formation in Polymer-Dispersed Liquid Crystals, *Phys. Rev. E,* 53(2):1805–1815 (1996)

25. Pflügl, W., and Titulär, U. M., The Droplet Size Distribution in the Late Stage of Phase Separation, *Physica A,* 214(1):52–67 (1995)

26. Tanaka, H., Elementary Process of Coalescence among Droplets in Phase-Separating Binary Polymer Mixtures: New Mechanism of Droplet Coalescence, Slow Dynamics in Condensed Matter, *AIP Conf. Proc.,* 256:240–241 (1992)

27. Hu, G., Krall, A. H., and Weitz, D. A., Thermal Fluctuations of the Shapes of Droplets in Dense and Compressed Emulsions, *Phys. Rev. E,* 52(6B):6289–6302 (1995)

28. Ueda, T., Gao, Q. Z., Yamaichi, E., Yamagishi, C., and Akiyama, M., Growth of GaAs Microcrystal by Ga Droplet Formation and Successive As Supply with Low-Pressure Metalorganic Chemical Vapor Deposition, *J. Crystal Growth,* 145(1–4):707–713 (1994)

29. Belmonte, T., Bonnetain, L., and Ginoux, J. L., Synthesis of Silicon Carbide Whiskers Using the Vapour-Liquid-Solid Mechanism in a Silicon-Rich Droplet, *J. Mater. Sci.,* 31(9):2367–2371 (1996)

30. Astakhova, T. Yu., Shaginyan, S. A., and Vinogradov, G. A., Droplet Mechanism of Fullerene Formation from Supersaturated Carbon Vapor, in: *Proc. of the Symp. on Recent Advances in the Chemistry and Physics of Fullerenes and Related Materials VII* (Ruoff, R. S., and Kadish, K. M., eds.), Electrochem. Soc., Pennington, NJ, USA, pp. 1253–1262 (1995)

31. Van de Riet, E., Biermann, U. K. P., and Dieleman, J., A New Method for the Reduction of Droplet Deposition onto Laser Deposited Thin Metal Films, *Thin Solid Films,* 241(1–2):134–137 (1994)

32. Schilling, M., Goerke, F., and Merkt, U., Laser Deposition of Epitaxial $YBa_2Cu_3O_7$ Films with Low Droplet Densities, *Thin Solid Films,* 235(1–2):202–209(1993); Blank, D. H. A., Ijsselsteijn, R. P. J., Out, P. G., Kuiper, H. J. H., Flokstra, J., and Rogalla, H., High T_c Thin Film Prepared by Laser Ablation: Material Distribution and Droplet Problem, *Mater. Sci. Eng. B,* 13(1):67–74 (1992)

33. Haraguchi, T., Shimada, R., and Takeyama, T., Microscope Observations of the Initial Droplet Formation Mechanism in Dropwise Condensation, *Heat Transfer - Jpn. Res.,* 21(6):573–585 (1992)

34. Meakin, P., Dropwise Condensation: The Deposition Growth and Coalescence of Fluid Droplets, *Physica Scripta, T*44:31–41 (1992)

35. Meakin, P., Droplet Deposition Growth and Coalescence, *Reports on Prog. in Phys.,* 55(2):157–240 (1992)

36. Kuz, V. A., Dynamics of the Condensation of a Saturated Vapor into Droplets, *Phys. Rev. E,* 49(2):1246–1250 (1994)

37. Rayleigh, Lord, On the Instability of Jets, *Proc. Lond. Math. Soc.,* 10:4–13 (1878)

38. Weber, G., Zum Zerfall eines Flüssigkeitsstrahles, *Z. Angew. Math. Mech.,* 11(2):136–154 (1931)

39. Haenlein, A., Über den Zerfall eines Flüssigkeitsstrahles, *Forsch. Geb. Ingenieurwesens A,* 2(4):139–149 (1931)

40. Ohnesorge, W. V., Die Bildung von Tropfen aus Düsen beim Zerfall flüssiger Strahlen, *Z. Angew. Math. Mech.,* 16:355–358 (1936)

41. Tanasawa, Y., and Toyoda, S., On the Atomization of Liquid Jet Issuing from a Cylindrical Nozzle, *The Technology Reports of the Tohoku University,* Jpn., No. 19-2, pp. 135–156 (1955)

42. Entezarian, M., Allaire, F., Tsantrizos, P., and Drew, R. A. L., Plasma Atomization: A New Process for the Production of Fine, Spherical Powders, *JOM,* 6:53–55 (1996)

43. Apelian, D., Paliwal, M., Smith, R. W., and Schilling, W. F., Melting and Solidification in Plasma Spray Deposition – Phenomenological Review, *Int. Metals Rev.,* 28(5):271–294 (1983)

44. Bragg, M. B., and Mohler, S. H., Predicting Droplet Impingement on Yawed Wings, *J. Aircraft,* 29(5):964–968 (1991)

45. Forney, L. J., Droplet Impaction on a Supersonic Wedge: Consideration of Similitude, *AIAA J.,* 28(4):650–654 (1990)

46. Savic, P., and Boult, G. T., The Fluid Flow Associated with the Impact of Liquid Drops with Solid Surfaces, in: *Proc. Heat Transfer Fluid Mech. Inst.*, Heat Transfer and Fluid Mechanics Institute, Stanford University Press, Stanford, CA, USA, pp. 43–84 (1957)

47. Zable, J. L., Splatter during Ink Jet Printing, *IBM J. Res. Develop.*, 21(4):315–320 (1977)

48. Heiniö, J., Kaski, K., and Abraham, D. B., Dynamics of a Microscopic Droplet on a Solid Surface: Theory and Simulation, *Phys. Rev. B*, 45(8):4409–4416 (1992)

49. Paliwal, M., and Holland, R. J., Production of Fine Powders via Microatomization, *Adv. in Powder Metall.*, 3:35–44 (1989)

50. Waldvogel, J. M., and Poulikakos, D., Solidification Phenomena in Picoliter Size Solder Droplet Deposition on a Composite Substrate, *Int. J. Heat Mass Transfer*, 40(2):295–309 (1997)

51. Liu, H., Sickinger, A., Mühlberger, E., Lavernia, E. J., and Rangel, R. H., Deformation and Interaction Behavior of Molten Droplets Impinging on a Flat Substrate in Plasma Spray Process, in: *Thermal Spray Coatings: Research, Design and Applications* (Berndt, C. C., and Bernecki, T. F., eds.), ASM International, Materials Park, OH, USA, pp. 457–462 (1993)

52. Liu, H., Lavernia, E. J., and Rangel, R. H., Numerical Investigation of Micro-Pore Formation during Substrate Impact of Molten Droplets in Plasma Spray Processes, *Atomization and Sprays*, 4(4):369–384 (1994)

53. Meakin, P., Steady State Behavior in a Model for Droplet Growth, Sliding and Coalescence: The Final Stage of Dropwise Condensation, *Physica A*, 183(4):422–438 (1992)

54. Young, J. B., The Condensation and Evaporation of Liquid Droplets at Arbitrary Knudsen Number in the Presence of an Inert Gas, *Int. J. Heat Mass Transfer*, 36(11):2941–2956 (1993)

55. Zhang, S., and Gogos, G., Film Evaporation of a Spherical Droplet over a Hot Surface: Fluid Mechanics and Heat/Mass Transfer Analysis, *J. Fluid Mech.*, 222:543–563 (1991)

56. Delplanque, J. P., and Sirignano, W. A., Numerical Study of the Transient Vaporization of an Oxygen Droplet at Sub- and Super-Critical Conditions, *Int. J. Heat Mass Transfer*, 36(2):303–314 (1993)

57. Elperin, T., and Krasovitov, B., Radiation, Thermal Diffusion and Kinetic Effects in Evaporation and Combustion of Large and Moderate Size Fuel Droplets, *Int. J. Heat Mass Transfer*, 38(3):409–418 (1995)

58. Chiang, C. H., and Sirignano, W. A., Interacting, Convecting, Vaporizing Fuel Droplets with Variable Properties, *Int. J. Heat Mass Transfer,* 36(4):875–886 (1993)

59. Bayazitoglu, Y., and Cerny, R., Electromagnetic Vaporization of Molten-Metal Drops, *Int. J. Heat Mass Transfer,* 36(2):277–286 (1993)

60. Paik, S., and Nguyen, H. D., Momentum and Heat Transfer of a Translating Liquid Droplet under a Thermal Plasma Environment, *Phys. Plasmas,* 1(7):2383–2397 (1994)

61. Lage, P. L. C., Rangel, R. H., and Hackenberg, C. M., Multicomponent Heat and Mass Transfer for Flow over a Droplet, *Int. J. Heat Mass Transfer,* 36(14):3573–3581 (1993)

62. Nguyen, H. D., Paik, S., and Chung, J. N., Unsteady Conjugate Heat Transfer Associated with a Translating Spherical Droplet: A Direct Numerical Simulation, *Numer. Heat Transfer A,* 24(2):161–180 (1993)

63. Oliver, D. L. R., and De Witt, K. J., High Peclet Number Heat Transfer from a Droplet Suspended in an Electric Field: Interior Problem, *Int. J. Heat Mass Transfer,* 36(12):3153–3155 (1993)

64. D'yakonov, S. G., Elizarov, V. I., Laptev, A. G., and Zaikova, O. V., Modeling of Mass Transfer in Continuous and Dispersed Phases during Droplet Motion in Extractors, *J. Eng. Phys. Thermophys.,* 64(1):8–13 (1993)

65. Berlemont, A., Grancher, M. S., and Gouesbet, G., Heat and Mass Transfer Coupling between Vaporizing Droplets and Turbulence Using a Lagrangian Approach, *Int. J. Heat Mass Transfer,* 38(16):3023–3034 (1995)

66. Hwang, T. H., and Lin, J. K., Combined Convection and Radiation Heat Transfer to Thermally Developing Laminar Droplet Flow in Concentric Annuli, *Wärme- und Stoffübertragung,* 27(6):385–391 (1992)

67. Anidjar, F., Tambour, Y., and Greenberg, J. B., Mass Exchange between Droplets during Head-on Collisions of Multisize Sprays, *Int. J. Heat Mass Transfer,* 38(18):3369–3383 (1995)

68. Pineda, A. C., and Ronis, D., Classical Model for Energy Transfer in Microspherical Droplets, *Phys. Rev. E,* 52(5B):5178–5194 (1995)

69. Fisenko, S. P., Mathematical Modeling of Heat and Mass Transfer in Transpiration Cooling of Water Droplets in a Cooling Tower, *J. Eng. Phys. Thermophys.,* 64(2):119–123 (1993)

70. Delale, C. F., and Meier, G. E. A., A Semiphenomenological Droplet Model of Homogeneous Nucleation from the Vapor Phase, *J. Chem. Phys.,* 98(12):9850–9858 (1993)

71. Kielkiewicz, M., Condensational Growth and Removal of Droplets, *J. Aerosol Sci.,* 24(2):227–235 (1993)

72. Pflügl, W., and Titulär, U.M., The Size Distribution of Liquid Droplets during Their Growth from a Vapor, *Physica A,* 198(3–4):410–422 (1993)

73. Fraser, R. P., Dombrowski, N., and Routley, J. H., The Production of Uniform Liquid Sheets from Spinning Cups; The Filming of Liquids by Spinning Cups; The Atomization of a Liquid Sheet by an Impinging Air Stream, *Chem. Eng. Sci.,* 18:315–321, 323–337, 339–353 (1963)

74. Karim, G. A., and Kumar, R., The Atomization of Liquids at Low Ambient Pressure Conditions, in: *Proc. of the First Int. Conf. on Liquid Atomization and Spray Systems,* Fuel Society of Japan, Tokyo, Jpn, pp. 151–155 (1978)

75. Wigg, L. D., Drop-Size Predictions for Twin Fluid Atomizers, *J. Inst. Fuel,* 27:500–505 (1964)

76. Sakai, T., Kito, M., Saito, M., and Kanbe, T., Characteristics of Internal Mixing Twin-Fluid Atomizer, in: *Proc. of the First Int. Conf. on Atomization and Spray Systems,* Fuel Society of Japan, Tokyo, Jpn, pp. 235–241 (1978)

77. Inamura, T., and Nagai, N., The Relative Performance of Externally and Internally-Mixed Twin-Fluid Atomizers, in: *Proc. of the Third Int. Conf. on Liquid Atomization and Spray Systems* (Yule, A. J., and Eisenklam, P., eds.), Institute of Energy, London, UK, pp. IIC/2/1–11 (1985)

78. Elkotb, M. M., Mahdy, M. A., and Montaser, M. E., Investigation of External-Mixing Air-Blast Atomizers, in: *Proc. of the Second Int. Conf. on Liquid Atomization and Spray Systems,* ILASS-Americas and Department of Chemical Engineering, University of Wisconsin-Madison, Madison, Wis., USA, pp. 107–115 (1982)

79. Nukiyama, S., and Tanasawa, Y., Experiments on the Atomization of Liquids in an Airstream, *Trans. Soc. Mech. Eng. Jpn.,* 5(18):62–75 (1939)

80. Lorenzetto, G. E., and Lefebvre, A. H., Measurements of Drop Size on a Plain Jet Airblast Atomizer, *AIAA J.,* 15(7):1006–1010 (1977)

81. Jasuja, A. K., Plain-Jet Airblast Atomization of Alternative Liquid Petroleum Fuels under High Ambient Air Pressure Conditions, *ASME Paper 82-GT-32* (1982)

82. Rizk, N. K., and Lefebvre, A. H., Spray Characteristics of Plain-Jet Airblast Atomizers, *Trans. ASME J. Eng. Gas Turbines Power,* 106:639–644 (1984)

83. Jasuja, A. K., Atomization of Crude and Residual Fuel Oils, *Trans. ASME J. Eng. Power,* 101(2):250–258 (1979)

84. Rizkalla, A. A., and Lefebvre, A. H., The Influence of Air and Liquid Properties on Airblast Atomization, *Trans. ASME J. Fluids Eng.*, 97(3):316–320 (1975)
85. El-Shanawany, M. S. M. R., and Lefebvre, A. H., Airblast Atomization: The Effect of Linear Scale on Mean Drop Size, *J. Energy,* 4(4):184–189 (1980)
86. Lefebvre, A. H., and Miller, D., The Development of an Air Blast Atomizer for Gas Turbine Application, *CoA-Report-AERO-193*, College of Aeronautics, Cranfield, Bedford, England, June (1966)
87. Lefebvre, A. H., Wang, X. F., and Martin, C. A., Spray Characteristics of Aerated Liquid Pressure Atomizers, *AIAA J. Propul. Power,* 4(4):293–298 (1988)
88. Lee, K. W., Putnam, A. A., Gieseke, J. A., Golovin, M. N., and Hale, J. A., Spray Nozzle Designs for Agricultural Aviation Applications, *NASA-CR*-159702 (1979)
89. Weast, R. C., Astle, M. J., and Beyer, W. H., *CRC Handbook of Chemistry and Physics*, Sixty-Sixth Edition, CRC Press, Boca Raton, FL, USA (1985)
90. Hiroyasu, H., and Kadota, T., Models for Combustion and Formation of Nitric Oxide and Soot in Direct-Injection Diesel Engines, *SAE Trans.*, 85:513–526 (1976)
91. Mock, F. C., and Ganger, D. R., Practical Conclusions on Gas Turbine Spray Nozzles, *SAE Q. Trans.*, 4(3):357–367 (1950)
92. Carey, F. H., The Development of the Spill Flow Burner and Its Control System for Gas Turbine Engines, *J. R. Aeronaut. Soc.*, 58(527):737–753 (1954)
93. Radcliffe, A., The Performance of a Type of Swirl Atomizer, *Proc. Inst. Mech. Eng.*, 169:93–106 (1955)
94. Dombrowski, N., and Munday, G., *Spray Drying, Biochemical and Biological Engineering Science*, Vol. 2, Chapter 16, Academic Press, New York, NY, USA, pp. 209–320 (1968)
95. Dombrowski, N., Hasson, D., and Ward, D. E., Some Aspects of Liquid Flow through Fan Spray Nozzles, *Chem. Eng. Sci.*, 12:35–50 (1960)
96. Ortman, J., and Lefebvre, A. H., Fuel Distributions from Pressure-Swirl Atomizers, *AIAA J. Propul. Power,* 1(1):11–15 (1985)
97. Carr, E., Further Applications of the Lucas Fan Spray Fuel Injection System, *ASME Paper 85-IGT-116* (1985)

98. Rizk, N. K., and Lefebvre, A. H., Influence of Liquid Film Thickness on Airblast Atomization, *Trans. ASME J. Eng. Power,* 102:706–710 (1980)

99. Mansour, A., and Chigier, N., Air-Blast Atomization of Non-Newtonian Liquids, *J. Non-Newtonian Fluid Mech.,* 58(2–3):161–194 (1995)

100. Tsai, S. C., and Viers, B., Airblast Atomization of Viscous Newtonian Liquids Using Twin-Fluid Jet Atomizers of Various Designs, *Trans. ASME J. Fluids Eng.,* 114(1):113–118 (1992)

101. Wigg, L. D., The Effect of Scale on Fine Sprays Produced by Large Airblast Atomizers, *National Gas Turbine Establishment Report* No. 236 (1959)

102. Gretzinger, J., and Marshall, W. R., Jr., Characteristics of Pneumatic Atomization, *AIChE J.*, 7(2):312–318 (1961)

103. Mullinger, P. J., and Chigier, N. A., The Design and Performance of Internal Mixing Multi-Jet Twin-Fluid Atomizers, *J. Inst. Fuel,* 47:251–261 (1974)

104. Bryce, W. B., Cox, N. W., and Joyce, W. I., Oil Droplet Production and Size Measurement from a Twin-Fluid Atomizer Using Real Fluids, in: *Proc. of the First Int. Conf. on Liquid Atomization and Spray Systems*, Fuel Society of Japan, Tokyo, Jpn, pp. 259–263 (1978)

105. Sargeant, M., Blast Atomizer Developments in the Central Electricity Generating Board, in: *Proc. of the Second Int. Conf. on Liquid Atomization and Spray Systems*, ILASS-Americas and Department of Chemical Engineering, University of Wisconsin-Madison, Madison, Wis., USA, pp. 131–135 (1982)

106. Hurley, J. F., and Doyle, B. W., Design of Two-Phase Atomizers for Use in Combustion Furnaces, in: *Proc. of the Third Int. Conf. on Liquid Atomization and Spray Systems* (Yule, A. J., and Eisenklam, P., eds.), Institute of Energy, London, UK, pp. 1A/3/1–13 (1985)

107. Stambuleanu, A., *Flame Combustion Processes in Industry*, Abacus Press, Turnbridge Wells, Kent, UK (1976)

108. Rizk, N. K., and Lefebvre, A. H., Influence of Airblast Atomizer Design Features on Mean Drop Size, *AIAA J.,* 21(8):1139–1142 (1983)

109. Matsumoto, S., Crosby, E. J., and Belcher, D. W., Rotary Atomizers: Performance Understanding and Prediction, in: *Proc. of the Third Int. Conf. on Liquid Atomization and Spray Systems* (Yule, A. J., and Eisenklam, P., eds.), Institute of Energy, London, UK, pp. 1A/1/1–21 (1985)

110. Kayano, A., and Kamiya, T., Calculation of the Mean Size of the Droplets Purged from the Rotating Disk, in: *Proc. of the First Int. Conf. on Liquid Atomization and Spray Systems*, Fuel Society of Japan, Tokyo, Jpn, pp. 133–143 (1978)

111. Tanasawa, Y., Miyasaka, Y., and Umehara, M., Effect of Shape of Rotating Disks and Cups on Liquid Atomization, in: *Proc. of the First Int. Conf. on Liquid Atomization and Spray Systems*, Fuel Society of Japan, Tokyo, Jpn, pp. 165–172 (1978)

112. Hinze, J. O., and Milborn, H., Atomization of Liquids by Means of a Rotating Cup, *Trans. ASME J. Appl. Mech.*, 17(2):145–153 (1950)

113. Christensen, L. S., and Steely, S. L., Monodisperse Atomizers for Agricultural Aviation Applications, *NASA CR* 159777 (1980)

114. Kitamura, Y., and Takahashi, T., Atomization of Highly Viscous Liquids by a Spinning Disk, *Atomization and Sprays,* 2(1):61–71 (1992)

115. Eisenklam, P., Recent Research and Development Work on Liquid Atomization in Europe and the USA, in: *Fifth Jpn Conf. on Liquid Atomization and Spray Systems*, The Japan Institute of Energy, Tokyo, Jpn, pp. 86–104 (1976)

116. Fraser, R. P., Eisenklam, P., Dombrowski, N., and Hasson, D., Drop Formation from Rapidly Moving Liquid Sheets, *AIChE J.,* 8(5):672–680 (1962)

117. Spillmann, J., and Sanderson, R., A Disc–Windmill Atomizer for the Aerial Application of Pesticides, in: *Proc. of the Second Int. Conf. on Liquid Atomization and Spray Systems*, ILASS-Americas and Department of Chemical Engineering, University of Wisconsin-Madison, Madison, Wis., USA, pp. 169–172 (1982)

118. Solomon, A. S. P., Rupprecht, S. D., Chen, L. D., and Faeth, G. M., Flow and Atomization in Flashing Injectors, *Atomization Spray Technol.*, 1(1):53–76 (1985)

119. Sato, M., Hatori, T., and Saito, M., Experimental Investigation of Droplet Formation Mechanisms by Electrostatic Dispersion in a Liquid-Liquid System, in: *1995 IEEE Industry Applications Conference, Thirtieth IAS Annual Meeting*, IEEE, New York, NY, USA, pp. 1485–1492 (1995)

120. Shiryaeva, S. O., and Grigor'ev, A. J., Some Regularities of Rayleigh Breakup of Drops in a Highly Nonuniform Electrostatic Field, *Sov. Phys. - Tech. Phys.,* 37(3):254–256 (1992)

121. Drozin, V. G., The Electrical Dispersion of Liquids as Aerosols, *J. Colloid Sci.*, 10(2):158–164 (1955)

122. Peskin, R. L., Raco, R. J., and Morehouse, J., A Study of Parameters Governing Electrostatic Atomization of Fuel Oil, in: *Proc. of API Res. Conf. on Distillate Fuel Combustion*, API Publication 704 (1965)

123. Kelly, A. J., The Electrostatic Atomization of Hydrocarbons, in: *Proc. of the Second Int. Conf. on Liquid Atomization and Spray Systems*, ILASS-Americas and Department of Chemical Engineering, University of Wisconsin-Madison, Madison, Wis., USA, pp. 57–65 (1982)

124. Bailey, A. G., The Theory and Practice of Electrostatic Spraying, *Atomization Spray Technol.*, 2:95–134 (1986)

125. Okuda, H., and Kelly, A. J., Electrostatic Atomization – Experiment, Theory and Industrial Applications, *Phys. Plasma,* 3(5 Pt. 2):2191–2206 (1996)

126. Sato, M., Kato, S., and Saito, M., Production of Oil/Water Type Uniformly Sized Droplets Using a Convergent AC Electric Field, *IEEE Trans. on Industry Applications,* 32(1):138–145 (1996)

127. Lang, R. J., Ultrasonic Atomization of Liquids, *J. Acoust. Soc. Am.*, 34(1):6–8 (1962)

128. Berger, H. L., Characterization of a Class of Widely Applicable Ultrasonic Nozzles, in: *Proc. of the Third Int. Conf. on Liquid Atomization and Spray Systems* (Yule, A. J., and Eisenklam, P., eds.), Institute of Energy, London, UK, pp. 1A/2/1–13 (1985)

129. Topp, M. N., and Eisenklam, P., Industrial and Medical Use of Ultrasonic Atomizers, *Ultrasonics,* 10(3):127–133 (1972)

130. Mochida, T., Ultrasonic Atomization of Liquids, in: *Proc. of the First Int. Conf. on Liquid Atomization and Spray Systems*, Fuel Society of Japan, Tokyo, Jpn, pp. 193–200 (1978)

131. Wilcox, R. L., and Tate, R. W., Liquid Atomization in a High Intensity Sound Field, *J. Am. Inst. Chem. Eng.*, 11(1):69–72 (1965)

132. Crawford, A. E., Production of Spray by High Power Magnetostriction Transducers, *J. Acoust. Soc. Am.,* 27(1):176–177 (1955)

133. Antonevich, J., Ultrasonic Atomization of Liquids, *IRE Trans. PGUE,* 7:6–15 (1959)

134. Bisa, K., Dirnagl, K., and Esche, R., Zerstäubung von Flüssigkeiten mit Ultraschall, *Siemens Z.,* 28(8):341–347 (1954)

135. McCubbin, T. K., Jr., The Particle Size Distribution in Fog Produced by Ultrasonic Radiation, *J. Acoust. Soc. Am.,* 25(5):1013–1014 (1953)

136. Topp, M. N., Ultrasonic Atomization–A Photographic Study of the Mechanism of Disintegration, *J. Aerosol Sci.,* 4(1):17–25 (1973)

137. Muromtsev, S. N., and Nenashev, V. P., The Study of Aerosols – III. An Ultrasonic Aerosol Atomizer, *J. Microbiol. Epidemiol. Immunobiol.,* 31(10):1840–1846 (1960)

138. Orme, M., Willis, K., and Nguyen, T. V., Droplet Patterns from Capillary Stream Breakup, *Phys. Fluids A,* 5(1):80–90 (1993)

139. Warnica, W. D., Van Reenen, M., Renksizbulut, M., and Strong, A. B., Charge Synchronization for a Piezoelectric Droplet Generator, *Rev. Scientific Instruments,* 64(8):2334–2339 (1993)

140. Giovannini, A., Guyomar, D., Gschwind, M., and Fonzes, G., Evaluation and Design of New Piezoelectrical Droplets Generator, in: *1994 IEEE Ultrasonics Symp. Proc.* (Levy, M., Schneider, S. C., and McAvoy, B. R., eds.), IEEE, New York, NY, USA, pp. 611–614 (1994)

141. Toda, K., and Ishii, J., Operation Performance of Self-Oscillation Ultrasonic Vibrating Device for Liquid Atomization, *Jpn. J. Appl. Phys., Pt. 1,* 34(9B):5332–5334 (1995)

142. Ruthardt, R., and Lierke, E. G., A New Ultrasonic Atomization Technique for the Production of Metal Powder, in: *Modern Developments in Powder Metallurgy* (Hausner, H. H., Antes, H. W., and Smith, G. D., eds.), MPIF, Princeton, NJ, USA, Vol. 12, pp. 105–111 (1981)

143. Bauckhage, K., Andersen, O., Hansmann, S., Reich, W., and Schreckenberg, P., Production of Fine Powders by Ultrasonic Standing Wave Atomization, *Powder Technol.,* 86(1):77–86 (1996)

144. Liu, H., and Dandy, D. S., Deformation and Solidification of Droplets Generated by a Piezo-Electric Transducer, *J. Mater. Sci. Lett.*, 14(3):185–187 (1995)

145. Gummeson, P. U., Modern Atomizing Techniques, *Powder Metall.*, 15(29):67–94 (1972)

146. Beddow, J. K., *The Production of Metal Powders by Atomization*, Heyden Press, Philadelphia, PA, USA (1978)

147. Klar, E., and Shafer, W. M., High-Pressure Gas Atomization of Metals, in: *Powder Metallurgy for High Performance Applications* (Burke, J. J., and Weiss, V., eds.), Syracuse University Press, Syracuse, NY, USA, pp. 57–68 (1972)

148. Klar, E., and Fesko, J. W., Gas and Water Atomization, in: *Metals Handbook*, Ninth Edition, Vol. 7, *Powder Metallurgy* (Klar, E., ed.), Am. Soc. for Metals, Metals Park, OH, USA, pp. 25–39 (1984)

149. Howells, R. I. L., Dunstan, G. R., and Moore, C., Production of Gas Atomized Metal Powders and Their Major Industrial Uses, *Powder Metall.,* 31(4):259–266 (1988)

150. Reinshagen, J. H., and Neupaver, A. J., Principles of Atomization, in: *Physical Chemistry of Powder Metals Production and Processing* (Small, W. M., ed.), TMS, Warrendale, PA, USA, pp. 3–18 (1989)

151. Hilsenrath, J., Hoge, H. J., Beckett, C. W., Masi, J. F., Benedict, W. S., Nuttall, R. L., Fano, L., Touloukian, Y. S., and Woolley, H. W., *Tables of Thermodynamic and Transport Properties of Air, Argon, Carbon Dioxide, Carbon Monoxide, Hydrogen, Nitrogen, Oxygen and Steam*, Pergamon Press, Oxford, UK (1960)

152. Iida, T., and Guthrie, R. I. L., *The Physical Properties of Liquid Metals*, Clarendon Press, Oxford, UK (1988)

153. Brandes, E. A., *Smithell's Metals Reference Book*, Seventh Edition, Butterworths, London, UK (1992)

154. Liu, H., Rangel, R. H., and Lavernia, E. J., Modeling of Reactive Atomization and Deposition Processing of Ni_3Al, *Acta Metall. Mater.*, 42(10):3277–3289 (1994)

155. Clyne, T. W., Numerical Treatment of Rapid Solidification, *Metall. Trans.*, 15B(6):369–381 (1984)

156. Liu, H., Rangel, R. H., and Lavernia, E. J., Modeling of Droplet-Gas Interactions in Spray Atomization of Ta-2.5W Alloy, *Mater. Sci. Eng. A*, 191(1–2):171–184 (1995)

157. San Marchi, C., Liu, H., Lavernia, E. J., Rangel, R. H., Sickinger, A., and Mühlberger, E., Numerical Analysis of the Deformation and Solidification of a Single Droplet Impinging onto a Flat Substrate, *J. Mater. Sci.*, 28(12):3313–3321 (1993)

158. Nachtrieb, N. H., *Bericht der Bunsen-Gesellschaft*, 80(8):678–688 (1976)

159. Sheikhaliev, S. M., and Dunkley, J. J., A Novel Internal Mixing Gas Atomiser for Fine Powder Production, in: *Advances in Powder Metallurgy & Particulate Materials* (Cadle, T. M., and Narasimhan, K. S., eds.), MPIF and APMI International, Princeton, NJ, USA, Vol. 1, Part 1, pp. 161–170 (1996)

160. Anderson, I. E., and Figliola, R. S., Observations of Gas Atomization Process Dynamics, in: *Modern Developments in Powder Metallurgy* (Gummeson, P. U., and Gustafson, D. A., eds.), MPIF, Princeton, NJ, USA, Vol. 20, pp. 205–223 (1988)

161. Mi, J., Figliola, R. S., and Anderson, I. E., A Numerical Simulation of Gas Flow Field Effects on High Pressure Gas Atomization due to Operating Pressure Variation, *Mater. Sci. Eng. A*, 208(1):20–29 (1996)

162. Ting, J., Terpstra, R., Anderson, I. E., Figliola, R. S., and Mi, J., A Novel High Pressure Gas Atomizing Nozzle for Liquid Metal Atomization, in: *Advances in Powder Metallurgy & Particulate Materials* (Cadle, T. M., and Narasimhan, K.S., eds.), MPIF and APMI International, Princeton, NJ, USA, Vol. 1, Part 1, pp. 97–108 (1996)

163. Ridder, S. D., Osella, S. A., Espina, P. I., and Biancaniello, F. S., Intelligent Control of Particle Size Distribution during Gas Atomization, *Int. J. Powder Metall.*, 28(2):133–147 (1992)

164. Biancaniello, F. S., Conway, J. J., Espina, P. I., Mattingly, G. E., and Ridder, S. D., Particle Size Measurement of Inert-Gas-Atomized Powder, *Mater. Sci. Eng. A*, 124:9–14 (1990)

165. Ünal, A., Rapid Solidification of Magnesium by Gas Atomization, *Mater. Manufacturing Processes*, 73:441–461 (1992)

166. Ünal, A., Effect of Processing Variables on Particle Size in Gas Atomization of Rapidly Solidified Aluminum Powders, *Mater. Sci. Technol.*, 3(12):1029–1039 (1987)

167. Ünal, A., Influence of Gas Flow on Performance of 'Confined' Atomization Nozzles, *Metall. Trans.*, 20B(12):833–843 (1989)

168. Ünal, A., Liquid Break-up in Gas Atomization of Fine Aluminum Powders, *Metall. Trans.*, 20B(2):61–69 (1989)

169. Ünal, A., Flow Separation and Liquid Rundown in a Gas-Atomization Process, *Metall. Trans.*, 20B(10):613–622 (1989)

170. Mates, S. P., and Settles, G. S., A Flow Visualization Study of the Gas Dynamics of Liquid Metal Atomization Nozzles, in: *Advances in Powder Metallurgy and Particulate Materials* (Phillips, M., and Porter, J., eds.), APMI International, Princeton, NJ, USA, Part I, pp. 15–29 (1995)

171. Schulz, G., Laminar Sonic and Supersonic Gas Flow Atomization – The NANOVAL Process, in: *Advances in Powder Metallurgy & Particulate Materials* (Cadle, T. M., and Narasimhan, K. S., eds.), MPIF and APMI International, Princeton, NJ, USA, Vol. 1, Part 1, pp. 43–54 (1996)

172. Grant, N. J., A Review of Various Atomization Processes, in: *Rapid Solidification Processing: Principles and Technologies* (Mehrabian, R., Kear, B. H., and Cohen, M., eds.), Claitor's Publishing Division, Baton Rouge, LA, USA, pp. 230–245 (1978)

173. Zeng, X., Liu, H., Chu, M., and Lavernia, E. J., An Experimental Investigation of Reactive Atomization and Deposition Processing of Ni_3Al/Y_2O_3 Using N_2-O_2 Atomization, *Metall. Trans.*, 23A(12):3394–3399 (1992)

174. Liu, H., Zeng, X., and Lavernia, E. J., Processing Maps for Reactive Atomization and Deposition Processing, *Scripta Metall. Mater.*, 29(10):1341–1344 (1993)

175. Donaldson, C. D., and Snedeker, R. S., A Study of Free Jet Impingement, Part 1. Mean Properties of Free and Impinging Jets, *J. Fluid Mech.*, 45(2):281–319 (1971)

176. Nilsson, E. O. F., Method and Device for Pulverizing Solid Materials, U.S. Patent 3.302.892 (1967)

177. Liu, J., Arnberg, L., Backstrom, N., Klang, H., and Savage, A., *Fluid Flow Behavior, Solidification Studies and Some Practical Experience from Ultrasonic Gas Atomization*, Swedish Institute for Metals Research, Stockholm, Sweden (1992)

178. Tsao, C. Y. A., and Grant, N. J., Modeling of the Liquid Dynamic Compaction Spray Process, *Int. J. Powder Metall.*, 30(3):323–333 (1994)

179. Tamura, E., Karasuno, I., Ichidate, M., and Kubo, T., Properties of Oil-Atomized 4100 Low Alloy Steel Powder, in: *Progress in Powder Metallurgy* (Sanderow, H. I., Giebelhausen, W. L., and Kulkarni, K. M., eds.), MPIF, Princeton, NJ, USA, Vol. 41, pp. 13–23 (1986)

180. Wentzell, J. M., Metal Powder Production by Vacuum Atomization, *J. Vac. Sci. Technol.*, 11(6):1035–1037 (1974)

181. Kaufmann, A. A., Method and Apparatus for Making Powder, U.S. Patent 3.099.041 (1963)

182. Kaufmann, A. A., Production of Pure Spherical Powders, U.S. Patent 3.802.816 (1974)

183. Friedman, G., and Kosinski, E., High Performance Material from a Hot-Worked Superalloy Powder, in: *Progress in Powder Metallurgy*, MPIF, Princeton, NJ, USA, Vol. 25, pp. 3–13 (1969)

184. Roberts, P. R., Rotating Electrode Process, in: *Metals Handbook*, Ninth Edition, Vol. 7, *Powder Metallurgy* (Klar, E., ed.), Am. Soc. for Metals, Metals Park, OH, USA, pp. 39–51 (1984)

185. Kear, B. H., Holiday, P. R., and Cox, A. R., On the Microstructure of Rapidly Solidified IN-100 Powders, *Metall. Trans., 10A*(2):191–197 (1979)

186. Anderson, R., Powder Metallurgy at Pratt and Whitney, *Int. J. Powder Metall.*, 26(2):171–178 (1990)

187. Pietsch, W., Stephan, H., Feuerstein, A., Heimerl, J., and Ruthardt, R., Atomization of Reactive and Refractory Metals by the Electron Beam Rotating Disk Process, *Powder Metall. Int.*, 15(2):77–83 (1983)

188. Singer, A. R. E., and Roche, A. D., Roller Atomization of Molten Metals for the Production of Powder, in: *Modern Developments in Powder Metallurgy* (Hausner, H. H., and Taubenblat, P. V., eds.), MPIF and APMI, Princeton, NJ, USA, Vol. 9, pp. 127–140 (1977)

189. Hodkin, D. J., Sutcliffe, P. W., Hardon, P. G., and Russel, L. E., Centrifugal Shot Casting: A New Atomization Process for the Preparation of High-Purity Alloy Powders, *Powder Metall.*, 16(32):278–313 (1973)

190. Liles, D. T., and Deleeuw, D. C., Rapid Solidification via Centrifugal Atomization with a Volatile Liquid Coolant, in: *Rapidly Solidified Crystalline Alloys* (Das, S. K., Kear, B. H., and Adam, C. M., eds.), The Metallurgical Society, Warrendale, PA, USA, pp. 285–290 (1985)

191. Yamauchi, I., Nomura, Y., Ohnaka, I., and Matsumoto, Y., Production of Rapidly Solidified Fe-C-Si Alloy Powder by the Rotating-Water-Atomization Process and its Structure, *Tetsu To Hagane,* 73(9):1138–1145 (1987)

192. Erich, D. L., and Patel, A. N., Battelle Plan Commercialization of Two-Stage Spinning Cup Atomization Process, *Met. Powder Report,* 42(10):698–700 (1987)

193. Konitzer, D. G., Walters, K. W., Heiser, E. L., and Fraser, H. L., Rapidly Solidified Prealloyed Powders by Laser Spin Atomization, *Metall. Trans.*, *15B*:149–153 (1984)

194. Westinghouse Electric Corporation, The Evaluation of Ti-6Al-2V-2Sn Pre-Alloyed Powder Processing, *Technical Report AFML-TR-76-65*, Air Force Materials Laboratory, Wright-Patterson Air Force Base, Dayton, OH, USA (1975)

195. Matei, G., Bicsak, E., Huppmann, W. J., and Claussen, N., Atomization of Metal Powders Using the Vibrating Electrode Method, in: *Modern Developments in Powder Metallurgy* (Hausner, H. H., and Taubenblat, P. V., eds.), MPIF and APMI, Princeton, NJ, USA, Vol. 9, pp. 153–159 (1977)

196. Panda, D., and Robertson, D., Simple Apparatus for Producing Single Liquid Drops, *Rev. Sci.Instrum.,* 65(7):2388–2394 (1994)

197. Sassa, K., Ninagawa, S., and Asai, S., Liquid Metal Atomization by the Use of Electromagnetic Force, *Magnetohydrodynamics,* 28(3):248–252 (1992)

198. Rulison, A. J., and Flagan, R. C., Scale-Up of Electrospray Atomization Using Linear Arrays of Taylor Cones, *Rev. Sci. Instrum.,* 64(3):683–686 (1993)

199. Lefebvre, A. H., *Gas Turbine Combustion*, Hemisphere, Washington, D.C., USA (1983)

200. Wu, J., Production of Spume Drops by the Wind Tearing of Wave Crests: The Search for Quantification, *J. Geophys. Res.,* 98(C10):18221–18227 (1993)

201. Spiel, D. E., On the Births of Jet Drops from Bubbles Bursting on Water Surfaces, *J. Geophys. Res.,* 100(C3):4995–5006 (1995)

202. Wu, J., Film Drops Produced by Air Bubbles Bursting at the Surface of Seawater, *J. Geophys. Res.,* 99(C8):16403–16407 (1994)

203. Spiel, D. E., The Number and Size of Jet Drops Produced by Air Bubbles Bursting on a Fresh Water Surface, *J. Geophys. Res.,* 99(C5):10289–10296 (1994)

204. Castleman, R. A., Jr., The Mechanism of the Atomization of Liquids, *J. Res. Natl. Bur. Standard,* 6(281):369–376 (1931)

205. Taylor, G. I., Generation of Ripples by Wind Blowing over a Viscous Fluid, in: *The Scientific Papers of Sir Geoffrey Ingram Taylor* (Batchelor, G. K., ed.), Vol. III, University Press, Cambridge, England, UK, pp. 244–254 (1963)

206. Taylor, G. I., The Instability of Liquid Surfaces when Accelerated in a Direction Perpendicular to their Planes, I, *Proc. R. Soc. Lond. A,* 201:192–196 (1950)

207. Fraser, R. P., and Eisenklam, P., Research into the Performance of Atomizers for Liquids, *Imp. Coll. Chem. Eng. Soc. J.,* 7:52–68 (1953)

208. Fraser, R. P., Liquid Fuel Atomization, in: *Sixth Symp. (Int.) on Combustion*, Combustion Institute, Reinhold Pub. Corp., NY, USA, pp. 687–701 (1957)

209. Mansour, A., and Chigier, N., Disintegration of Liquid Sheets, *Phys. Fluids,* 2(5):706–719 (1990)

210. Faragó, Z., and Chigier, N., Morphological Classification of Disintegration of Round Liquid Jets in a Coaxial Air Stream, *Atomization and Sprays,* 2(2):137–153 (1992)

211. Chigier, N., The Physics of Atomization, in: *Proc. of the Fifth Int. Conf. on Liquid Atomization and Spray Systems* (Semerjian, H. G., ed.), ILASS-Americas and NIST, Gaithersburg, MD, USA, pp. 49–64 (1991)

212. Hashimoto, H., Suzuki, T., and Matsuya, T., Interaction between Two Thin Liquid Jets with Disintegration by Gas Flow, *Atomization and Sprays,* 1:47–62 (1991)

213. Giffen, E., and Muraszew, A., *The Atomization of Liquid Fuels*, John Wiley & Sons, New York, NY, USA (1953)

214. Chigier, N., *Energy, Combustion and Environment*, McGraw-Hill, New York, NY, USA (1981)

215. Nemchinsky, V. A., Size and Shape of the Liquid Droplet at the Molten Tip of an Arc Electrode, *J. Phys. D: Appl. Phys.,* 27(7):1433–1442 (1994)

216. Iliev, S. D., Iterative Method for the Shape of Static Drops, *Comp. Meth. Appl. Mech. Eng.,* 126(3–4):251–265 (1995)

217. Basaran, O. A., and DePaoli, D. W., Nonlinear Oscillations of Pendant Drops, *Phys. Fluids,* 6(9):2923–2943 (1994)

218. Wohlhuter, F. K., and Basaran, O. A., Shapes and Stability of Pendant and Sessile Dielectric Drops in an Electric Field, *J. Fluid Mech.,* 235:481–510 (1992)

219. Meister, B. J., and Scheele, G. F., Drop Formation from Cylindrical Jets in Immiscible Liquid System, *AIChE J.,* 15(5):700–706 (1969)

220. Miesse, C. C., Correlation of Experimental Data on the Disintegration of Liquid Jets, *Ind. Eng. Chem.,* 47(9):1690–1701 (1955)

221. Krzywoblocki, M. A., Jets – Review of Literature, *Jet Propul.*, 26:760–779 (1957)

222. Phinney, R. E., The Breakup of a Turbulent Liquid Jet in a Gaseous Atmosphere, *J. Fluid Mech.*, 60:689–701 (1973)

223. Kocamustafaogullari, G., Chen, I. Y., and Ishii, M., Unified Theory for Predicting Maximum Fluid Particle Size for Drops and Bubbles, *Argonne National Laboratory Report,* NUREG/CR-4028 (1984)

224. Yi, S., Xie, M., and Chen, B., The Breakup and Atomization of a Viscous Liquid Jet, *Acta Mech. Sinica,* 12(2):124–134 (1996)

225. Kowalewski, T. A., On the Separation of Droplets from a Liquid Jet, *Fluid Dynamics Res.,* 17(3):121–145 (1996)

226. Mayer, W. O. H., Coaxial Atomization of a Round Liquid Jet in a High Speed Gas Stream: A Phenomenological Study, *Exp. in Fluids,* 16(6):401–410 (1994)

227. Reitz, R. D., *Atomization and Other Breakup Regimes of a Liquid Jet,* Ph.D. Thesis, Princeton University, Princeton, NJ, USA, September (1978)

228. Martinelli, L., Reitz, R. D., and Bracco, F. V., Comparison of Computed and Measured Dense Spray Jets, in: *Ninth Int. Colloquium on Dynamics of Explosions and Reactive Systems* (Bowen, J. R., Manson, N., Oppenheim, A. K., and Soloukhin, R. I., eds.), Poitiers, France, pp. 484–512 (1983)

229. Reitz, R. D., and Bracco, F. V., Mechanisms of Breakup of Round Liquid Jets, in: *The Encyclopedia of Fluid Mechanics* (Cheremisinoff, N. P., ed.), Gulf Publishing, Houston, TX, USA, 3(10):233–249 (1986)

230. McCarthy, M. J., and Molloy, N. A., Review of Stability of Liquid Jets and the Influence of Nozzle Design, *Chem. Eng. J.,* 7:1–20 (1974)

231. Tyler, F., Instability of Liquid Jets, *Phil. Mag. Lond.,* 16:504–518 (1933)

232. Reitz, R. D., Modeling Atomization Processes in High-Pressure Vaporizing Sprays, *Atomization and Spray Technol.,* 3:309–337 (1987)

233. Shavit, U., and Chigier, N., Development and Evaluation of a New Turbulence Generator for Atomization Research, *Exp. in Fluids,* 20(4):291–301 (1996)

234. Hiroyasu, H., Shimizu, M., and Arai, M., The Breakup of High Speed Jet in a High Pressure Gaseous Atmosphere, in: *Proc. of the Second Int. Conf. on Liquid Atomization and Spray Systems,* ILASS-Americas and Department of Chemical Engineering, University of Wisconsin-Madison, Madison, Wis., USA, pp. 69–74 (1982)

235. Crapper, G. D., Dombrowski, N., Jepson, W. P., and Pyott, G. A. D., A Note on the Growth of Kelvin-Helmholtz Waves on Thin Liquid Sheets, *J. Fluid Mech.,* 57(4):671–672 (1973); Crapper, G. D., Dombrowski, N., and Pyott, G. A. D., Kelvin-Helmholtz Wave Growth on Cylindrical Sheets, *J. Fluid Mech.,* 68(3):497–502 (1975)

236. Crapper, G. D., Dombrowski, N., and Pyott, G. A. D., Large Amplitude Kelvin-Helmholtz Waves on Thin Liquid Sheets, *Proc. R. Soc. Lond. A,* 342(1629):209–224 (1975)

237. Rangel, R. H., and Sirignano, W. A., Nonlinear Growth of Kelvin-Helmholtz Instability, Effect of Surface Tension and Density Ratio, *Phys. Fluids A,* 31(7):1845–1855 (1988)

238. Dombrowski, N., and Johns, W. R., The Aerodynamic Instability and Disintegration of Viscous Liquid Sheets, *Chem. Eng. Sci.*, 18:203–214 (1963)

239. Kitamura, Y., and Takahashi, T., Stability of a Liquid Jet in Air Flow Normal to the Jet Axis, *J. Chem. Eng. Jpn.,* 9(4):282–286 (1976)

240. Shavit, U., and Chigier, N., The Role of Dynamic Surface Tension in Air Assist Atomization, *Phys. Fluids,* 7(1):24–33 (1995)

241. Ibrahim, E. A., Effect of Swirl on Jet Atomization, *AIAA J.,* 31(12):2376–2377 (1993)

242. Lian, Z. W., and Lin, S. P., Breakup of a Liquid Jet in a Swirling Gas, *Phys. Fluids A,* 2(12):2134–2139 (1990)

243. Lian, Z. W., and Reitz, R. D., The Effect of Vaporization and Gas Compressibility on Liquid Jet Atomization, *Atomization and Sprays,* 3:249–264 (1993)

244. Issac, K., Missoum, A., Drallmeier, J., and Johnston, A., Atomization Experiments in a Coaxial Coflowing Mach 1.5 Flow, *AIAA J.,* 32(8):1640–1646 (1994)

245. Taylor, G. I., The Shape and Acceleration of a Drop in a High Speed Air Stream, in: *The Scientific Papers of G. I. Taylor* (Batchelor, G. K., ed.), Vol. III, University Press, Cambridge, England, UK, pp. 457–464 (1963)

246. Ranger, A. A., and Nicholls, J. A., Aerodynamic Shattering of Liquid Drops, *AIAA J.,* 7(2):285–290 (1969)

247. Grant, R. P., and Middleman, S., Newtonian Jet Stability, *AIChE J.,* 12(4):669–678 (1966)

248. Mahoney, T. J., and Sterling, M. A., The Breakup Length of Laminar Newtonian Liquid Jets in Air, in: *Proc. of the First Int. Conf. on Liquid Atomization and Spray Systems*, Fuel Society of Japan, Tokyo, Jpn., pp. 9–12 (1978)

249. Wu, P. K., Ruff, G. A., and Faeth, G. M., Primary Breakup in Liquid-Gas Mixing Layers, *Atomization and Sprays,* 1:421–440 (1991)

250. Arai, M., Shimizu, M., and Hiroyasu, H., Breakup Length and Spray Angle of High Speed Jet, in: *Proc. of the Third Int. Conf. on Liquid Atomization and Spray Systems* (Yule, A. J., and Eisenklam, P., eds.), Institute of Energy, London, UK, pp. 1B/4/1–10 (1985)

251. Shavit, U., and Chigier, N., Fractal Geometry of Liquid Jets under Breakup, *Fractals,* 2(2):277–281 (1994)

252. Li, X., and Tankin, R. S., On the Prediction of Droplet Size and Velocity Distributions in Sprays through Maximum Entropy Principle, *Part. Part. Syst. Charact.,* 9(3):195–201 (1992); Li, X., and Tankin, R. S., On the Temporal Instability of a Two-Dimensional Viscous Liquid Sheet, *J. Fluid Mech.*, 226:425–443 (1991)

253. Taylor, G. I., The Dynamics of Thin Sheets of Fluid I. Water Bells, *Proc. R. Soc. Lond. A,* 253:289–295 (1959); Taylor, G. I., The Dynamics of Thin Sheets of Fluid III. Disintegration of Fluid Sheets, *Proc. R. Soc. Lond. A,* 253:313–321 (1959); Taylor, G. I., Formation of Thin Flat Sheet of Water, *Proc. R. Soc. Lond. A,* 259:1–17 (1960); Taylor, G. I., Oblique Impact of a Jet on a Plane Surface, *Philos. Trans. R. Soc. Lond. A,* 260:96–100 (1966)

254. Taylor, G. I., and Michael, D. H., On Making Holes in a Sheet of Fluid, *J. Fluid Mech.,* 58(4):625–639 (1973)

255. York, J. L., Stubbs, H. F., and Tek, M. R., The Mechanism of Disintegration of Liquid Sheet, *Trans. ASME,* 75:1279–1286 (1953)

256. Hagerty, W. W., and Shea, J. F., A Study of the Stability of Plane Fluid Sheets, *J. Appl. Mech.,* 22:509–514 (1955)

257. Weihs, D., Stability of Thin, Radially Moving Liquid Sheets, *J. Fluid Mech.*, 87(2):289–298 (1978)

258. Squire, H. B., Investigation of the Instability of a Moving Liquid Film, *Br. J. Appl. Phys.*, 4:167–169 (1953)

259. Weiss, M. A., and Worsham, C. H., Atomization in High Velocity Air-Streams, *J. Am. Rocket Soc.*, 29(4):252–259 (1959)

260. Tokuoka, N., Nagaosa, S., Hora, S., and Sato, G. T., Study on the Disintegration of a Liquid Film by Air Impingement, *Atomization and Spray Technol.*, 1:103–123 (1985)

261. Arai, T., and Hashimoto, H., Disintegration of a Thin Liquid Sheet in a Cocurrent Gas Stream, in: *Proc. of the Third Int. Conf. on Liquid Atomization and Spray Systems* (Yule, A. J., and Eisenklam, P., eds.), Institute of Energy, London, UK, pp. 1B/1/1–7 (1985)

262. Eroglu, H., Chigier, N., and Faragó, Z., Coaxial Atomizer Liquid Intact Lengths, *Phys. Fluids A,* 3(2):303–308 (1991)

263. Knoll, K. E., and Sojka, P. E., Flat-Sheet Twin-Fluid Atomization of High-Viscosity Fluids, Part I: Newtonian Liquids, *Atomization and Sprays,* 2:17–36 (1992)

264. Simmons, H. C., The Correlation of Drop-Size Distributions in Fuel Nozzle Sprays, *Trans. ASME J. Eng. Power,* 99:309–319 (1977)

265. Nam Hong, U., Capillary Wave Instability of a Spherical Liquid Shell with Pulsating Cavity: Formation of Drops, *Bull. Russian Academy of Sci. Phys.*, 57(suppl.2):87–94 (1993)

266. Lefebvre, A. H., The Prediction of Sauter Mean Diameter for Simplex Pressure-Swirl Atomizers, *Atomization Spray Technol.*, 3(1):37–51 (1987)

267. Lenard, P., Über Regen, *Meteorol. Z.,* 21:249–262 (1904)

268. Hinze, J. O., Critical Speeds and Sizes of Liquid Globules, *Appl. Sci. Res. A,* 1(4):273–288 (1949)

269. Hochschwender, E., *Über das Zerblasen von Wassertropfen im Luftstrom und die Wasserfalltheorie der Gewitter*, Ph.D. Thesis, Universität Heidelberg, Rössler and Herbert, Heidelberg, Germany (1919)

270. Hinze, J. O., Fundamentals of the Hydrodynamic Mechanism of Splitting in Dispersion Processes, *AIChE J.,* 1(3):289–295 (1955)

271. Taylor, G. I., The Function of Emulsion in Definable Field Flow, *Proc. R. Soc. Lond. A,* 146:501–523 (1934)

272. Sauter, J., Determining Size of Drops in Fuel Mixture of Internal Combustion Engines, *NACA-TM*-390 (1926)

273. Scheubel, F. N., Über die Zerstäubung in Vergasern, *Jahrbuch d. Wiss. Ges. f. Luftfahrt*, R. Oldenburg, München and Berlin, Germany, pp. 140–146 (1927)

274. Castleman, R. A., Jr., The Mechanism of Atomization Accompanying Solid Injection, *NACA-TR*-440 (1933)

275. Lane, W. R., Shatter of Drops in Streams of Air, *Ind. Eng. Chem.,* 43(6):1312–1317 (1951)

276. Haas, F. C., Stability of Droplets Suddenly Exposed to a High Velocity Gas Stream, *AIChE J.,* 10:920–924 (1964)

277. Hanson, A. R., Domich, E. G., and Adams, H. S., Shock Tube Investigation of the Breakup of Drops by Air Blasts, *Phys. Fluids,* 6:1070–1080 (1963)

278. Sleicher, C. A., Maximum Drop Size in Turbulent Flow, *AIChE J.,* 8:471–477 (1962)

279. Sevik, M., and Park, S. H., The Splitting of Drops and Bubbles by Turbulent Fluid Flow, *Trans. ASME J. Fluids Eng.,* 95:53–60 (1973)

280. Kolmogorov, A. N., On the Disintegration of Drops in a Turbulent Flow, *Dokl. Akad. Nat. USSR,* 66:825–828 (1949)

281. Clay, P. H., The Mechanism of Emulsion Formation in Turbulent Flow, *Proc. R. Acad. Sci.* (Amsterdam), 43(7, Pt. 2):852–865 (1940)

282. Tomotika, S., Breaking up of a Drop of Viscous Liquid Immersed in another Viscous Fluid which is Extending at a Uniform Rate, *Proc. R. Soc. Lond. A,* 153:302–320 (1936)

283. Rumscheidt, F. D., and Mason S. G., Particle Motion in Sheared Suspensions, Deformation and Burst of Fluid Drops in Shear and Hyperbolic Flows, *J. Colloid Sci.,* 16:238–261 (1967)

284. Hsiang, L. P., and Faeth, G. M., Drop Deformation and Breakup due to Shock Wave and Steady Disturbances, *Int. J. Multiphase Flow,* 214:545–560 (1995)

285. Hsiang, L. P., and Faeth, G. M., Near-Limit Drop Deformation and Secondary Breakup, *Int. J. Multiphase Flow,* 18(5):635–652 (1992)

286. Kennedy, J. B., and Roberts, J., Rain Ingestion in a Gas Turbine Engine, in: *Proc. of the Fourth Ann. Conf. ILASS* (Simmons, H., Kennedy, J. B., Reitz, R. D., and Zimmer, D., eds.), ILASS-Americas, Hartford, CT, USA, pp. 154–162 (1990)

287. Goulian, M., and Milner, S. T., Shape Fluctuations of a Droplet Containing a Polymer, *J. de Physique II,* 6(4):543–550 (1996)

288. Delaby, I., Ernst, B., and Muller, R., Drop Deformation during Elongational Flow in Blends of Viscoelastic Fluids, Small Deformation Theory and Comparison with Experimental Results, *Rheologica Acta,* 34(6):525–533 (1995)

289. Fortelny, I., and Zivny, A., Theory of Competition between Breakup and Coalescence of Droplets in Flowing Polymer Blends, *Polymer Eng. Sci.,* 35(23):1872–1877 (1995)

290. Brenn, G., and Frohn, A., An Experimental Method for the Investigation of Droplet Oscillations in a Gaseous Medium, *Exp. in Fluids,* 15(2):85–90 (1993)

291. Deng, Z. T., and Jeng, S. M., Numerical Simulation of Droplet Deformation in Convective Flows, *AIAA J.*, 30(5):1290–1297 (1992)

292. Sheth, K. S., and Pozrikidis, C., Effects of Inertia on the Deformation of Liquid Drops in Simple Shear Flow, *Computers & Fluids,* 24(2):101–119 (1995)

293. Podvysotskiy, A. M., and Dubrovskiy, V. V., Experimental Study of Time-Varying Aerodynamic Breakup of Liquid Drops and of Drops Containing Solids, *Fluid Mech. Res.,* 21(5):34–38 (1992)

294. Olbricht, W. L., and Kung, D. M., The Deformation and Breakup of Liquid Drops in Low Reynolds Number Flow through a Capillary, *Phys. Fluids A,* 4(7):1347–1354 (1992)

295. Kelbaliev, G. I., Rzaev, A. G., and Kasymov, A. A., Application of a Fokker-Planck Equation to Describe Processes of Coagulation and Disintegration of Droplets in a Turbulent Flow, *J. Eng. Phys. Thermophys.,* 64(2):115–118 (1993)

296. Inculet, I. I., Floryan, J. M., and Haywood, R. J., Dynamics of Water Droplets Breakup in Electric Fields, *IEEE Transactions on Industry Applications,* 28(5):1203–1204 (1992)

297. Abdella, K., Rasmussen, H., and Inculet, I. I., Interfacial Deformation of Liquid Drops by Electric Fields at Zero Gravity, *Computers & Mathematics with Applications,* 31(10):67–82 (1996)

298. Vladimirov, V. V., Gabovich, M. D., and Nazarenko, O. K., Disintegration of Charged Drops in the Supercritical Rayleigh Regime, *Tech. Phys. Lett.,* 20(5):343–344 (1994)

299. Shiryaeva, S. O., and Grigor'eva, I. D., Growth Rate of the Instability of a Highly Charged Droplet, *Tech. Phys. Lett.,* 20(3):214–215 (1994)

300. Kandaurova, N. V., Toroptsev, E. L., and Cheunkova, I. Yu., Deformation of a Magnetic Fluid Droplet in Electric and Magnetic Fields, *Magnetohydrodynamics,* 28(3):310–312 (1992)

301. Tao Shi, W., and Apfel, R. E., Deformation and Position of Acoustically Levitated Liquid Drops, *J. Acoust. Soc. Am.,* 99(4 Pt. 1):1977–1984 (1996)

302. Danilov, S. D., and Mironov, M. A., Breakup of a Droplet in a High-Intensity Sound Field, *J. Acoust. Soc. Am.,* 92(5):2747–2755 (1992)

303. Utheza, F., Saurel, R., Daniel, E., and Loraud, J. C., Droplet Break-up through an Oblique Shock Wave, *Shock Waves,* 5(5):265–273 (1996)

304. Hu, G., Krall, A. H., and Weitz, D. A., Shape Fluctuations of Interacting Fluid Droplets, *Phys. Rev. Lett.,* 73(25):3435–3438 (1994)

305. He, X., and Dembo, M., Numerical Simulation of Oil-Droplet Cleavage by Surfactant – Cell Division Mechanics Application, *J. Biomech. Eng.,* 118(2):201–209 (1996)

306. Janssen, J. M. H., and Meijer, H. E. H., Droplet Breakup Mechanisms: Stepwise Equilibrium versus Transient Dispersion, *J. Rheology,* 37(4):597–608 (1993)

307. Sotolongo-Costa, O., Moreno-Vega, Y., Lloveras-Gonzalez, J. J., and Antoranz, J. C., Criticality in Droplet Fragmentation, *Phys. Rev. Lett.,* 76(1):42–45 (1996)

308. Brenner, M. P., Shi, X. D., and Nagel, S. R., Iterated Instabilities during Droplet Fission, *Phys. Rev. Lett.,* 73(25):3391–3394 (1994)

309. Andreani, M., and Yadigaraoglu, G., Parametric Studies on Droplet Size Distribution and Break-up Processes in Dispersed Flow Film Boiling during Reflooding, in: *Proc. of the Int. Topical Meeting on Nuclear Thermal Hydraulics* (Courtland, M., and Delhaye, J. M., eds.), Grenoble, France, Vol. 1, pp. 129–138 (1993)

310. O'Rourke, P. J., and Amsden, A. A., The Tab Method for Numerical Calculation of Spray Droplet Breakup, *SAE Paper 872089*:1–10 (1987)

311. Krzeczkowski, S. A., Measurement of Liquid Droplet Disintegration Mechanisms, *Int. J. Multiphase Flow,* 6:227–239 (1980)

312. Reinecke, W. G., and Waldman, G. D., A Study of Drop Breakup behind Strong Shocks with Application to Flight, *AVCO Report,* AVSD-0110-70–77 (1970)

313. Gel'fand, B. E., Gubin, S. A., and Kogarko, S. M., Various Forms of Drop Fractionation in Shock Waves and Their Special Characteristics, *Inzh. Fiz. Zh.*, 27:119–126 (1974)

314. Wierzba, A., and Takayama, K., Experimental Investigation of the Aerodynamic Breakup of Liquid Drops, *AIAA J.*, 26:1329–1335 (1988)

315. Hwang, S. S., and Reitz, R. D., A Photographic Study of the Breakup of High Speed Vaporizing Drops, in: *Proc. of Int. Conf. on Liquid Atomization and Spray Systems* (Bellan, J., ed.), ILASS, Troy, Michigan, USA, pp. 164–188 (1995)

316. Reitz, R. D., and Diwakar, R., Effect of Drop Breakup on Fuel Sprays, *SAE Paper 860469* (1986); Structure of High-Pressure Fuel Sprays, *SAE Paper 870598* (1987)

317. Hiroyasu, H., and Kadota, T., Fuel Droplet Size Distribution in Diesel Combustion Chamber, *SAE Paper 740715*:2615–2624 (1974)

318. See, J. B., Runkle, J. C., and King, T. B., The Disintegration of Liquid Lead Streams by Nitrogen Jets, *Metall. Trans.,* 4(11):2669–2673 (1973)

319. See, J. B., and Johnston, G. H., Interactions between Nitrogen Jets and Liquid Lead and Tin Streams, *Powder Technol.,* 21:119–133 (1978)

320. Champagne, B., and Angers, R., Fabrication of Powders by the Rotating Electrode Process, *Int. J. Powder Metall. Powder Technol.*, 16(4):359–367 (1980); Champagne, B., and Angers, R., Size Distributions of Powders Atomized by the Rotating Electrode Process, in: *Modern Developments in Powder Metallurgy* (Hausner, H. H., Antes, H. W., and Smith, G. D., eds.), MPIF, Princeton, NJ, USA, Vol. 12, pp. 83–104 (1981); Champagne, B., and Angers, R., REP Atomization Mechanisms, *Powder Metall. Int.,* 16(3):125–128 (1984)

321. Mehrotra, S. P., Mathematical Modeling of Gas Atomization Process for Metal Powder Production, Part 1, *Powder Metall. Int.,* 13(2):80–84 (1981)

322. Mehrotra, S. P., Mathematical Modeling of Gas Atomization Process for Metal Powder Production, Part 2, *Powder Metall. Int.,* 13(3):132–135 (1981)

323. Lubanska, H., Correlation of Spray Ring Data for Gas Atomization of Liquid Metals, *J. Met.,* 22(2):45–49 (1970)

324. Kuntz, D. W., and Payne, J. L., Simulation of Powder Metal Fabrication with High Pressure Gas Atomization, in: *Advances in Powder Metallurgy and Particulate Materials* (Phillips, M., and Porter, J., eds.), APMI International, Princeton, NJ, USA, Part I, pp. 63–77 (1995)

325. Liu, H., Numerical Modeling of Gas Atomization in Spray Forming Process, *J. Mater. Synth. Proc.,* 5(1):11–17 (1997)

326. Bauckhage, K., Universität Bremen, Bremen, Germany, Private Communication (1998)

327. Miller, S. A., and Miller, R. S., Real Time Visualization of Close-Coupled Gas Atomization, in: *Advances in Powder Metallurgy and Particulate Materials* (Capus, J. M., and German, R. M., eds.), APMI International, Princeton, NJ, USA, Vol. 1, pp. 113–125 (1992)

328. Hespel, J. M., Brunet, Y., and Dyment, A., Liquid Sheet and Film Atomization: A Comparative Experimental Study, *Exp. in Fluids,* 19(6):388–396 (1995)

329. Bradley, D., On the Atomization of Liquids by High-Velocity Gases, *J. Phys. D.: Appl. Phys.,* 6:1724–1736 (1973)

330. Bradley, D., On the Atomization of a Liquid by High-Velocity Gases: II, *J. Phys. D.: Appl. Phys.*, 6:2267–2272 (1973)

331. Rao, K. P., and Mehrotra, S. P., Effect of Process Variables on Atomization of Metals and Alloys, in: *Modern Developments in Powder Metallurgy, Principles and Processes* (Hausner, H. H., Antes, H. W., and Smith, G. D., eds.), MPIF and APMI International, Princeton, NJ, USA, Vol. 12, pp. 113–130 (1981)

332. Lewis, D. J., The Instability of Liquid Surfaces when Accelerated in a Direction Perpendicular to Their Planes, II, *Proc. R. Soc. Lond. A,* 202(1068):81–96 (1950)

333. Gerretsen, J. H., and Visser, A., Inhibitor Performance under Liquid Droplet Impingement Conditions in CO_2-Containing Environment, *Corrosion Sci.*, 34(8):1299–1310 (1993)

334. Tsurutani, K., Yao, M., Senda, J., and Fujimoto, H., Numerical Analysis of the Deformation Process of a Droplet Impinging upon a Wall, *JSME Int. J.*, II 33(3):555–561 (1990)

335. Senda, J., Yamada, K., Fujimoto, H., and Miki, H., The Heat Transfer Characteristics of a Small Droplet Impinging upon a Hot Surface, *JSME Int. J.*, II 31(1):105–111 (1988)

336. Shi, M. H., Bai, T. C., and Yu, J., Dynamic Behavior and Heat Transfer of a Liquid Droplet Impinging on a Solid Surface, *Exp. Thermal Fluid Sci.*, 6:202–207 (1993)

337. Worthington, A. M., On the Forms Assumed by Drops of Liquids Falling Vertically on a Horizontal Plate, *Proc. R. Soc. Lond. A,* 25:261–271 (1876); Worthington, A. M., A Second Paper on the Forms Assumed by Drops of Liquids Falling Vertically on a Horizontal Plate, *Proc. R. Soc. Lond. A,* 25:498–503 (1877); Worthington, A. M., *A Study of Splashes*, Longmans, Green and Co., London, UK (1908)

338. Liu, H., Lavernia, E. J., and Rangel, R. H., Modeling of Molten Droplet Impingement on a Non-Flat Surface, in: *ASME Winter Annual Meeting, Symp. on Multiphase Flow and Heat Transfer in Materials Processing* (Chen, M., ed.), Chicago, IL, USA, pp. 1–5 (1994)

339. Liu, H., Influence of Droplet/Substrate Interactions in Thermal Spraying on Microstructure of Thermal Coatings, in: *95' Beijing Int. Conf. for Surf. Sci., and Eng.* (Zhu, R., ed.), International Academic Publishers, Beijing, China, pp. 130–134 (1995)

340. Hatta, N., Fujimoto, H., and Takuda, H., Deformation Process of a Water Droplet Impinging on a Solid Surface, *Trans. ASME J. Fluids Eng.*, 117(3):394–401 (1995)

341. Fukai, J., Zhao, Z., Poulikakos, D., Megaridis, C. M., and Miyatake, O., Modeling of the Deformation of a Liquid Droplet Impinging upon a Flat Surface, *Phys. Fluids A,* 5(11):2588–2599 (1993)

342. Cheng, E., and Ebner, C., Dynamics of Liquid-Droplet Spreading: A Monte Carlo Study, *Phys. Rev. B,* 47(20):13808–13811 (1993)

343. Uijttewaal, W. S. J., Nijhof, E. J., and Heethaar, R. M., Droplet Migration, Deformation, and Orientation in the Presence of a Plane Wall: A Numerical Study Compared with Analytical Theories, *Phys. Fluids A,* 5(4):819–825 (1993)

344. Healy, W. M., Hartley, J. G., and Abdel-Khalik, S. I., Comparison between Theoretical Models and Experimental Data for the Spreading of Liquid Droplets Impacting a Solid Surface, *Int. J. Heat Mass Transfer,* 39(14):3079–3082 (1996)

345. McHale, G., Rowan, S. M., and Newton, M. I., Frenkel's Method and the Spreading of Small Spherical Droplets, *J. Phys. D: Appl. Phys.,* 27(12):2619–2623 (1994)

346. Venalainen, O., Ala-Nissila, T., and Kaski, K., Morphological Changes of Small Viscous Droplets under Spreading, *Europhysics Lett.,* 25(2):125–130 (1994)

347. Anders, K., Roth, N., and Frohn, A., The Velocity Change of Ethanol Droplets during Collision with a Wall Analyzed by Image Processing, *Exp. in Fluids,* 15(2):91–96 (1993)

348. Law, B. M., and Pak, H. K., Ellipsometric Imaging of Surface Drops, *J. Opt. Soc. Am. A,* 13(2):379–384 (1996)

349. Hocking, L. M., The Spreading of Drops with Intermolecular Forces, *Phys. Fluids,* 6(10):3224–3228 (1994)

350. Kholodenko, A. L., Statistical Mechanics of the Deformable Droplets on Flat Surfaces, *J. Math. Phys.,* 37(3):1287–1313 (1996)

351. Kholodenko, A. L., Statistical Mechanics of the Deformable Droplets on Riemannian Surfaces: Applications to Reptation and Related Problems, *J. Math. Phys.,* 37(3):1314–1335 (1996)

352. Venalainen, O., Ala-Nissila, T., and Kaski, K., Simulations of Solid-on-Solid Models of Spreading of Viscous Droplets, *Physica A,* 210(3–4):362–375 (1994)

353. Brenner, M., and Bertozzi, A., Spreading of Droplets on a Solid Surface, *Phys. Rev. Lett.,* 71(4):593–596 (1993)

354. Dos Santos, F. D., and Ondarcuhu, T., Free-Running Droplets, *Phys. Rev. Lett.,* 75(16):2972–2975 (1995)

355. Fujimoto, H., and Hatta, N., Deformation and Rebounding Processes of a Water Droplet Impinging on a Flat Surface above Leidenfrost Temperature, *Trans. ASME J. Fluids Eng.,* 118(1):142–149 (1996)

356. Mundo, C., Sommerfeld, M., and Tropea, C., Droplet-Wall Collisions: Experimental Studies of the Deformation and Breakup Process, *Int. J. Multiphase Flow,* 21(2):151–173 (1995)

357. Yarin, A. L., and Weiss, D. A., Impact of Drops on Solid Surfaces: Self-Similar Capillary Waves, and Splashing as a New Type of Kinematic Discontinuity, *J. Fluid Mech.,* 283:141–173 (1995)

358. Rein, M., The Transitional Regime between Coalescing and Splashing Drops, *J. Fluid Mech.,* 306:145–165 (1996)

359. Brailovsky, A. B., Gaponov, S. V., and Luchin, V. I., Mechanisms of Melt Droplets and Solid-Particle Ejection from a Target Surface by Pulsed Laser Action, *Appl. Phys. A,* D61(1):81–86 (1995)

360. Chate, D. M., and Kamra, A. K., Charge Separation Associated with Splashing of Water Droplets on Solid Surfaces, *Atmospheric Res.,* 29(1–2):115–128 (1993)

361. Zapalowicz, Z., Influence of Air Velocity on Droplet's Wetting and Evaporation Conditions on a Flat Surface, *Int. Comm. Heat Mass Transfer,* 21(3):411–420 (1994); Zapalowicz, Z., Analysis of the Process of Wetting a Horizontal Heated Surface Area by a Single Water Droplet, *Int. Comm. Heat Mass Transfer,* 22(5):713–720 (1995)

362. Fukai, J., Shiiba, Y., Yamamoto, T., Miyatake, O., Poulikakos, D., Megaridis, C. M., and Zhao, Z., Wetting Effects on the Spreading of a Liquid Droplet Colliding with a Flat Surface: Experiment and Modeling, *Phys. Fluids,* 7(2):236–247 (1995)

363. Chandra, S., di Marzo, M., Qiao, Y. M., and Tartarini, P., Effect of Liquid-Solid Contact Angle on Droplet Evaporation, *Fire Safety J.,* 27:141–158 (1996)

364. Popov, V. G., Effect of Real Wetting Conditions on the Hysteresis of the Contact Angle of Drops, *High Temperature,* 30(5):753–760 (1992); Bausch, R., and Blossey, R., Critical Droplets on a Wall Near a First-Order Wetting Transition, *Phys. Rev. E,* 48(2):1131–1135 (1993); Zosel, A., Studies of the Wetting Kinetics of Liquid Drops on Solid Surfaces, *Colloid & Polymer Sci.,* 271(7):680–687 (1993)

365. Hocking, L. M., Rival Contact-Angle Models and the Spreading of Drops, *J. Fluid Mech.,* 239:671–681 (1992)

366. Pasandideh-Fard, M., Qiao, Y. M., Chandra, S., and Mostaghimi, J., Capillary Effects during Droplet Impact on a Solid Surface, *Phys. Fluids,* 8(3):650–659 (1996)

367. Pasandideh-Fard, M., Bhola, R., Chandra, S., and Mostaghimi, J., Deposition of Tin Droplets on a Steel Plate: Simulations and Experiments, *Int. J. Heat Mass Transfer,* 41:2929–2945 (1998)

368. Zhao, Z., Poulikakos, D., and Fukai, J., Heat Transfer and Fluid Dynamics during the Collision of a Liquid Droplet on a Substrate – I. Modeling, *Int. J. Heat Mass Transfer,* 39(13):2771–2789 (1996); Zhao, Z., Poulikakos, D., and Fukai, J., Heat Transfer and Fluid Dynamics during the Collision of a Liquid Droplet on a Substrate – II. Experiments, *Int. J. Heat Mass Transfer,* 39(13):2791–2802 (1996)

369. Inada, S., and Yang, W., Film Boiling Heat Transfer for Saturated Drops Impinging on a Heating Surface, *Int. J. Heat Mass Transfer,* 37(16):2588–2591 (1994)

370. Anderson, D. M., and Davis, S. H., The Spreading of Volatile Liquid Droplets on Heated Surfaces, *Phys. Fluids,* 7(2):248–265 (1995)

371. Trapaga, G., Matthys, E. F., Valencia, J. J., and Szekely, J., Fluid Flow, Heat Transfer, and Solidification of Molten Metal Droplets Impinging on Substrates: Comparison of Numerical and Experimental Results, *Metall. Trans., 23B*(6):701–718 (1992)

372. Matida, E. A., and Torii, K., Droplet Deposition and Heat Transfer Simulations of Turbulent Air-Water Dispersed Flow in a Vertical Tube, *JSME Int. J. B,* 38(4):628–636 (1995)

373. Venalainen, O., Computer Simulations of Evaporation and Spreading of Stranski-Krastanov Islands and Viscous Fluid Droplets, *Physica Scripta,* 48(4):487–493 (1993)

374. Halvorson, P. J., Carson, R. J., Jeter, S. M., and Abdel-Khalik, S. I., Critical Heat Flux Limits for a Heated Surface Impacted by a Stream of Liquid Droplets, *J. Heat Transfer,* 116(3):679–685 (1994)

375. Vasil'ev, N. I., Guguchkin, V. V., and Nigmatulin, B. I., The Interaction between Liquid Droplets and a Heated Surface, *High Temperature,* 31(5):738–743 (1993)

376. Shrayber, A. A., Podvysotskiy, A. M., and Khelemskiy, S. L., Effect of Viscosity on the Rate of Mass Transfer in Impingement of Droplets on a Wetted Wall, *Fluid Mech. Res.,* 21(2):6–10 (1992)

377. Takano, T., Fujita, T., and Kobayashi, K., Vaporization Behavior of a Single Droplet Impinging on a Hot Surface with a Flame-Sprayed Ceramic Coating and a Pressurized Atmosphere, *Heat Transfer - Jpn. Res.,* 24(1):80–97 (1995)

378. Liu, H., Sickinger, A., Mühlberger, E., Lavernia, E. J., and Rangel, R. H., Numerical Investigation of Micro-Pore Formation during Substrate Impact of Molten Droplets in Spraying Processes, in: *1994 Thermal Spray Industrial Applications* (Berndt, C. C., and Sampath, S., eds.), ASM International, Materials Park, OH, USA, pp. 375–380 (1994)

379. Liu, H., Cai, W., Rangel, R. H., and Lavernia, E. J., Numerical and Experimental Study of Porosity Evolution during Plasma Spray Deposition of Tungsten, in: *NATO Workshop on Sci. Technol. of Rapid Solidification and Processing* (Otooni, M.A., ed.), West Point, NY, USA, pp. 73–107 (1994)

380. Bennett, T., and Poulikakos, D., Splat-Quench Solidification: Estimating the Maximum Spreading of a Droplet Impacting a Solid Surface, *J. Mater. Sci.,* 28(4):963–970 (1993)

381. Kang, B., Zhao, Z., and Poulikakos, D., Solidification of Liquid Metal Droplets Impacting Sequentially on a Solid Surface, *J. Heat Transfer,* 116(2):436–445 (1994)

382. Shue, K., and Yeh, J., Formation of Highly Elongated Flakes by Splat Quenching Droplets on a Moving Substrate at a Small Angle, *Mater. Sci. Eng. A,* 189(1–2):241–245 (1994)

383. Rein, M., Shock Waves Formed during the Collision of a Droplet of High Velocity with a Liquid Surface, *Z. Angew. Math. Mech.,* 75(1):315–316 (1995)

384. Wagner, L., Effect of Substrate Corrugation on the Spreading of Polymer Droplets, *Phys. Rev. E,* 52(3A):2797–2800 (1995)

385. Chandra, S., and Avedisian, C. T., Observations of Droplet Impingement on a Ceramic Porous Surface, *Int. J. Heat Mass Transfer,* 35(10):2377–2388 (1992)

386. Brochard-Wyart, F., Debrégeas, G., and De Gennes, P. G., Spreading of Viscous Droplets on a Non Viscous Liquid, *Colloid & Polymer Sci.,* 274(1):70–72 (1996)

387. Rein, M., Phenomena of Liquid Drop Impact on Solid and Liquid Surfaces, *Fluid Dynamics Res.,* 12:61–93 (1993)

388. Liu, H., Lavernia, E. J., and Rangel, R. H., Numerical Simulation of Substrate Impact and Freezing of Droplets in Plasma Spray Processes, *J. Phys. D: Appl. Phys.,* 26:1900–1908 (1993)

389. Liu, H., Lavernia, E. J., and Rangel, R. H., Modeling of Molten Droplet Impingement on a Non-Flat Surface, *Acta Metall. Mater.*, 43(5):2053–2072 (1995)

390. Dykhuizen, R. C., Review of Impact and Solidification of Molten Thermal Spray Droplets, *J. Thermal Spray Technol.*, 3(4):351–361 (1994)

391. Dussan, E. B., On the Spreading of Liquids on Solid Surfaces: Static and Dynamic Contact Lines, *Ann. Rev. Fluid Mech.*, 11:371–400 (1979); Davis, S. H., Contact-Line Problems in Fluid Mechanics, *Trans. ASME: J. Appl. Mech.*, 50:977–982 (1983); Gennes, P. G. De, Wetting: Statics and Dynamics, *Rev. Mod. Phys.*, 57:827–863 (1985)

392. Hinkle, S. E., Water Drop Kinetic Energy and Momentum Measurement Considerations, *Appl. Eng. Agric.*, 5(3):386–391 (1979)

393. Stow, C. D., and Stainer, R. D., The Physical Products of a Splashing Water Drop, *J. Meteorol. Soc. Jpn.*, 55:518–532 (1977)

394. Levin, Z., and Hobbs, P. V., Splashing of Water Drops on Solid and Wetted Surfaces: Hydrodynamics and Charge Separation, *Philos. Trans. R. Soc. Lond. A,* 269:555–585 (1971)

395. Engel, O. G., Water Drop Collisions with Solid Surfaces, *J. Res. Nat. Bur. Stand.*, 54(5):281–298 (1955)

396. Stow, C. D., and Hadfield, M. G., An Experimental Investigation of Fluid Flow Resulting from the Impact of a Water Drop with an Unyielding Dry Surface, *Proc. R. Soc. Lond. A,* 373:419–441 (1981)

397. Harlow, F. H., and Shannon, J. P., The Splash of a Liquid Drop, *J. Appl. Phys.*, 38(10):3855–3866 (1967).

398. Walzel, P., Zerteilgrenze beim Tropfenaufprall, *Chem. Ing. Tech.*, 52(4):338–339 (1980)

399. Schmidt, P., and Knauss, G., Prallzerstäubung von Flüssigkeiten bei Nichtbenetzung, *Chem. Ing. Tech.*, 48(7):659–659 (1976)

400. Berg, M., and Ulrich, J., Experimental Based Detection of the Splash Limits for the Normal and Oblique Impact of Molten Metal Particles on Different Surfaces, *J. Mater. Synth. Proc.*, 5(1):45–49 (1997)

401. Madejski, J., Solidification of Droplets on a Cold Surface, *Int. J. Heat Mass Transfer,* 19:1009–1013 (1976)

402. Hasui, A., Kitahara, S., and Fukushima, T., On Relation between Properties of Coating and Spraying Angle in Plasma Jet Spraying, *Trans. Natl. Res. Inst. Met.*, 12(1):9–20 (1970)

403. Kuroda, S., Fukushima, T., and Kitahara, S., Significance of Quenching Stress in the Cohesion and Adhesion of Thermally Sprayed Coatings, *J. Thermal Spray Technol.*, 1(4):325–332 (1992)

404. Pech, J., Hannoyer, B., Bianchi, L., Fauchais, P., and Denoirjean, A., Study of Oxide Layers Obtained onto 304L Substrate Heated by a DC Plasma Jet, in: *Thermal Spray: A United Forum for Scientific and Technological Advances* (Berndt, C. C., ed.), ASM International, Materials Park, OH, USA, pp. 775–782 (1997)

405. Bianchi, L., Blein, F., Baradel, N., and Jeandin, M., In-Situ Measurements of Residual Stress within Coatings Plasma-Sprayed under Industrial Conditions, in: *Thermal Spray: A United Forum for Scientific and Technological Advances* (Berndt, C. C., ed.), ASM International, Materials Park, OH, USA, pp. 831–838 (1997)

406. Fantassi, S., Vardelle, M., Vardelle, A., and Fauchais, P., Influence of the Velocity of Plasma Sprayed Particles on the Splat Formation, in: *Thermal Spray Coatings: Research, Design and Applications* (Berndt, C. C., and Bernecki, T. F., eds.), ASM International, Materials Park, OH, USA, pp. 1–6 (1993)

407. Houben, J. M., *Relation of the Adhesion of Plasma Sprayed Coatings to the Process Parameters Size, Velocity and Heat Content of the Spray Particles*, Ph.D. Thesis, Technische Universiteit Eindhoven, Eindhoven, The Netherlands (1988)

408. Sobolev, V. V., Guilemany, J. M., and Nutting, J., Analysis of Different Mechanisms of Coating-Substrate Adhesion in Thermal Spraying, in: *Thermal Spray: A United Forum for Scientific and Technological Advances* (Berndt, C.C., ed.), ASM International, Materials Park, OH, USA, pp. 797–804 (1997)

409. Matson, D. M., Rolland, M., and Flemings, M. C., Impingement Behavior of a Spray Cast Aluminum Alloy, in: *Solidification* (Marsh, S. P., Dantzig, J. A., Trivedi, R., Hofmeister, W., Chu, M. G., Lavernia, E. J., and Chun, J. H., eds.), TMS, Warrendale, PA, USA, pp. 389–400 (1998)

410. Hofmeister, W., Bayuzick, R. J., Trapaga, G., Matson, D. M., and Flemings, M. C., Observation of Thermal Profiles during Impact and Solidification of Nickel Droplets, in: *Solidification* (Marsh, S. P., Dantzig, J. A., Trivedi, R., Hofmeister, W., Chu, M. G., Lavernia, E. J., and Chun, J. H., eds.), TMS, Warrendale, PA, USA, pp. 375–387 (1998)

411. Chandra, S., and Avedisian, C. T., On the Collision of a Droplet with a Solid Surface, *Proc. R. Soc. Lond. A,* 432:13–41 (1991)

412. Bernardin, J. D., Stebbins, C. J., and Mudawar, I., Mapping of Impact and Heat Transfer Regimes of Water Drops Impinging on a Polished Surface, *Int. J. Heat Mass Transfer,* 40(2):247–267 (1997)

413. Bernardin, J. D., Stebbins, C. J., and Mudawar, I., Effects of Surface Roughness on Water Droplet Impact History and Heat Transfer Regimes, *Int. J. Heat Mass Transfer,* 40(1):73–88 (1997)

414. Tio, K. K., and Sadhal, S. S., Thermal Analysis of Droplet Spray Evaporation from a Heated Solid Surface, *Trans. ASME J. Heat Transfer,* 114:220–226 (1992)

415. Takeuchi, K., Senda, J., and Sato, Y., Experimental Studies on the Behavior of a Small Droplet Impinging upon a Hot Surface, in: *Proc. of the Second Int. Conf. on Liquid Atomization and Spray Systems*, ILASS-Americas and Department of Chemical Engineering, University of Wisconsin-Madison, Madison, Wis., USA, pp. 397–404 (1982)

416. Mortensen, A., Drevet, B., and Eustathopoulos, N., Kinetics of Diffusion-Limited Spreading of Sessile Drops in Reactive Wetting, *Scripta Mater.*, 36(6):645–651 (1997)

417. Carles, P., Cazabat, A. M., and Heslot, F., Droplet Spreading in Presence of Contamination, *Colloids and Surfaces,* 58:287–292 (1991)

418. Thompson, J., On Certain Curious Motions Observable at the Surfaces of Wine and other Alcoholic Liquors, *Philos. Mag.*, 4(10):330–333 (1855); Marangoni, C., Sul Principio Della Viscosita Superficiale Dei Liquidi Stabilito Dal, *Nuovo Cimento,* 5:239–273 (1871); Scriven, L. E., and Sternling, C. V., The Marangoni Effects, *Nature,* 187:186–188 (1960); Levich, V. G., and Krylov, V. S., Surface-Tension-Driven Phenomena, *Ann. Rev. Fluid Mech.*, 1:293–316 (1969)

419. Braun, R. J., Murray, B. T., Boettinger, W. J., and McFadden, G. B., Lubrication Theory for Reactive Spreading of a Thin Drop, *Phys. Fluids,* 7(8):1797–1810 (1995)

420. Choo, R. T., Mukai, K., and Toguri, J. M., Marangoni Interaction of a Liquid Droplet Falling onto a Liquid Pool, *Welding Res. Suppl.*, 139–146 (1992)

421. Al-Roub, M., Farrell, P. V., and Senda J., Near Wall Interaction in Spray Impingement, *SAE Paper 960863*:171–181 (1996)

422. Senda, J., Characteristics of Diesel Spray Impinging on a Flat Wall, in: *Int. Symp. Adv. Spray Combust.*, Hiroshima, Jpn., pp. 55–69 (1994); Senda, J., Kobayashi, M., Iwashita, S., and Fujimoto, H., Modeling of Diesel Spray Impingement on a Flat Wall, *SAE Paper 941894*: pp. 1–14 (1994); Senda, J., Fujimoto, H., Kobayashi, M., Yamamoto, K., and Enomoto, Y., Heat Transfer Characteristics of a Diesel Spray Impinging on a Wall, *Bull. M.E.S.J.*, 23(2):67–76 (1995); Senda, J., Kanda, T., Al-Roub, M., Farrell, P. V., Fukami, T., and Fujimoto, H., Modeling Spray Impingement Considering Fuel Film Formation on the Wall, *SAE Paper* 970047, pp. 37–51 (1997)

423. Mugele, R. A., and Evans, H. D., Droplet Size Distributions in Sprays, *Ind. Eng. Chem.*, 43(6):1317–1324 (1951)

424. Fraser, R. P., and Eisenklam, P., Liquid Atomization and the Drop Size of Sprays, *Trans. Inst. Chem. Eng.*, 34:294–319 (1956)

425. Miesse, C. C., and Putnam, A. A., Mathematical Expressions for Drop-Size Distributions, *Injection and Combustion of Liquid Fuels*, Sec. II, WADC Technical Report 56–344, Battelle Memorial Institute, Columbus, OH, USA (1957)

426. Brodkey, R. A., *The Phenomena of Fluid Motions*, Addison-Wesley, Reading, MA, USA (1967)

427. Marshall, W. R., Jr., Mathematical Representations of Drop-Size Distributions of Sprays, in: *Atomization and Spray Drying*, Chapter VI, *Chem. Eng. Prog. Monogr. Ser.* 50(2), AIChE, New York, NY, USA (1954)

428. Rosin, P., and Rammler, E., The Laws Governing the Fineness of Powdered Coal, *J. Inst. Fuel,* 7(31):29–36 (1933)

429. Kim, K. Y., and Marshall, W. R., Jr., Drop-Size Distributions from Pneumatic Atomizers, *AIChE J.,* 17(3):575–584 (1971)

430. Licht, W., Maximum Drop Size Produced by Pneumatic Atomization, *AIChE J.*, 20(3):595–596 (1974)

431. Troesch, H. A., Die Zerstäubung von Flüssigkeiten, *Chem. Ing. Tech.*, 26(6):311–320 (1954)

432. Shellens, R. W., and Brzustowski, T. A., A Prediction of the Drop Size Distribution in a Spray from First Principles, *Atomization and Spray Technol.*, 1:89–102 (1985)

433. Tate, R. W., Some Problems Associated with the Accurate Representation of Drop-Size Distributions, in: *Proc. of the Second Int. Conf. on Liquid Atomization and Spray Systems*, ILASS-Americas and Department of Chemical Engineering, University of Wisconsin-Madison, Madison, Wis., USA, pp. 341–351 (1982)

434. Merrington, A. C., and Richardson, E. G., The Breakup of Liquid Jets, *Proc. Phys. Soc. Lond.*, 59(331):1–13 (1947)

435. Panasenkov, N. J., *Zh. Tekh. Fiz.*, 21(2):160–166 (1951)

436. Harmon, D. R., Jr., Drop Sizes from Low Speed Jets, *J. Franklin Inst.*, 259(5):519–522 (1955)

437. Cheremisinoff, N. P., *Gas-Liquid-Flows*, in: *The Encyclopedia of Fluid Mechanics*, Gulf Publishing, Houston, TX, USA, Vol. 3 (1986)

438. Elkotb, M. M., Fuel Atomization for Spray Modeling, *Prog. Energy Combust. Sci.*, 8(1):61–91 (1982)

439. Yamane, Y., Yokota, H., and Kamimoto, T., Atomization and Air-Entrainment Characteristics of Unsteady Dense Sprays, *JSME Int. J. B*, 37(3):604–610 (1994)

440. Giffen, E., and Lamb, T. A J., The Effect of Air Density on Spray Atomization, *Motor Industry Research Association Report*, 1953/5 (1953)

441. Razumovskii, N. A., Shape of Drops and Satellite Droplets Formed as a Result of Forced Capillary Breakup of a Liquid Jet, *Tech. Phys.*, 38(9):752–761 (1993)

442. Teng, H., Kinoshita, C. M., and Masutani, S. M., Prediction of Droplet Size from the Breakup of Cylindrical Liquid Jets, *Int. J. Multiphase Flow*, 21(1):129–136 (1995)

443. Radcliffe, A., Fuel Injection, in: *High Speed Aerodynamics and Jet Propulsion*, Vol. XI, Sec. D, Princeton University Press, Princeton, NJ, USA (1960)

444. Babu, K. R., Narasimhan, M. V., and Narayanaswamy, K., Prediction of Mean Drop Size of Fuel Sprays from Swirl Spray Atomizers, in: *Proc. of the Second Int. Conf. on Liquid Atomization and Spray Systems*, ILASS-Americas and Department of Chemical Engineering, University of Wisconsin-Madison, Madison, Wis., USA, pp. 99–106 (1982)

445. Kennedy, J. B., High Weber Number SMD correlations for Pressure Atomizers, *ASME Paper 85-GT-37* (1985)

446. Jones, A. R., Design Optimization of a Large Pressure-Jet Atomizer for Power Plant, in: *Proc. of the Second Int. Conf. on Liquid Atomization and Spray Systems*, ILASS-Americas and Department of Chemical Engineering, University of Wisconsin-Madison, Madison, Wis., USA, pp. 181–185 (1982)

447. Abou-Ellail, M. M. M., and Rafat, N. M., Effect of Fuel Pressure, Air Pressure and Air Temperature on Droplet Size Distribution in Hollow-Cone Kerosine Sprays, in: *Proc. of the First Int. Conf. on Liquid Atomization and Spray Systems*, Fuel Society of Japan, Tokyo, Jpn, pp. 85–92 (1978)

448. Suyari, M., and Lefebvre, A. H., Film Thickness Measurements in a Simplex Swirl Atomizer, *AIAA J. Propul. Power,* 2(6):528–533 (1986)

449. Wang, X. F., and Lefebvre, A. H., Mean Drop Sizes from Pressure-Swirl Nozzles, *AIAA J. Propul. Power,* 3(1):11–18 (1987)

450. Hasson, D., and Mizrahi, J., The Drop Size of Fan Spray Nozzles: Measurements by the Solidifying Wax Method Compared with Those Obtained by Other Sizing Techniques, *Trans. Inst. Chem. Eng.*, 39(6):415–422 (1961)

451. Simmons, H. C., The Prediction of Sauter Mean Diameter for Gas Turbine Fuel Nozzles of Different Types, *ASME Paper 79-WA/GT-5* (1979)

452. Dodge, L. G., and Biaglow, J. A., Effect of Elevated Temperature and Pressure on Sprays from Simplex Swirl Atomizers, *ASME Paper 85-GT-58* (1985)

453. Simmons, H. C., and Harding, C. F., Some Effects on Using Water as a Test Fluid in Fuel Nozzle Spray Analysis, *ASME Paper 80-GT-90* (1980)

454. Rizk, N. K., and Lefebvre, A. H., Spray Characteristics of Simplex Swirl Atomizers, *Prog. Astronaut. Aeronaut.,* 95:563–580 (1985)

455. De Corso, S. M., Effect of Ambient and Fuel Pressure on Spray Drop Size, *Trans. ASME J. Eng. Power,* 82(1):10–18 (1960)

456. Neya, K., and Sato, S., Effect of Ambient Air Pressure on the Spray Characteristics of Swirl Atomizers, *Paper 27*, *Ship Res. Inst.*, Tokyo, Jpn. (1968)

457. De Corso, S. M., and Kemeny, G. A., Effect of Ambient and Fuel Pressure on Nozzle Spray Angle, *Trans. ASME,* 79(3):607–615 (1957)

458. Rizk, N. K., and Lefebvre, A. H., Spray Characteristics of Spill-Return Atomizer, *AIAA J. Propul. Power,* 1(3):200–204 (1985)

459. Elkotb, M. M., Rafat, N. M., and Hanna, M. A., The Influence of Swirl Atomizer Geometry on the Atomization Performance, in: *Proc. of the First Int. Conf. on Liquid Atomization and Spray Systems*, Fuel Society of Japan, Tokyo, Jpn., pp. 109–115 (1978)

460. Chin, J. S., Nickolaus, D., and Lefebvre, A. H., Influence of Downstream Distance on the Spray Characteristics of Pressure-Swirl Atomizers, *Trans. ASME J. Eng. Gas Turbines Power,* 106(1):219–224 (1984)

461. Clare, H., and Radcliffe, A., An Airblast Atomizer for Use with Viscous Fuels, *J. Inst. Fuel,* 27(165):510–515 (1954)

462. Mayer, E., Theory of Liquid Atomization in High Velocity Gas Streams, *Am. Rocket Soc. J.,* 31:1783–1785 (1961)

463. Suyari, M., and Lefebvre, A. H., Drop-Size Measurements in Air-Assist Swirl Atomizer Sprays, in: *Central States Combustion Institute Spring Meeting*, NASA-Lewis Research Center, Cleveland, OH, USA (1986)

464. Bryan, R., Godbole, P. S., and Norster, E. R., Characteristics of Airblast Atomizers, in: *Combustion and Heat Transfer in Gas Turbine Systems, Cranfield Int. Symp. Ser. 11* (Norster, E. R., ed.), Pergamon, New York, NY, USA, pp. 343–359 (1971)

465. Ingebo, R. D., Atomization of Liquid Sheets in High Pressure Airflow, *ASME Paper HT-WA/HT-27* (1984)

466. Ingebo, R. D., and Foster, H. H., Drop-Size Distribution for Cross-Current Breakup of Liquid Jets in Airstreams, *NACA-TN*-4087 (1957)

467. Ingebo, R. D., Capillary and Acceleration Wave Breakup of Liquid Jets in Axial Flow Air-Streams, *NASA Technical Paper, 1791* (1981)

468. Aigner, M., and Wittig, S., Performance and Optimization of an Airblast Nozzle, Drop-Size Distribution and Volumetric Air Flow, in: *Proc. of the Third Int. Conf. on Liquid Atomization and Spray Systems* (Yule, A. J., and Eisenklam, P., eds.), Institute of Energy, London, UK, pp. HC/3/1–8 (1985)

469. Lefebvre, A. H., Airblast Atomization, *Prog. Energy Combust. Sci.,* 6:233–261 (1980)

470. Matsumoto, S., Saito, K., and Takashima, Y., Phenomenal Transition of Liquid Atomization from Disk, *J. Chem. Eng. Jpn.,* 7(1):13–19 (1974)

471. Walton, W. H., and Prewett, W.C., The Production of Sprays and Mists of Uniform Drop Size by Means of Spinning Disc Type Sprayers, *Proc. Phys. Soc. Lond. B.,* 62(6):341–350 (1949)

472. Oyama, Y., and Endou, K., On the Centrifugal Disk Atomization; Studies on the Atomization of Water Droplets, *Kagaku Kogaku,* 17:256–260, 269–275 (1953)

473. Matsumoto, S., and Takashima, Y., Droplet Size Distribution in Spray, *Kagaku Kogaku,* 33(4):357–361 (1969)

474. Hege, H., Die Auflösung von Flüssigkeiten in Tropfen, *Aufbereit.-Tech.,* 10(3):142–147 (1969)

475. Friedman, S. J., Gluckert, F. A., and Marshall, W.R., Jr., Centrifugal Disk Atomization, *Chem. Eng. Prog.,* 48(4):181–191 (1952)

476. Herring, W. M., Jr., and Marshall, W. R., Jr., Performance of Vaned-Disk Atomizers, *J. Am. Inst. Chem. Eng.,* 1(2):200–209 (1955)

477. Fraser, R. P., Eisenklam, P., and Dombrowski, N., Liquid Atomization in Chemical Engineering – Liquid Disintegration and Dispersion Process and Study of Various Types of Atomizers; Rotary Atomizers, Their Applications and Design, Drop Formation and Sizes; Types of Available Pressure Nozzles, Energy Requirements to Effect Atomization; Twin Fluid Atomizers and Their Operating Principles, *Brit. Chem. Eng.,* 2(8):414–417; 2(9):496–501; 2(10):536–543; 2(11):610–613 (1957)

478. Scott, M. W., Robinson, M. J., Pauls, J. F., and Lantz, R. J., Spray Congealing: Particle Size Relationships Using a Centrifugal Wheel Atomizer, *J. Pharm. Sci.,* 53(6):670–675 (1964)

479. Dombrowski, N., and Fraser, R. P., A Photographic Investigation into the Disintegration of Liquid Sheets, *Philos. Trans. R. Soc. Lond. A,* 247(924):101–130 (1954)

480. Tanasawa, Y., Miyasaka, Y., and Umehara, M., On the Filamentation of Liquid by Means of Rotating Discs, *Trans. Jpn. Soc. Mech. Eng.,* 25(156):879–897 (1959)

481. Mori, Y., Hijikata, K., and Nagasaki, T., Electrostatic Atomization for Small Droplets of Uniform Diameter, *Trans. Jpn. Soc. Mech. Eng. B,* 47(421):1881–1890 (1981)

482. Mizutani, Y., Uga, Y., and Nishimoto, T., Investigation on Ultrasonic Atomization, *Trans. Jpn. Soc. Mech. Eng.,* 37(301):1746–1753 (1971)

483. Peskin, R. I., and Raco, R. J., Ultrasonic Atomization of Liquids, *J. Acoust. Soc. Am.,* 35(9):1378–1381 (1963)

484. Lobdell, D. D., Particle Size-Amplitude Relation for the Ultrasonic Atomizer, *J. Acoust. Soc. Am.,* 43(2):229–231 (1967)

485. Aller, A. J., and Losada, A., Models in the Metal Powders Atomization, *Powder Metall. Sci. Technol.,* 2(2):13–37 (1991)

486. Tornberg, C., Particle Size Prediction in an Atomization System, in: *Advances in Powder Metallurgy & Particulate Materials* (Capus, J. M., and German, R. M., eds.), APMI International, Princeton, NJ, USA, Vol. 1, pp. 137–150 (1992)

487. Dunkley, J. J., Evaluating the Performance of Atomizers, in: *Advances in Powder Metallurgy* (Gasbarre, T. G., and Jandeska, W. F., eds.), MPIF, Princeton, NJ, USA, Vol. 2, pp. 1–13 (1989)

488. Nichiporenko, O. S., Investigation of the Process of Bronze Atomization by Air, *Sov. Powder Metall. Met. Ceram.,* 7(60):947–949 (1967); Nichiporenko, O. S., and Naida, Yu. I., Heat Exchange between Metal Particles and Gas in the Atomization Process, *Sov. Powder Metall. Met. Ceram.,* 7(67):509–512 (1968)

489. Schmitt, H., Mathematical-Physical Considerations Regarding the Production of Metal Powders for Powder Metallurgy (Part I), *Powder Metall. Int.*, 11(1):17–21 (1979)

490. Tamura, K., and Takeda, T., On Manufacture of Irregular Particle Shaped Copper Powders by Gas Atomization, *J. Jpn. Soc. Powder Metall.,* 10(4):153–159 (1963)

491. Thompson, J. S., A Study of Process Variables in the Production of Aluminum Powder by Atomization, *J. Inst. Met.*, 74:101–132 (1948)

492. Antipas, G., Lekakou, C., and Tsakiropoulos, P., The Breakup of Melt Streams by High Pressure Gases in Spray Forming, in: *Proc. of the Second Int. Conf. on Spray Forming*, Osprey Metals Ltd., Swansea, UK, pp. 15–24 (1993)

493. Liu, H., A Comparison of Predictions by Various Correlations for Mean Droplet Sizes, Unpublished Work (1998)

494. Date, U. R., Tendolkar, G. S., and Vartak, M.N., Statistical Interpretation of Metal Atomization, *Int. J. Powder Metall.,* 3(2):49–58 (1967)

495. Bauckhage, K., Liu, H., and Fritsching, U., Models for the Transport Phenomena in a New Spray Compacting Process, in: *Proc. of the Fourth Int. Conf. on Liquid Atomization and Spray Systems*, Fuel Society of Japan, Sendai, Jpn, pp. 425–430 (1988)

496. Fritsching, U., Liu, H., and Bauckhage, K., Numerical Modeling in the Metal Spray Compacting Process, in: *Proc. of the Fifth Int. Conf. on Liquid Atomization and Spray Systems* (Semerjian, H.G., ed.), ILASS-Americas and NIST, Gaithersburg, MD, USA, pp. 491–498 (1991)

497. Mathur, P., Apelian, D., and Lawley, A., Analysis of the Spray Deposition Process, *Acta Metall. Mater.*, 37(2):429–443 (1989)

498. Small, S., and Bruce, T. J., The Comparison of Characteristics of Water and Inert Gas Atomized Powders, *Int. J. Powder Metall.*, 4(3):7–17 (1968)

499. Putimtsev, B. N., Effect of the Thermophysical Properties of Gases and Molten Metals on the Properties of Atomized Powders, *Sov. Powder Metall. Met. Ceram.*, 11:171–175 (1972)

500. Zeng, X., Liu, H., and Lavernia, E. J., Reactive Spray Processing of Metal Matrix Composites, in: *Proc. of the Ninth Int. Conf. on Composite Materials* (Miravete, A., ed.), Madrid, Spain, pp. 731–738 (1993)

501. Grandzol, R. J., and Tallmadge, J. A., Water Jet Atomization of Molten Steel, *AIChE J.*, 19(6):1149–1158 (1973); Grandzol, R. J., and Tallmadge, J. A., Effect of Jet Angle on Water Atomization, *Int. J. Powder Metall. Powder Technol.*, 11(2):103–114 (1975)

502. Seki, Y., Okamoto, S., Takigawa, H., and Kawai, N., Effect of Atomization Variables on Powder Characteristics in the High-Pressured Water Atomization Process, *Met. Powder Report,* 45(1):38–40 (1990)

503. Kato, T., Tezuka, H., Kusaka, K., and Suzuki, K., Study of Stainless Steel Powder, *Denki Seiko* (*Electric Furnace Steel*), 46:4–10 (1975)

504. Kishidaka, H., Theory and Manufacturing of Atomized Steel Powder, in: *Proc. Multidisciplinary Meeting on Sintered Metals and Magnetic Properties*, Japan Society for Power and Powder Metallurgy, Vol. 19, p. 72 (1972)

505. Ranz, W. E., and Marshall, W. R., Jr., Evaporation from Drops, Parts I and II, *Chem. Eng. Prog.*, 48(3):141–146 and 173–180 (1952)

506. Rao, P., and Tallmadge, J. A., Change of Shape of Metal Droplets in Quench Atomization, in: *Proc. 1971 Fall Powder Metall. Conf.* (Mocarski, S., ed.), MPIF, Princeton, NJ, USA, pp. 251–266 (1972)

507. Sundaram, A., Sundaram, K. M. S., and Leung, J. W., Droplet Spreading and Penetration of Non-Aqueous Pesticide Formulations and Spray Diluents in Kromekote® Cards, *Trans. ASAE,* 34(5):1941–1951 (1991)

508. Jones, H., Cooling, Freezing and Substrate Impact of Droplets Formed by Rotary Atomization, *J. Phys. D: Appl. Phys.*, 4:1657–1660 (1971)

509. Cheng, L., Dynamic Spreading of Drops Impacting onto a Solid Surface, *Ind. Eng. Chem. Process Des. Dev.*, 16(2):192–197 (1977)

510. Fukanuma, H., and Ohmori, A., Behavior of Molten Droplets Impinging on Flat Surfaces, in: *1994 Thermal Spray Industrial Applications* (Berndt, C. C., and Sampath, S., eds.), ASM International, Materials Park, OH, USA, pp. 563–568 (1994)

511. Sobolev, V. V., and Guilemany, J. M., Flattening of Thermally Sprayed Particles, *Mater. Lett.*, 22:209–213 (1995); Sobolev, V. V., Guilemany, J. M., and Martin, A. J., Engineering Formulas for Flattening of Composite Powder Particles during Thermal Spraying, in: *Thermal Spray: A United Forum for Scientific and Technological Advances* (Berndt, C. C., ed.), ASM International, Materials Park, OH, USA, pp. 653–656 (1997)

512. Solonenko, O. P., Advanced Thermophysical Fundamentals of the Melt Microdroplet Flattening and Solidification on a Substrate, in: *Thermal Spray Science and Technology* (Berndt, C.C., and Sampath, S., eds.), ASM International, Materials Park, OH, USA, pp. 237–242 (1995)

513. Tan, N. T., Klevtsov, A. V., and Chilikina, N. V., The Theory of Deformation of a Droplet on a Horizontal Surface, *Thermal Eng.*, 36(10):572–574 (1989)

514. Collings, E. W., Markworth, A. J., McCoy, J.K., and Saunders, J. H., Splat-Quench Solidification of Freely Falling Liquid-Metal Drops by Impact on a Planar Substrate, *J. Mater. Sci.*, 25:3677–3682 (1990)

515. Trapaga, G., and Szekely, J., Mathematical Modeling of the Isothermal Impingement of Liquid Droplets in Spraying Processes, *Metall. Trans.*, *22B*:901–914 (1991)

516. Hamatani, H., Okada, T., and Yoshida, T., in: *Proc. Int. Symp. on Plasma Chemistry*, Bochum, Germany, pp. 1527–1532 (1989)

517. Watanabe, T., Kuribayashi, I., Honda, T., and Kanzawa, A., Deformation and Solidification of a Droplet on a Cold Substrate, *Chem. Eng. Sci.*, 47:3059–3065 (1992)

518. Bertagnolli, M., Marchese, M., Jacucci, G., Noelting, S., and Doltsinis, I., Thermomechanical Simulation of the Splashing Process of a Ceramic Droplet on a Rigid Substrate, in: *1994 Int. Energy Technol. Conf., Material, Design and Analysis Symp.*, New Orleans, LA, USA (1994)

519. Ueda, T., Enomoto, T., and Kanetsuki, M., Heat Transfer Characteristics and Dynamic Behavior of Saturated Droplets Impinging on a Heated Vertical Surface, *Bull. Jpn. Soc. Mech. Eng.*, 22(167):724–732 (1979)

520. Madejski, J., Droplets on Impact with a Solid Surface, *Int. J. Heat Mass Transfer,* 26(7):1095–1098 (1983)

521. Fiedler, R., and Naber, J., Estimating Film Size due to Drop Impact, in: *Proc. Combustion Institute Spring Technical Meeting on Combustion Fundamentals and Applications* (Adamczyk, A.A., ed.), Combustion Institute, Pittsburgh, PA, USA, pp. 269–274 (1989)

522. Markworth, A. J., and Saunders, J. H., An Improved Velocity Field For the Madejski Splat-Quench Solidification Model, *Int. J. Heat Mass Transfer,* 35(7):1836–1837 (1992)

523. Haley, P. J., and Miksis, M. J., The Effect of the Contact Line on Droplet Spreading, *J. Fluid Mech.,* 223:57–81 (1991)

524. Moreau, C., Lamontagne, M., and Cielo, P., Influence of the Coating Thickness on the Cooling Rate of Plasma-Sprayed Particles Impinging on a Substrate, in: *Thermal Spray Coatings: Properties, Processes and Applications* (Bernecki, T.F., ed.), ASM International, Materials Park, OH, USA, pp. 237–243 (1992)

525. Reitz, R. D., Advances in Internal Combustion Engine Modeling, in: *Int. Conf. on Liquid Atomization and Spray Systems-Americas-93* (Chigier, N., ed.), ILASS-Americas, Worcester, MA, USA, pp. 81–88 (1993)

526. Chigier, N., Spray Combustion, *ASME HTD-321/FED,* 233:1-30 (1995)

527. Bogy, D. B., Drop Formation in a Circular Liquid Jet, *Ann. Rev. Fluid Mech.*, 11:207–228 (1979)

528. Spangler, C. A., Hilbing, J. H., and Heister, S. D., Nonlinear Modeling of Jet Atomization in the Wind-Induced Regime, *Phys. Fluids,* 7(5):964–971 (1995)

529. Hilbing, J. H., and Heister, S. D., Droplet Size Control in Liquid Jet Breakup, *Phys. Fluids,* 8(6):1574–1581 (1996)

530. Kothe, D. B., Mjolsness, R. C., and Torrey, M. D., *RIPPLE: A Computer Program for Incompressible Flows with Free Surfaces*, LA-12007-MS, Los Alamos National Laboratory, Los Alamos, NM, USA (1989)

531. Ponstein, J., Instability of Rotating Cylindrical Jets, *Appl. Sci. Res.*, 8A:425–456 (1959)

532. Kang, D. J. and Lin, S. P., Breakup of a Swirling Liquid Jet, *Int. J. Eng. Fluid Mech.*, 2(1):47–62 (1989)

533. Shokoohi, F., and Elrod, H. G., Numerical Investigation of the Disintegration of Liquid Jets, *J. Comp. Phys.*, 71(2):324–342 (1987)

534. Green, A. E., On the Nonlinear Behavior of Fluid Jets, *Int. J. Eng. Sci.*, 14(1):49–63 (1976)

535. Ibrahim, E. A., and Lin, S. P., Weakly Nonlinear Instability of a Liquid Jet in a Viscous Gas, *J. Appl. Mech.*, 59(Pt. 2):S291–S296 (1992)

536. Lee, H. C., Drop Formation in a Liquid Jet, *IBM J. Res. Dev.*, 18:364–369 (1974)

537. Chuech, S. G., Przekwas, A. J., and Yang, H. Q., Direct Simulation for the Instability and Breakup of Laminar Liquid Jets, *AIAA Paper 90-2066* (1990)

538. Ibrahim, E. A., Satellite Drop Formation in a Swirling Liquid Jet, *Mech. Res. Comm.,* 23(4):409–412 (1996)

539. Li, X., On the Instability of Plane Liquid Sheets in Two Gas Streams of Unequal Velocities, *Acta Mech.,* 106:137–156 (1994)

540. Clark, C. J., and Dombrowski, N., Aerodynamic Instability and Disintegration of Inviscid Liquid Sheets, *Proc. R. Soc. Lond. A,* 329:467–478 (1972)

541. Williams, F. A., Spray Combustion and Atomization, *Phys. Fluids,* 1(6):541–545 (1958); Williams, F. A., Spray Combustion Theory, *Combust. Flame,* 3(2):215–228 (1959); Williams, F. A., *Combustion Theory*, Second Edition, Addison-Wesley Publishing Co., Reading, MA, USA (1985)

542. Kennard, E. H., *Kinetic Theory of Gases*, McGraw-Hill Book Company Inc., New York, USA (1938)

543. Gupta, H. C., and Bracco, F.V., Numerical Computations of Two-Dimensional Unsteady Sprays for Application to Engines, *AIAA J.*, 16(10):1053–1061 (1978)

544. Hasselman, L. C., and Westbrook, C. K., A Theoretical Model for Fuel Injection in Stratified Charge Engines, *SAE Paper 780138* (1978)

545. Roache, P. J., *Computational Fluid Dynamics*, Hermosa Publishers, Albuquerque, NM, USA (1982)

546. Borman, G. L., and Johnson, J. H., Unsteady Vaporization Histories and Trajectories of Fuel Drops Injected in Swirling Air, *SAE Paper 598C* (1962)

547. Lambiris, S., Combes, L. P., and Levine, R. S., Stable Combustion Processes in Liquid Propellant Rocket Engines, in: *Fifth Colloquium of the Combustion and Propulsion Panel*, AGARD (1962)

548. Crowe, C. T., Sharma, M. P., and Stock, D. E., The Particle-Source-in-Cell Method for Gas Droplet Flow, *Trans. ASME J. Fluids Eng.*, 99:325–332 (1977)

549. Dukowicz, J. K., A Particle-Fluid Numerical Model for Liquid Sprays, *J. Comp. Phys.*, 35(2):229–253 (1980)

550. O'Rourke, P. J., *Collective Drop Effects on Vaporizing Liquid Sprays*, LA-9069-T, Los Alamos National Laboratory, Los Alamos, NM, USA (1981)

551. Bracco, F. V., Modeling of Engine Sprays, *SAE Paper 850394* (1985)

552. Reitz, R. D., and Bracco, F. V., On the Dependence of Spray Angle and other Spray Parameters on Nozzle Design and Operating Conditions, *SAE Paper 790494* (1979)

553. Chehroudi, B., Chen, S. H., Bracco, F. V., and Onuma, Y., On the Intact Core of Full-Cone Sprays, *SAE Paper 850126* (1985)

554. Lamb, H., *Hydrodynamics*, Dover Publications, New York, NY, USA (1932)

555. Amsden, A. A., Ramshaw, J. D., O'Rourke, P. J., and Dukowicz, J. K., *KIVA: A Computer Program for Two- and Three-Dimensional Fluid Flow with Chemical Reactions and Fuel Sprays*, LA-10254-MS, Los Alamos National Laboratory, Los Alamos, NM, USA (1985); Amsden, A. A., O'Rourke, P. J., and Butler, T. D., *KIVA-II – A Computer Program for Chemically Reactive Flows with Sprays*, LA-11560-MS, Los Alamos National Laboratory, Los Alamos, NM, USA (1989); Amsden, A. A., *KIVA-3: A KIVA Program with Block-Structured Mesh for Complex Geometries*, LA-12503-MS, Los Alamos National Laboratory, Los Alamos, NM, USA (1993)

556. Ibrahim, E. A., Yang, H. Q., and Przekwas, A. J., Modeling of Spray Droplets Deformation and Breakup, *AIAA J. Propul. Power,* 9:651–654 (1993)

557. Aslanov, S. K., and Shamshev, K. N., On the Theory of Drop Atomization in a Spray, *Phys. Doklady,* 39(6):444–446 (1994)

558. Ashgriz, N., and Poo, J. Y., Coalescence and Separation in Binary Collisions of Liquid Drops, *J. Fluid Mech.*, 221:183–204 (1990)

559. O'Rourke, P. J., and Bracco, F. V., Modeling Drop Interactions in Thick Sprays and a Comparison with Experiments, in: *Stratified Charge Automotive Engines*, I. Mech. E. Conference Publications, Automobile Division of the Institution of Mechanical Engineers, Westminster, London, UK, pp. 101–116 (1980)

560. Langmuir, I., The Production of Rain by a Chain Reaction in Cumulous Clouds at Temperatures above Freezing, *J. Meteorol.*, 5:175–192 (1948)

561. Tsouris, C., and Tavlarides, L. L., Breakage and Coalescence Models for Drops in Turbulent Dispersions, *AIChE J.,* 40(3):395–406 (1994)

562. Tambour, Y., A Lagrangian Sectional Approach for Simulating Droplet Size Distributions of Vaporizing Fuel Sprays in a Turbulent Jet, *Combust. Flame,* 60:15–28 (1985)

563. Faeth, G. M., Mixing Transport and Combustion in Sprays, *Prog. Energy Combust. Sci.*, 13(4):293–345 (1987)

564. Ishii, M., *Thermo-Fluid Dynamic Theory for Two-Phase Flows*, Eyrolles, Paris, France (1975)

565. Raju, M. S., and Sirignano, W. A., Interaction between Two Vaporizing Droplets in an Intermediate Reynolds Number Flow, *Phys. Fluids A,* 2(10):1780–1796 (1990)

566. Dandy, D. S., and Dwyer, H., A Sphere in Shear Flow at Finite Reynolds Number: Effect of Shear on Particle Lift, Drag and Heat Transfer, *J. Fluid Mech.*, 216:381–410 (1990)

567. Yung, C. N., De Witt, K. J., Brockwell, J. L., and Chai, A. T., The Transient Motion of a Spherical Fluid Droplet, *Chem. Eng. Comm.*, 110:163–186 (1991)

568. Soo, S. L., *Fluid Dynamics of Multiphase Systems*, Blaisdell, Waltham, MA, USA (1967)

569. Carlson, D. J., and Hoglund, R. F., Particle Drag and Heat Transfer in Rocket Nozzles, *AIAA J.,* 2:1980–1984 (1964)

570. Crowe, C. T., Drag Coefficient of Particles in a Rocket Nozzle, *AIAA J.,* 5:1021–1022 (1967)

571. Korkan, K. D. Petrie, S. L., and Bodonyi, R. J., Particle Concentrations in High Mach Number, Two-Phase Flows, *TR 740102*, Aerospace Research Laboratories, Wright-Patterson Air Force Base, Dayton, OH, USA (1974)

572. Liu, A. B., Mather, D., and Reitz, R. D., Effects of Drop Drag and Breakup on Fuel Sprays, *SAE Paper 930072* (1993); Liu, Z., Obokata, T., and Reitz, R. D., Modeling Drop Drag Effects on Fuel Spray Impingement in Direct Injection Diesel Engines, *SAE Paper 971256*, pp. 9–22 (1997)

573. Renksizbulut, M., and Yuen, M. C., Experimental Study of Droplet Evaporation in a High Temperature Air Stream, *Trans. ASME J. Heat Transfer,* 105:384–388 (1983)

574. Gonzalez, M., Lian, Z., and Reitz, R. D., Modeling Diesel Engine Spray Vaporization and Combustion, *SAE Paper 920579* (1992)

575. Henderson, C. B., Drag Coefficients of Spheres in Continuum and Rarefied Flows, *AIAA J.,* 14(6):707–708 (1976)

576. Crowe, C. T., Challenges in Numerical Simulation of Metal Sprays in Spray Forming Processes, *Spray Forming*, Band 2:1–16 (1997)

577. Torobin, L. B., and Gauvin, W. H., The Drag Coefficients of Single Spheres Moving in Steady and Accelerated Motion in a Turbulent Fluid, *AIChE J.,* 7:615–619 (1961)

578. Warnica, W. D., Renksizbulut, M., and Strong, A. B., Drag Coefficients of Spherical Liquid Droplets, Part 2, Turbulent Gaseous Fields, *Exp. Fluids,* 18:265–276 (1994)

579. Thomas, P. J., On the Influence of the Basset History Force on the Motion of a Particle through a Fluid, *Phys. Fluids A,* 4(9):2090–2093 (1992)

580. Baranovskii, S. I., Nadvorskii, A. S., and Turishchev, A. I., Method of Calculating Supersonic Gas-Droplet Jets, *J. Eng. Phys. Thermophys.,* 62(4):412–415 (1992)

581. Smolders, H. J., and van Dongen, M. E. H., Shock Wave Structure in a Mixture of Gas, Vapour and Droplets, *Shock Waves,* 2(4):255–267 (1992)

582. Surov, V. S., Numerical Modeling of the Interaction of a Strong Shock Wave with Liquid Drops, *J. Appl. Mech. Tech. Phys.,* 36(3):354–359 (1995)

583. Higdon, J. J. L., and Muldowney, G. P., Resistance Functions for Spherical Particles, Droplets and Bubbles in Cylindrical Tubes, *J. Fluid Mech.,* 298:193–210 (1995)

584. Whitaker, S., Forced Convection Heat Transfer Correlations for Flow in Pipes, Past Flat Plates, Single Cylinders, Single Spheres, and for Flow in Packed Beds and Tube Bundles, *AIChE J.,* 18(2):361–371 (1972)

585. Yao, S. C., and Schrock, V. E., Heat and Mass Transfer from Freely Falling Drops, *Trans. ASME J. Heat Transfer,* 98(1):120–126 (1976)

586. Priem, R. J., Borman, G. L., El Wakil, M. M., Uyehara, O. A., and Myers, P. S., Experimental and Calculated Histories of Vaporizing Fuel Drops, *NACA-TN*-3988 (1957)

587. Yearling, P. R., and Gould, R. D., Convective Heat and Mass Transfer from Single, Evaporating Water, Methanol and Ethanol Droplets, *ASME FED,* 233:33–38 (1995)

588. Silverman, I., and Sirignano, W. A., Multi-Droplet Interaction Effects in Dense Sprays, *Int. J. Multiphase Flow,* 20(1):99–116 (1994)

589. Faeth, G. M., Evaporation and Combustion of Sprays, *Prog. Energy Combust. Sci.,* 9:1–76 (1983)

590. Crowe, C. T., Modeling Turbulence in Multiphase Flows, in: *Engineering Turbulence Modeling and Experiments* 2 (Rodi, W., and Martelli, F., eds.), Elsevier Science Publishers B.V., Amsterdam, The Netherlands, pp. 899–913 (1993)

591. Crowe, C. T., Chung, J. N., and Troutt, T. R., Particle Dispersion in Turbulent Shear Flows, *Prog. Energy Combust. Sci.,* 14:171–194 (1988)

592. Piomelli, U., Ferziger, J. H., and Moin, P., Models for Large Eddy Simulation of Turbulent Channel Flows Including Transpiration, *Stanford University Department of Mechanical Engineering Report*, TF-32 (1988)

593. Leonard, A., Computing Three-Dimensional Incompressible Flows with Vortex Elements, *Ann. Rev. Fluid Mech.*, 17:523–559 (1985)

594. Young, J. B., The Fundamental Equations of Gas-Droplet Multiphase Flow, *Int. J. Multiphase Flow,* 21(2):175–191 (1995)

595. Creismeas, P., An Eulerian/Lagrangian Model to Calculate the Evolution of a Water Droplet Spray, *Int. J. Numer. Meth. Fluids,* 20(2):135–155 (1995)

596. Haselman, L. C., and Westbrook, C. K., A Theoretical Model for Two-Phase Fuel Injection in Stratified Charge Engines, *SAE Paper 780318* (1978)

597. Sinnamon, J. F., Lancaster, D. R., and Steiner, J. C., An Experimental and Analytical Study of Engine Fuel Spray Trajectories, *SAE Trans.*, 89(1):765–783 (1980)

598. Kong, S. C., and Reitz, R. D., Spray Combustion Processes in Internal Combustion Engines, in: *Recent Advances in Spray Combustion*, Chapter 9, *Progress in Astronautics and Aeronautics*, Series (Kuo, K., ed.), AIAA, Reston, VA, USA, pp. 395–423 (1993)

599. Naber, J. D., and Reitz, R. D., Modeling Engine Spray/Wall Impingement, *SAE Paper 881316*:1–23 (1988)

600. Halstead, M., Kirsh, L., and Quinn, C., The Auto Ignition of Hydrocarbon Fuels at High Temperatures and Pressures – Fitting of a Mathematical Model, *Combust. Flame,* 30:45–60 (1977)

601. Magnussen, B. F., and Hjertager, B. H., On Mathematical Modeling of Turbulent Combustion with Special Emphasis on Soot Formation and Combustion, in: *Sixteenth Symp. (Int.) on Combustion*, Combustion Institute, Pittsburgh, PA, USA, pp. 719–729 (1977)

602. Marble, F. E., and Broadwell, J. E., The Coherent Flame Model for Turbulent Chemical Reactions, TRW-9-PU (1977)

603. Heywood, J. B., Pollutant Formation and Control in Spark-Ignition Engines, *Prog. Energy Combust. Sci.*, 1:135–164 (1976)

604. Han, Z., Uludogan, A., Hampson, G. J., and Reitz, R. D., Mechanism of Soot and NOx Emission Reduction Using Multiple-Injection in a Diesel Engine, *SAE Paper 960633* (1996)

605. Liu, H., *Application of Osprey Process to Spray Deposition of Steels*, Internal Report, Verfahrenstechnik/FB4, Universität Bremen, Bremen, Germany, July (1986)

606. Liu, H., Lavernia, E. J., and Rangel, R. H., An Analysis of Freeze-Up Phenomena during Gas Atomization of Metals, *Int. J. Heat Mass Transfer,* 38(12):2183–2193 (1995)

607. Liu, H., and Dandy, D. S., Modeling of Liquid Metal Flow and Heat Transfer in Delivery Tube during Gas Atomization, *Mater. Sci. Eng. A,* 197(2):199–208 (1995)

608. Fritsching, U., Uhlenwinkel, V., Bauckhage, K., and Urlau, U., Gas und Partikelströmungen im Düsennahbereich einer Zweistoffdüse - Modelluntersuchungen zur Zerstäubung von Metallschmelzen, *Chem. Ing. Tech.*, 62(2):146–147 (1990); Uhlenwinkel, V., Fritsching, U., Bauckhage, K., and Urlau, U., Strömungsuntersuchungen im Düsennahbereich einer Zweistoffdüse - Modelluntersuchungen für die Zerstäubung von Metallschmelzen, *Chem. Ing. Tech.*, 62(3):228–229 (1990)

609. Bewlay, B. P., and Cantor, B., Modeling of Spray Deposition: Measurements of Particle Size, Gas Velocity, Particle Velocity, and Spray Temperature in Gas-Atomized Sprays, *Metall. Trans., 21B*:899–912 (1990)

610. Bewlay, B. P., and Cantor, B., The Relationship between Thermal History and Microstructure in Spray-Deposited Tin-Lead Alloys, *J. Mater. Res.*, 6(7):1433–1454 (1991)

611. Espina, P. I., Ridder, S. D., Biancaniello, F. S., and Mattingly, G. E., Aerodynamic Analysis of the Aspiration Phenomena in a Close-Coupled Inert Gas Atomizer, in: *Synthesis and Analysis in Materials Processing: Advances in Characterization and Diagnostics of Ceramics and Metal Particulate Processing* (Lavernia, E. J., Henein, H., and Anderson, I., eds.), TMS, Warrendale, PA, USA, pp. 49–61 (1989)

612. Liu, H., and Dax, F. R., Effects of Atomizer Geometry and Configuration on Gas Flow Field in Gas Atomization, in: *Advances in Powder Metallurgy and Particulate Materials* (McKotch, R. A. and Webb, R., eds.), MPIF and APMI International, Princeton, NJ, USA, Vol. 1, pp.5/3–5/11 (1997)

613. Power, G. D., Cooper, G. K., and Sirbaugh, J. R., NPARC 2.2 - Features and Capabilities, *AIAA* 95–2609:1–12 (1995)

614. Liu, H., and Seuren, B., Spray Atomization and Deposition of Steel, in: *Second National Symposium*, Society for the Advancement of Material and Process Engineering, e.V., Wuppetal, Germany, November 4 (1988)

615. Liu, H., *Berechnungsmodelle für die Geschwindigkeiten und die Abkühlung von Tropfen im Sprühkegel einer Stahl-Zerstäubungsanlage*, Ph.D. Thesis, Universität Bremen, Bremen, Germany, February (1990)

616. Cai, W., Liu, H., Sickinger, A., Mühlberger, E., Bailey, D., and Lavernia, E. J., Low-Pressure Plasma Deposition of Tungsten, *J. Thermal Spray Technol.,* 3(2):135–141 (1994)

617. Bewlay, B. P., and Cantor, B., Gas Velocity Measurements from a Close-Coupled Spray Deposition Atomizer, *Mater. Sci. Eng. A,* 118:207–222 (1989)

618. Trapaga, G. M., *Gas-Particle-Deposit Interactions during Plasma Spraying and Spray Forming Processes*, Sc.D. Thesis, Massachusetts Institute of Technology, Cambridge, MA, USA, September (1990)

619. Grant, P. S., Cantor, B., and Katgerman, L., Modeling of Droplet Dynamic and Thermal Histories during Spray Forming – I. Individual Droplet Behavior and II. Effect of Process Parameters, *Acta Metall. Mater.*, 41(11):3097–3118 (1993)

620. Bergmann, D., Fritsching, U., and Bauckhage, K., Modellierung der Abkühlung und raschen Erstarrung von Metalltropfen im Fluge während des Sprühkompaktierens, *Spray Forming*, Band 3:175–195 (1998)

621. Bauckhage, K., Liu, H., Seuren, B., and Uhlenwinkel, V., Spray Forming of Liquid Steel, Local Size and Velocity Distribution of Particle in the Spray Cone and Their Reference to Varying Process Parameters, in: *Proc. Powder Metall. Conf., PM'90*, Institute of Metals, London, UK, pp. 207–215 (1990)

622. Liu, H., Seuren, B., and Uhlenwinkel, V., On-Line and In-Line Measurements of Local Size and Velocity Distributions of Particles by LDVS in the Spray Cone during Spray Deposition of Liquid Steel, in: *Int. Conf. on Liquid Atomization and Spray Systems-Europe-89*, ILASS-Europe and Universität Bremen, Bremen, Germany, July 1–4 (1989)

623. Johnson, H. A., and Rubesin, M. W., Aerodynamic Heating and Convective Heat Transfer – Summary of Literature Survey, *Trans. ASME,* 71(5):447–456 (1949)

624. Hirth, J. P., Nucleation, Undercooling and Homogeneous Structures in Rapidly Solidified Powders, *Metall. Trans., 9A*:401–404 (1978)

625. Turnbull, D., and Fisher, J. C., Rate of Nucleation in Condensed Systems, *J. Chem. Phys.,* 17(1):71–73 (1949)

626. Levi, C. G., and Mehrabian, R., Heat Flow during Rapid Solidification of Undercooled Metal Droplets, *Metall. Trans., 13A*:221–234 (1982)

627. Zhang, X., and Atrens, A., A Thermokinetic Model for Rapid Solidification, *Int. J. Rapid Solidification,* 7:83–107 (1992)

628. Clyne, T. W., and Kurz, W., Solute Redistribution during Solidification with Rapid Solid State Diffusion, *Metall. Trans., 12A*:965–971 (1981)

629. Liang, X., Earthman, J. C., and Lavernia, E. J., On the Mechanism of Grain Formation during Spray Atomization and Deposition, *Acta Metall. Mater.,* 40(11):3003–3016 (1992)

630. Ojha, S. N., Tripathi, A. K., and Singh, S. N., Spray Atomization and Deposition of an Al-4Cu-20Pb Alloy, *Powder Metall. Int.,* 25(2):65–69 (1993)

631. Uhlenwinkel, V., Kramer, C., Bauckhage, K., and Ulrich, J., Experimental Investigation of the Sticking Efficiency during Spray Forming, in: *Advances in Powder Metallurgy & Particulate Materials* (Cadle, T. M., and Narasimhan, K. S., eds.), MPIF and APMI International, Princeton, NJ, USA, Vol. 1, Part 9, pp. 29–40 (1996); Kramer, C., Uhlenwinkel, V., and Bauckhage, K., Spray Forming: The Sticking Efficiency and its Dependency on Affecting Parameters, in: *Solidification* (Marsh, S. P., Dantzig, J. A., Trivedi, R., Hofmeister, W., Chu, M. G., Lavernia, E. J., and Chun, J. H., eds.), TMS, Warrendale, PA, USA, pp. 401–413 (1998)

632. Uhlenwinkel, V., and Bauckhage, K., Mass Flux Profile and Local Particle Size in the Spray Cone during Spray Forming of Steel, Copper and Tin, in: *Proc. of the Second Int. Conf. on Spray Forming*, Osprey Metals Ltd., Swansea, UK, pp. 25–34 (1993)

633. Welch, J. E., Harlow, F. H., Shannon, J. P., and Daly, B. J., *The MAC Method*, LA-3425, UC-32, Mathematics and Computers, TID-4500, Los Alamos National Laboratory, Los Alamos, NM, USA (1966)

634. Montavon, G., Feng, Z. G., Coddet, C., Feng, Z. Q., and Domaszewski, M., Influence of the Spray Parameters on the Transient Pressure within a Molten Particle Impacting onto a Flat Substrate, in: *Thermal Spray: A United Forum for Scientific and Technological Advances* (Berndt, C. C., ed.), ASM International, Materials Park, OH, USA, pp. 627–633 (1997)

635. Samsonov, V. M., and Shcherbakov, L. M., Thermodynamic Theory of Droplet Evolution after Contact with a Solid Surface, *Kolloidnyi Zhurnal,* 52(6):1122–1128 (1990)

636. Nieminen, J. A., Abraham, D. B., Karttunen, M., and Kaski, K., Molecular Dynamics of a Microscopic Droplet on Solid Surface, *Phys. Rev. Lett.,* 69(1):124–127 (1992)

637. Shi, H. H., and Dear, J. P., Oblique High-Speed Liquid-Solid Impact, *JSME Int. J., I* 35(3):285–295 (1992)

638. Pasandideh-Fard, M., and Mostaghimi, J., Droplet Impact and Solidification in a Thermal Spray Process: Droplet-Substrate Interaction, in: *Thermal Spray: Practical Solution for Engineering Problems* (Berndt, C. C., ed.), ASM International, Materials Park, OH, USA, pp. 637–646 (1996)

639. Hirt, C. W., and Nichols, B. D., Volume of Fluid (VOF) Method for the Dynamics of Free Boundaries, *J. Comp. Phys.,* 39:201–225 (1981)

640. Itoh, A., Hirata, M., and Ayagaki, M., Effects of Substrate Temperature During Spraying on the Properties of Sprayed Coatings, in: *Thermal Spray Coatings: Research, Design and Applications* (Berndt, C. C., and Bernecki, T. F., eds.), ASM International, Materials Park, OH, USA, pp. 593–600 (1993)

641. Wachters, L. H. J., and Westerling, N. A. J., The Heat Transfer from a Hot Wall to Impinging Water Drops in the Spheroidal State, *Chem. Eng. Sci.,* 21:1047–1056 (1966)

642. Akao, F. Araki, K., Mori, S., and Moriyama, A., Deformation Behaviors of a Liquid Droplet Impinging onto a Hot Metal Surface, *Trans. I.S.I.J.,* 20:737–743 (1980)

643. Nagaoka, M., Kawazoe, H., and Nomura, N., Modeling Fuel Spray Impingement on a Hot Wall for Gasoline Engines, *SAE Paper 940525* (1994)

644. Kuniyoshi, H., Yamamoto, H., Fujimoto, K., and Sato, G. T., Investigation of the Characteristics of Diesel Fuel Spray Impinging upon a Flat Plate, *J. M.E.S.J.,* 15:57–64 (1980)

645. Naber, J. D., and Farrell, P. V., Hydrodynamics of Droplet Impingement on a Heated Surface, *SAE Paper 930919* (1993)

646. Eckhause, J. E., and Reitz, R. D., Modeling Heat Transfer to Impinging Fuel Sprays in Direct Injection Engines, *Atomization and Sprays,* 5(2):213–242 (1995)

647. Chigier, N., *Combustion Measurements*, Hemisphere Publishing Corporation, Washington, DC, USA (1991)

648. Israel, F., Taylor, A. M. K. P., and Whitelaw, J. H., Simultaneous Measurement of Droplet Velocity and Size and Flame Mantle Temperature by Phase Doppler Anemometry and Two-Colour Pyrometry, *Measurement Sci. & Technol.*, 6(6):727–741 (1995)

649. Yule, A., Chigier, N., Atakan, S., and Ungut, A., Particle Size and Velocity Measurement by Laser Anemometry, *AIAA Paper 77–214*, in: *Fifteenth Aerospace Sciences Meeting*, Los Angles, CA, USA (1977)

650. Murakami, T., and Ishikawa, M., Laser Holographic Study on Atomization Processes, in: *Proc. of the First Int. Conf. on Liquid Atomization and Spray Systems*, Fuel Society of Japan, Tokyo, Jpn., pp. 317–324 (1978)

651. Ante, A., Schaller, J. K., Fiedler, D. K., and Stojanoff, C. G., Digital Image Processing of Double Pulse Holograms: Investigation of Velocity and Number Density of Droplets in Sprays, in: *Proc. of the SPIE - The Int. Soc. for Opt. Eng.*, 2248:275–283 (1994)

652. Chen, B. T., Cheng, Y. S., and Yeh, H. C., A Study of Density Effect and Droplet Deformation in the TSI Aerodynamic Particle Sizer, *Aerosol Sci. Technol.*, 12:278–285 (1990)

653. Chigier, N. A., Drop Size and Velocity Instrumentation, *Prog. Energy Combust. Sci.*, 9:155–177 (1983)

654. Bachalo, W. D., Droplet Analysis Techniques: Their Selection and Applications, in: *Liquid Particle Size Measurement Techniques*, ASTM STP 848 (Tishkoff, J. M., Ingebo, R. D., and Kennedy, J. B., eds.), ASTM, West Conshohocken, PA, USA, pp. 5–21 (1984)

655. Bachalo, W. D., Experimental Methods in Multiphase Flows, *Int. J. Multiphase Flow*, 20:261–295 (1994)

656. Hirleman, E. D., Particle Sizing by Optical, Nonimaging Techniques, in: *Liquid Particle Size Measurement Techniques*, ASTM STP 848 (Tishkoff, J. M., Ingebo, R. D., and Kennedy, J. B., eds.), ASTM, West Conshohocken, PA, USA, pp. 35–60 (1984)

657. Jones, A. R., A Review of Drop Size Measurement - The Application of Techniques to Dense Fuel Sprays, *Prog. Energy Combust. Sci.*, 3:225–234 (1977)

658. Dodge, L. G., Comparison of Performance of Drop-Sizing Instruments, *Appl. Opt.*, 26(7):1328–1341 (1987)

659. May, K. R., The Measurement of Airborne Droplets by the Magnesium Oxide Method, *J. Sci. Instrum.*, 27:128–130 (1950)

660. Hausser, F., and Strobl, G. M., Method of Catching Drops on a Surface and Defining the Drop Size Distributions by Curves, *Z. Tech. Phys.*, 5(4):154–157 (1924)

661. Rupe, J. H., A Technique for the Investigation of Spray Characteristics of Constant Flow Nozzles, in: *Third Symp. on Combustion, Flame, and Explosion Phenomena*, Williams & Wilkins, Baltimore, MD, USA, pp. 680–694 (1949)

662. Zhang, G. J., and Ishii, M., Isokinetic Sampling Probe and Image Processing System for Droplet Size Measurement in Two-Phase Flow, *Int. J. Heat Mass Transfer*, 38(11):2019–2027 (1995)

663. Joyce, J. R., The Atomization of Liquid Fuels for Combustion, *J. Inst. Fuel,* 22(124):150–156 (1949)

664. Taylor, E. H., and Harmon, D. B., Jr., Measuring Drop Sizes in Sprays, *Ind. Eng. Chem.,* 46(7):1455–1457 (1954)

665. Choudhury, A. P. R., Lamb, G. G., and Stevens, W. F., A New Technique for Drop-Size Distribution Determination, *Trans. Indian Inst. Chem. Eng.,* 10:21–24 (1957)

666. Wicks, M., and Dukler, A. E., in: *Proc. of ASME Heat Transfer Conf.*, Vol. V, Chicago, USA, p. 39 (1966)

667. Gardiner, J. A., Measurement of the Drop Size Distribution in Water Sprays by an Electrical Method, *Instrum. Pract.,* 18(4):353–356 (1964)

668. Mahler, D. S., and Magnus, D. E., Hot-Wire Technique for Droplet Measurements, in: *Liquid Particle Size Measurement Techniques*, ASTM STP 848 (Tishkoff, J. M., Ingebo, R. D., and Kennedy, J. B., eds.), ASTM, West Conshohocken, PA, USA, pp. 153–165 (1984)

669. Chigier, N., Optical Imaging of Sprays, *Prog. Energy Combust. Sci.*, 17:211–262 (1991)

670. Bachalo, W. D., Method for Measuring the Size and Velocity of Spheres by Dual-Beam Light Scatter Interferometry, *Appl. Opt.*, 19(3):363–370 (1980)

671. Chigier, N. A., The Atomization and Burning of Liquid Fuel Sprays, *Prog. Energy Combust. Sci.*, 2:97–114 (1976)

672. Mellor, R., Chigier, N. A., and Beer, J. M., Pressure Jet Spray in Airstreams, *ASME Paper 70-GT-101*, in: *ASME Gas Turbine Conference*, Brussels (1970)

673. Simmons, H. C., and Lapera, D. L., A High-Speed Spray Analyzer for Gas Turbine Fuel Nozzles, in: *ASME Gas Turbine Conference*, Cleveland, OH, USA, March (1969)

674. MacLoughlin, P. F., and Walsh, J. J., A Holographic Study of Interacting Liquid Sprays, in: *Proc. of the First Int. Conf. on Liquid Atomization and Spray Systems*, Fuel Society of Japan, Tokyo, Jpn, pp. 325–332 (1978)

675. Thompson, B. J., Droplet Characteristics with Conventional and Holographic Imaging Techniques, in: *Liquid Particle Size Measurement Techniques*, ASTM STP 848 (Tishkoff, J. M., Ingebo, R. D., and Kennedy, J. B., eds.), ASTM, West Conshohocken, PA, USA, pp. 111–122 (1984)

676. Ridder, S. D., Osella, S. A., Espina, P. I., and Biancaniello, F. S., Process Modeling and Control of Inert Gas Atomization, in: *Thermal Structures and Materials for High-Speed Flight*, Vol. 140, *Progress in Astronautics and Aeronautics* (Thornton, E.A., ed.), AIAA, Washington, DC, USA, pp. 499–521 (1992)

677. Ridder, S. D., Biancaniello, F. S., Espina, P. I., and Miller, S. A., Laser Holographic Imaging of the Two Fluid Atomization Process, in: *Synthesis and Analysis in Materials Processing: Advances in Characterization and Diagnostics of Ceramics and Metal Particulate Processing* (Lavernia, E. J., Henein, H., and Anderson, I., eds.), TMS, Warrendale, PA, USA, pp. 73–86 (1989)

678. Durst, F., Melling, A., and Whitelaw, J. H., *Principles and Practice of Laser-Doppler Anemometry*, Academic Press, New York, NY, USA (1976)

679. Farmer, W. M., *The Interferometric Observation of Dynamic Particle Size, Velocity, and Number Density*, Ph.D. Thesis, University of Tennessee, Knoxville, TN, USA (1973)

680. Farmer, W. M., Sample Space for Particle Size and Velocity Measuring Interferometers, *Appl. Opt.*, 15:1984–1989 (1976)

681. Bachalo, W. D., Hess, C. F., and Hartwell, C. A., An Instrument for Spray Droplet Size and Velocity Measurements, *ASME Paper 79-WA/GT-13*, in: *ASME Winter Annual Meeting* (1979)

682. Hess, C. F., A Technique Combining the Visibility of a Doppler Signal with the Peak Intensity of the Pedestal to Measure the Size and Velocity of Droplets in a Spray, in: *AIAA Twenty-Second Aerospace Sciences Meeting*, Reno, NV, USA, January 9–12 (1984)

683. Jackson, T. A., and Samuelson, G. S., Spatially Resolved Droplet Size Measurements, *ASME Paper 85-GT-38* (1985)

684. Glover, A. R., Skippon, S. M., and Boyle, R. D., Interferometric Laser Imaging for Droplet Sizing: A Method for Droplet-Size Measurement in Sparse Spray Systems, *Appl. Opt.*, 34(36):8409–8421 (1995)

685. Bachalo, W. D., and Houser, M. J., Phase Doppler Spray Analyzer for Simultaneous Measurements of Drop Size and Velocity Distributions, *Opt. Eng.*, 23(5):583–590 (1984)

686. Bauckhage, K., The Phase-Doppler-Difference Method, A New Laser-Doppler Technique for Simultaneous Size and Velocity Measurement, *Part. Part. Syst. Charact.*, 5:16–22 and 66–71 (1988)

687. Bauckhage, K., and Flögel, H. H., Simultaneous Measurement of Droplet Size and Velocity in Nozzle Sprays, in: *Proc. of the Second Int. Symp. on Applications of Laser Anemometry to Fluid Mechanics*, pp. 18.1/1–6, Lisbon, Portugal, July 2–4 (1984)

688. Martin, W. W., Adbelmessih, A. H., Liska, J. J., and Durst, F., Characteristics of Laser-Doppler Signal from Bubbles, *Int. J. Multiphase Flow,* 7:433–460 (1981)

689. Manasse, U., Wriedt, Th., and Bauckhage, K., Phase Doppler Sizing of Optically Absorbent Single- and Multicomponent Liquid Droplets Using Semiconductor Devices, *Measurement Sci. & Technol.*, 4(3):369–377 (1993)

690. Manasse, U., Wriedt, Th., and Bauckhage, K., Phase-Doppler Sizing of Optically Absorbing Liquid Droplets: Comparison between Mie Theory and Experiment, *Part. Part. Syst. Charact.*, 9(3):176–185 (1992)

691. Holve, D. J., and Self, S. A, Optical Particle Sizing Counter for in Situ Measurements, Parts I and II, *Appl. Opt.*, 18(10):1632–1645 (1979)

692. Holve, D. J., and Annen, K., Optical Particle Counting and Sizing Using Intensity Deconvolution, *Opt. Eng.*, 23(5):591–603 (1984)

693. Holve, D. J., Transit Timing Velocimetry (TTV) for Two Phase Reacting Flows, *Combust. Flame,* 48:105–108 (1982)

694. Dobbins, R. A., Crocco, L., and Glassman, I., Measurement of Mean Particle Sizes of Sprays from Diffractively Scattered Light, *AIAA J.*, 1(8):1882–1886 (1963)

695. Rizk, N. K., and Lefebvre, A. H., Measurement of Drop-Size Distribution by a Light-Scattering Technique, in: *Liquid Particle Size Measurement Techniques*, ASTM STP 848 (Tishkoff, J. M., Ingebo, R. D., and Kennedy, J. B., eds.), ASTM, West Conshohocken, PA, USA, pp. 61–71 (1984)

696. Huckaby, J. L., Ray, A. K., and Das, B., Determination of Size, Refractive Index, and Dispersion of Single Droplets from Wavelength-Dependent Scattering Spectra, *Appl. Opt.*, 33(30):7112–7125 (1994)

697. Swithenbank, J., Beer, J. M., Abbott, D., and McCreath, C. G., A Laser Diagnostic Technique for the Measurement of Droplet and Particle Size Distribution, *AIAA Paper 76–69*, in: *Fourteenth Aerospace Sciences Meeting*, Washington, D.C., January 26–28, AIAA, Reston, VA, USA (1976)

698. Dodge, L. G., Calibration of Malvern Particle Sizer, *Appl. Opt.*, 23:2415–2419 (1984)

699. Dodge, L. G., and Cerwin, S.A., Extending the Applicability of Diffraction-Based Drop Sizing Instruments, in: *Liquid Particle Size Measurement Techniques*, ASTM STP 848 (Tishkoff, J.M., Ingebo, R.D., and Kennedy, J.B., eds.), ASTM, West Conshohocken, PA, USA, pp. 72–81 (1984)

700. Felton, P. G., Hamidi, A. A., and Aigal, A. K., Measurement of Drop Size Distribution in Dense Sprays by Laser Diffraction, in: *Proc. of the Third Int. Conf. on Liquid Atomization and Spray Systems* (Yule, A. J., and Eisenklam, P., eds.), Institute of Energy, London, UK, pp. IVA/4/1–11 (1985); Hamidi, A. A., and Swithenbank, J., Treatment of Multiple Scattering of Light in Laser Diffraction Measurement Techniques in Dense Sprays and Particle Fields, *J. Inst. Energy,* 59:101–105 (1986)

701. Naqwi, A. A., Liu, X. Z., and Durst, F., Evaluation of the Dual-Cylindrical Wave Laser Technique for Sizing of Liquid Droplets, *Part. Part. Syst. Charact.,* 9(1):44–51 (1992)

702. Hodkinson, J., Particle Sizing by Means of the Forward Scattering Lobe, *Appl. Opt.,* 5:839–844 (1966); Gravatt, C., Real Time Measurement of the Size Distribution of Particulate Matter by a Light Scattering Method, *J. Air Pollut. Control Assoc.*, 23(12):1035–1038 (1973)

703. Hirleman, E. D., and Wittig, S. L. K., Uncertainties in Particle Size Distributions Measured with Ratio-Type Single Particle Counters, in: *Conf. on Laser and Electro-Optical Systems*, San Diego, CA, USA, May 25–27 (1976); Wuerer, J.E., Oeding, R. G., Poon, C. C., and Hess, C. F., The Application of Nonintrusive Optical Methods to Physical Measurements in Combustion, *Paper AIAA-82-0236*, in: *AIAA 20th Aerospace Sciences Meeting*, NASA, Scientific and Technical Information Branch, Washington, D.C., USA (1982)

704. Zemlianskii, V. M., and Yanovsky, F. J., Phase Optical-Microwave Method for Sizing Droplets in Air, Theory and Numerical Treatment, in: *Conf. Proc. of the Second Topical Symp. on Combined Optical-Microwave Earth and Atmosphere Sensing,* Atlanta, GA, 3–6 April 1995, IEEE, New York, NY, USA, pp. 174–174 (1995)

705. Korobochka, A. N., and Pavlenko, A. M., Determination of Exact Droplet Size Distribution Generated by an Air Driven Nozzle, *Fluid Mech. Res.,* 21(2):11–16 (1992)

706. Chigier, N. A., Instrumentation Techniques for Studying Heterogeneous Combustion, *Prog. Energy Combust. Sci.*, 3:175–189 (1977)

707. Adrian, R. J., Particle Imaging Techniques for Experimental Fluid Mechanics, *Ann. Rev. Fluid Mech.*, 23:261–304 (1991)

708. Yurteri, C. U., Kadambi, J. R., and Arik, E., Spray Characterization and Droplet Interactions Study Using Particle Dynamic Analyzer, in: *Proc. of the SPIE - The Int. Soc. for Opt. Eng.,* 2052:145–152 (1993)

709. Tilleman, M. M., and Grasso, R., A Practical Method for Diameter, Number Density and Material Characterization of 40-90 ìm Size Droplets by Stimulated Raman Scattering, *Appl. Phys. B,* 61(6):581–584 (1995)

710. Massoli, P., Beretta, F., D'Alessio, A., and Lazzaro, M., Temperature and Size of Single Transparent Droplets by Light Scattering in the Forward and Rainbow Regions, *Appl. Opt.,* 32(18):3295–3301 (1993)

711. Naudin, N., Farre, J., Bissieres, D., and Lavergne, G., IR Instrumentation Dedicated to the Experimental Validation of Evaporating and Burning Droplets' Temperature Models, in: *Proc. of the SPIE - The Int. Soc. for Opt. Eng.,* 2552(Pt. 2):669–676 (1995); Naudin, N., Farre, J., Liousse, F., and Ravel, O., Temperature Measurement of Small Moving Droplets by a Non Invasive Infrared Method, *Atti della Fondazione Giorgio Ronchi,* 51(1–2):305–318 (1996).

712. Vardelle, M., Vardelle, A., Leger, A. C., and Fauchais, P., Dynamics of Splat Formation and Solidification in Thermal Spraying Processes, in: *1994 Thermal Spray Industrial Applications* (Berndt, C. C., and Sampath, S., eds.), ASM International, Materials Park, OH, USA, pp. 555–562 (1994)

713. Lawrynowicz, D. E., and Lavernia, E. J., Sensors and Techniques Used to Monitor Processing Parameters during Spray Atomization and Deposition, *J. Mater. Sci.,* 30(5):1125–1138 (1995)

714. Jiang, G., Henein, H., and Siegel, M., Overview: Intelligent Sensors for Atomization, *Int. J. Powder Metall.,* 26(3):253–268 (1990)

715. Henein, H., Process Diagnostics to Monitor and Control Gas Atomization Operations, in: *Advances in Powder Metallurgy & Particulate Materials* (Cadle, T. M., and Narasimhan, K. S., eds.), MPIF and APMI International, Princeton, NJ, USA, Vol. 1, Part 1, pp. 199–219 (1996)

Index

B

C

D

H

I

J

M

N

S

T